IAC INFORMATION CENTER/LIBRARY AZ15-1W10
001019

AF615388

CHLORINE PRODUCTION PROCESSES

CHLORINE PRODUCTION PROCESSES

Recent and Energy Saving Developments

Edited by J.S. Robinson

NOYES DATA CORPORATION
Park Ridge, New Jersey, U.S.A.
1981

Library of Congress Catalog Card Number: 81-2361
ISBN: 0-8155-0842-5
Printed in the United States

Published in the United States of America by
Noyes Data Corporation
Noyes Building, Park Ridge, New Jersey 07656

Library of Congress Cataloging in Publication Data

Main entry under title:

Chlorine production processes.

(Chemical technology review ; no. 185) (Energy
technology review ; no. 64)
Includes index.
1. Chlorine--Patents. I. Robinson, J. S.,
1936- II. Series. III. Series: Energy
technology review ; no. 64.
TP245.C5C46 665.8'3 81-2361
ISBN 0-8155-0842-5 AACR2

FOREWORD

The detailed, descriptive information in this book is based on over 200 patents issued since January 1978 that deal with chlorine production. In addition, the first chapter discusses new or potential chlorine processes, as well as methods of improving energy efficiencies in chlorine production.

This book is a data-based publication, providing information retrieved and made available from the U.S. patent literature. It thus serves a double purpose in that it supplies detailed technical information and can be used as a guide to the patent literature in this field. By indicating all the information that is significant, and eliminating legal jargon and juristic phraseology, this book presents an advanced commercially oriented review of chlorine production.

The U.S. patent literature is the largest and most comprehensive collection of technical information in the world. There is more practical, commercial, timely process information assembled here than is available from any other source. The technical information obtained from a patent is extremely reliable and comprehensive; sufficient information must be included to avoid rejection for "insufficient disclosure." These patents include practically all of those issued on the subject in the United States during the period under review; there has been no bias in the selection of patents for inclusion.

The patent literature covers a substantial amount of information not available in the journal literature. The patent literature is a prime source of basic commercially useful information. This information is overlooked by those who rely primarily on the periodical journal literature. It is realized that there is a lag between a patent application on a new process development and the granting of a patent, but it is felt that this may roughly parallel or even anticipate the lag in putting that development into commercial practice.

Many of these patents are being utilized commercially. Whether used or not, they offer opportunities for technological transfer. Also, a major purpose of this book is to describe the number of technical possibilities available, which may open up profitable areas of research and development. The information contained in this book will allow you to establish a sound background before launching into research in this field.

Advanced composition and production methods developed by Noyes Data are employed to bring these durably bound books to you in a minimum of time. Special techniques are used to close the gap between "manuscript" and "completed book." Industrial technology is progressing so rapidly that time-honored, conventional typesetting, binding and shipping methods are no longer suitable. We have bypassed the delays in the conventional book publishing cycle and provide the user with an effective and convenient means of reviewing up-to-date information in depth.

The table of contents is organized in such a way as to serve as a subject index. Other indexes by company, inventor and patent number help in providing easy access to the information contained in this book.

16 Reasons Why the U.S. Patent Office Literature Is Important to You

1. The U.S. patent literature is the largest and most comprehensive collection of technical information in the world. There is more practical commercial process information assembled here than is available from any other source. Most important technological advances are described in the patent literature.

2. The technical information obtained from the patent literature is extremely comprehensive; sufficient information must be included to avoid rejection for "insufficient disclosure."

3. The patent literature is a prime source of basic commercially utilizable information. This information is overlooked by those who rely primarily on the periodical journal literature.

4. An important feature of the patent literature is that it can serve to avoid duplication of research and development.

5. Patents, unlike periodical literature, are bound by definition to contain new information, data and ideas.

6. It can serve as a source of new ideas in a different but related field, and may be outside the patent protection offered the original invention.

7. Since claims are narrowly defined, much valuable information is included that may be outside the legal protection afforded by the claims.

8. Patents discuss the difficulties associated with previous research, development or production techniques, and offer a specific method of overcoming problems. This gives clues to current process information that has not been published in periodicals or books.

9. Can aid in process design by providing a selection of alternate techniques. A powerful research and engineering tool.

10. Obtain licenses–many U.S. chemical patents have not been developed commercially.

11. Patents provide an excellent starting point for the next investigator.

12. Frequently, innovations derived from research are first disclosed in the patent literature, prior to coverage in the periodical literature.

13. Patents offer a most valuable method of keeping abreast of latest technologies, serving an individual's own "current awareness" program.

14. Identifying potential new competitors.

15. It is a creative source of ideas for those with imagination.

16. Scrutiny of the patent literature has important profit-making potential.

CONTENTS AND SUBJECT INDEX

A SURVEY OF POTENTIAL CHLORINE PRODUCTION PROCESSES

The information in this chapter is based on:

> *A Survey of Potential Chlorine Production Processes*, ANL/OEPM-79-1, prepared by Versar, Inc. for the Office for Electrochemical Project Management, Argonne National Laboratory, April 1979.

INTRODUCTION

The chlorine industry is the second largest consumer of electricity among all electrochemical industries. In 1977 alone it consumed almost 35 billion kWh of fuel energy, 92% of which was used to power diaphragm and mercury cells. These cells have an overall energy efficiency of only 12% (3), and current research indicates that improvements in this situation are not expected in the near future.

Any significant reduction in the chlorine industry's consumption of electricity would require large scale introduction of either electrolytic or nonelectrolytic processes with an overall energy efficiency of 15 to 20%. Processes of this nature have been investigated over the years with varying degrees of success.

Diaphragm Cell and Mercury Cell

Conventional electrolytic technology is based on the diaphragm cell or the mercury cell. The fact that innovations in these processes are still being made is evidenced by the information in the next two chapters.

In the diaphragm cell, a concentrated brine is electrolyzed with the anode and cathode sections separated by a diaphragm which acts to retard the diffusion of hydrogen and caustic from the cathode, where they are produced, to the anode. If such a separator is not present, hydrogen and chlorine could mix, forming explosive mixtures, and hydroxide ions could react with the chlorine to yield undesired chlorates and perchlorates via competing electrochemical mechanisms. In the diaphragm cell process, the chlorine liberated at the anode is recovered, dried by passage through concentrated sulfuric acid, compressed, liquefied and readied for sale. The coproduct hydrogen is also compressed and either sold or used on-site for its chemical or fuel value. The caustic formed at the cathode in a diaphragm cell operation is 8 to 11% in strength. It normally must be concentrated by evaporation of water, first to 50% to effect precipitation of salt values present and, if desired, to higher values to recover solid products.

In the mercury cell process, the cells produce chlorine gas and sodium amalgam. The chlorine is pure but must be dried with sulfuric acid and compressed before liquefaction. The sodium amalgam is conveyed to a denuder section where it is washed with water, producing caustic, at a concentration of 50% by weight, and hydrogen. The mercury is recycled to the cell. A recent concern with mercury cells is the control of mercury emissions.

This chapter presents a survey of potential chlorine production processes. The study was undertaken for the Department of Energy (DOE). Its purpose is to identify chlorine production processes which are economically viable and more energy efficient than the brine electrolysis systems in use today.

Processes were grouped into 20 basic categories and evaluated according to four preselected criteria: raw material availability; type and amount of energy required; by-product salability/disposability; status of development.

Raw material availability was of special importance since DOE is primarily interested in chlorine production processes which are both energy- and cost-efficient. Supplies of raw material plentiful enough to produce 500,000 metric tons per year (5% of annual chlorine production) or more were considered acceptable to meet this first criterion.

The type and amount of energy used constitutes a necessary criterion if the principal objective of achieving energy efficient production is to be met. Nonelectrolytic, low temperature processes were given highest consideration; electrolytic and high temperature processes (600°C) were judged the least desirable.

The third criterion, salability or disposability of by-products, reflects environmental concern about those processes which produce liquid or solid wastes. Although there are penalties for improper disposal of wastes, this does little to alleviate the environmental problems created by their presence. Therefore, processes which do not produce waste products were given a higher numerical rating than those with major solid or liquid waste streams.

Status of development, the fourth criterion, was chosen because of the need to introduce energy efficiency processes as soon as possible. Processes which have been demonstrated only in the laboratory or on bench-scale receive a low numerical rating, since they would require considerable time and extensive funding before they could become operational and commercialized.

Based on these four criteria and preliminary economic analyses, the most promising chlorine production processes appear to be:

- The membrane cell (with and without hydrogen and oxygen electrodes);

- Kel-Chlor (homogenous catalytic oxidation of hydrogen chloride);
- Mobay (direct electrolysis of hydrogen chloride);
- The Shell Process (catalytic oxidation of hydrogen chloride); and
- Oxidation of ammonium chloride.

Patents dealing with these processes and related aspects are reviewed in later chapters.

The Membrane Cell

The membrane cell appears to be the most economic and energy efficient chlor-alkali process of the sodium chloride based processes examined. Developmental efforts on this process are underway at several laboratories worldwide and it is considered by many people to be the "chlorine process of the future." This process offers an additional advantage in that it is free of environmental problems plaguing current chlor-alkali production methods.

This chapter identifies a number of second generation process improvements for the membrane cell area, including generation of coproducts other than caustic soda, use of more efficient electrode materials and development of more efficient cell designs. It is recommended that further developments in the three areas be closely monitored and supported, as needed, since significant advances are expected within the next few years.

Conversion of Waste Hydrogen Chloride to Chlorine

Three processes have been identified which convert waste hydrogen chloride to chlorine: Shell, Kel-Chlor and Mobay. The latter two are already operational and have a proven capacity for considerable energy savings. The Shell process, however, is not currently used in the United States and its potential and feasibility should be investigated. Kel-Chlor also requires additional research. Previous economic studies indicate that Kel-Chlor has many economy-of-scale advantages over the other two processes but no R&D work has been accomplished to date on its application to small production plants (100 to 200 tons per day).

Oxidation of Ammonium Chloride

Conversion of waste ammonium chloride to chlorine was identified as the most energy-efficient chlorine production method which is also economically viable. The ability to implement this conceptualized process depends upon four factors which require further investigation:

(a) the availability of ammonium chloride itself or means to produce the chemical compounds economically;

(b) the conversion efficiency of nitrosyl chloride to chlorine and nitrogen dioxide at low temperatures catalytically. The catalyst is needed to maintain minimal energy requirements;

(c) the kinetics of the initial reaction between ammonium chloride and nitrogen dioxide. Neither the reaction rates nor the conversion efficiency were sufficiently characterized by earlier investigators; and

(d) the separation of ammonium nitrate from ammonium chloride by fractional crystallization if less than 98% conversion efficiency is attained in the initial reaction. The 98% conversion efficiency is a requirement of the fertilizer industry.

THE MEMBRANE CELL

In this section, we discuss three emerging areas in chlorine technology:

(a) The membrane cell process itself.

(b) Use of the membrane cell with improved cell cathodes employing either the oxygen electrode or a catalytic hydrogen evolution electrode.

(c) Use of the membrane cell to produce coproducts other than caustic soda.

Membrane Cell Process

Process History: Development of the membrane cell technology began in the mid-1960's with concurrent efforts undertaken at Asahi Chemical Company, in Osaka, Japan, and at the E.I. Du Pont de Nemours Corporate Laboratories in the United States. Asahi Chemical Company developed a membrane constructed of a polyperfluorocarboxylic acid, the exact composition of which has not been disclosed. The Du Pont efforts centered around development of a polymeric membrane using polyperfluorosulfuric acid and were first reported by W.G. Grot, U.S. Patent 3,718,627, February 1973 (assigned to Du Pont) (1). Since then, Du Pont has made a number of further improvements in their membrane based upon ethylenediamine modified material.

Testing of these two membrane materials for chlor-alkali applications began in the early 1970's in Japan and the U.S. Membrane type chlor-alkali cells were developed and tested, using these polymeric materials. In the U.S., Du Pont entered into a development program with the Diamond Shamrock Corporation to evaluate and develop membrane type cells using the perfluorosulfuric acid polymer (called Nafion) membranes.

The membrane cell is currently in use at several facilities in the U.S. and Canada. Other U.S. organizations including Hooker Chemical, Allied Chemical, and B.F. Goodrich have begun evaluation and developmental efforts based on the use of Nafion-type membranes for this type of cell.

The Asahi Chemical membranes have undergone similar development and evaluation in Japan (2). Two other Japanese firms, the Asahi Glass Company (not affiliated with Asahi Chemical) of Tokyo and Tokoyama Soda Company have entered the membrane field, using fluorocarboxylic acid-type membranes. Thus, there are at least six potential producers of membrane cells, and developmental efforts are underway to utilize this technology fully.

Scientific Basis: The membrane cell differs from the diaphragm cell in that the membrane allows passage of only positively charged ions (i.e., Na^+ or H_3O^+). This effectively prevents diffusion of either hydroxide or hydrogen and allows caustic or other solutions to be used as the catholyte. As hydroxide ions are formed at the cathode, overall electrical neutrality of the catholyte is maintained by permitting sodium ions to move from the anolyte compartment through the membrane. This permits direct production of caustic solutions of strengths up to 30%, significantly reducing the amount of evaporation required when compared to conventional display cells. Also, since chloride ions do not diffuse through the membrane, a relatively chloride-free caustic is generated.

In the operation of a membrane-type cell, concentrated, fresh sodium chloride brine is fed to the anolyte compartment of the cell, while a 20 to 30% caustic solution is circulated in the catholyte compartment. With application of electric power to this system, chlorine is liberated at the anode, and sodium ions diffuse through the membrane to the catholytic compartment, where simultaneously, hydrogen is liberated via the electrochemical reaction: $2H_2O + 2e^- \rightarrow H_2 + 2OH^-$.

In the catholyte chamber, sodium ions enter via the membrane and combine with hydroxyl ions to generate caustic: $Na^+ + OH^- \rightarrow NaOH$. The caustic is drawn off for further concentration of sodium hydroxide as desired.

Engineering: Figure 1.1 shows an overall membrane cell process diagram. The process consists of the following six operations:

(a) Brine Makeup–Salt is dissolved in water to produce a concentrated brine solution. This solution is filtered to remove any undissolved material and pumped to the brine purification system.

(b) Brine Purification–Brine purification is required because the membrane cell is sensitive to impurities which foul the membrane. An existing commercial chlor-alkali plant that uses membrane cells without brine purification must wash the membrane daily with hydrochloric acid to remove impurities that interfere with process operations. For brine purification, soda ash and strontium or barium salts are added to the makeup to remove calcium, magnesium and sulfate ions via precipitation (i.e., softening process). The precipitates, calcium carbonate and magnesium hydroxide, are removed from the brine by filtration and purified brine is fed to the anolyte compartment of the membrane cell. Asahi Chemical Industry Company sells an ion exchange unit for brine purification.

(c) Electrolysis–A direct current is passed through the cell between the anode and cathode, and the following chemical activities occur:

- chlorine is liberated at the anode,
- sodium ions enter the membrane and diffuse into the catholyte compartment, and
- at the cathode, hydrogen is liberated.

All three processes operate at comparable rates to maintain overall electrical neutrality of the different cell compartments. In the catholyte chamber, a 20 to 30% caustic soda solution circulates. As additional caustic is formed, part of the solution is withdrawn for further processing, and fresh makeup water is added to maintain the caustic concentration in the 20 to 30% range. This is required as the membrane cell performance declines if the caustic concentration rises above 30% or falls below 15%.

Figure 1.1: Membrane Cell

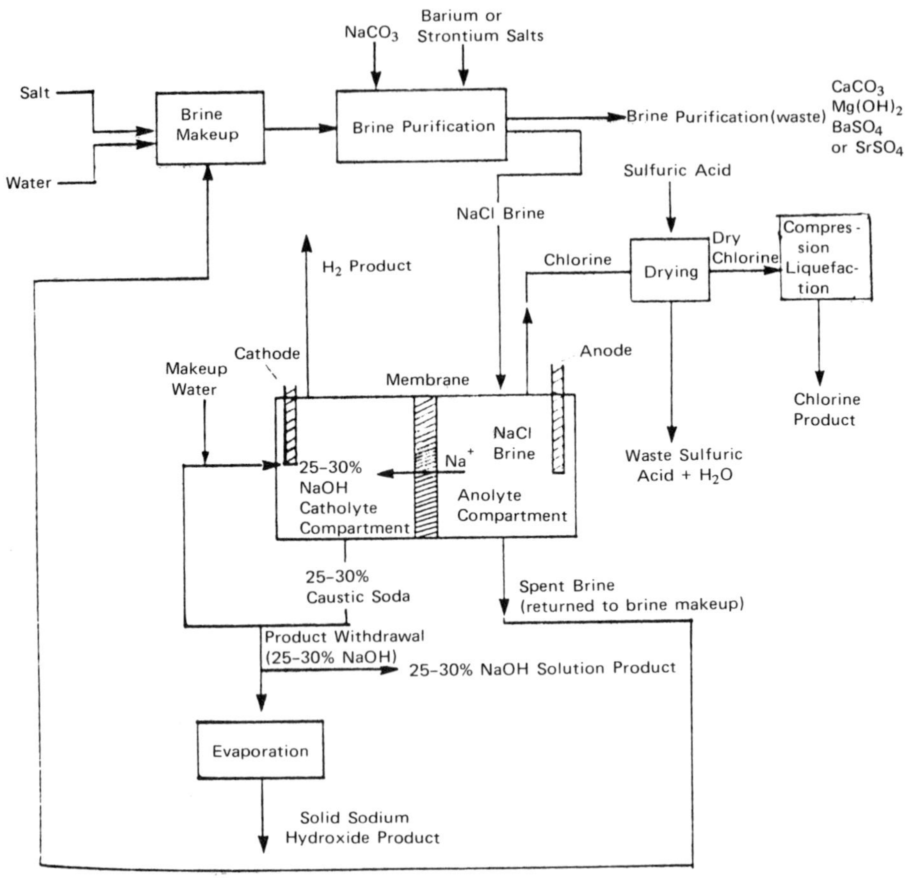

Source: ANL/OEPM-79-1

(d) Chlorine Purification—The chlorine liberated at the anode contains water vapor and 2 to 4% oxygen. This is removed by passing the chlorine through concentrated sulfuric acid (>80%). The dried chlorine is then compressed, liquefied, and prepared for shipment or use.

(e) Caustic Purification—The caustic withdrawn from the catholyte section is 20 to 30% strength. If this is adequate for the customer, the caustic is directly packaged for sale. For other purposes requiring a higher strength or solid caustic, the desired strength is readily prepared by evaporation of the solution. Evaporation is performed using a standard multiple effect steam evaporator or a heat recovery evaporator patented by Asahi Chemical.

(f) Brine Recycle–The depleted brines from the anolyte section of the membrane cell are recycled to the initial brine makeup step for refortification with salt.

While any concentration of caustic could be directly produced by this method, the overall process has been found to be most efficient when generating caustic solutions in the 20 to 30% range.

For a well-operated membrane plant producing caustic soda in the optimum concentration range, current efficiencies of 93% have been quoted for the Asahi Chemical polyperfluorinated carboxylic acid membrane (5)(6). For the Du Pont Nafion membranes, current efficiencies of over 90% have been reported (4)(7)(8)(9). Operating data for cells in continuous operation over a nine-month period show that, because of impurities present, cell efficiencies are about 88%. Efficiency drops produce corresponding increases in cell voltages and power consumption.

One advantage of the membrane cell is high purity caustic production. Analyses of caustic soda product generated by this method are listed below.

Table 1.1: Caustic Purity Using Membrane Cells

Parameter	Nafion Membrane Cell* (Diamond Shamrock)	Asahi Chemical Cell**
Caustic strength, wt %	28	21.6
NaCl, ppm	50	20
Ca, ppm	1	0.3
Mg, ppm	0.5	no data
Fe, ppm	1	1.3
SO_4, ppm	15	none
ClO_3, ppm	<50	no data
SiO_2, ppm	no data	4.4

*Diamond Shamrock data (E.J. Peters and D.R. Pulver paper presented at ECS Meeting, Atlanta, October 1977).

**Asahi Chemical data (Maomi Seko, paper presented at 20th Chlorine Plant Managers Meeting, New Orleans, February 9, 1977).

Similar comparisons are available for current efficiency versus concentration between the two types of cells. Asahi Chemical claims that their membrane cell will operate at current efficiencies of over 90% while producing caustic in the 10 to 40% range (5). These data were provided by Asahi Chemical Industry Company to show that its membrane was superior to the Du Pont membrane (2). However, an independent comparison of results by E. Hermana et al, showed power usages under identical conditions were 3,054 kWh/ton of chlorine for the Asahi membrane versus only 2,744 kWh/ton of chlorine for Nafion (10). The results, however, were obtained using a modified mercury cell with a membrane placed between the anode and a modified vertical mercury cathode. For the Asahi membrane cell, a cell voltage of 3.68 V and a dc power consumption of 2,650 kWh (dc)/metric ton of sodium hydroxide has been reported as compared with a Nafion membrane in a Diamond Shamrock cell of 2,765 kWh (dc)/metric ton (5).

Published estimates reveal that these values are expected to drop to about 2,500 kWh/metric ton of product in the near future.

Data have been released on the total energy requirements for the mercury diaphragm and membrane cells. Approximate total energy requirements, including cells, evaporation, and other energy needs to produce 50% caustic solutions were reported by G.L. Fish of Diamond Shamrock as follows:

> Membrane cell ~4,100 eq kWh/metric ton of 100% sodium hydroxide;
>
> Diaphragm cell ~5,500 eq kWh/metric ton of 100% sodium hydroxide; and
>
> Mercury cell 3,500 eq kWh/metric ton of 100% sodium hydroxide.

If, however, 20 to 30% caustic soda is acceptable as a finished product, as is the case in some industries, membrane cell energy requirements are reduced to ~3,200 kWh/metric ton of sodium hydroxide. Thus, the membrane cell is competitive with current processes in both product quality and energy needs. Ongoing development work is expected to reduce power requirements further.

Economic Considerations: A comparison of costs involved in the operation for membrane cell plant with the diaphragm cell and mercury cell plants has been released by Asahi Chemical (5) and is shown in Table 1.2.

Table 1.2: Cost of Raw Materials and Utilities of Membrane Cell, Mercury Cell and Diaphragm Cell*

		Asahi Chemical's Membrane Process		Mercury Process** . . .		Diaphragm Process*** . . .	
Raw Material	**Unit Price**	**Consumption**	**Amount ($/MT)**	**Consumption**	**Amount ($/MT)**	**Consumption**	**Amount ($/MT)**
Salt	21.34$/MT	1.495	31.90	1.495	31.90	1.585	33.82
Electrolysis power	2.88¢/ac-kWh	2,786	80.24	3,211	92.48	2,371	68.28
Other chemicals	-	-	4.00	-	7.46	-	2.72
Power for motor	2.88¢/kWh	95	2.74	86	2.48	210	6.05
Steam	7.72$/MT	0.7	5.40	0.1	0.77	3.6	27.79
Other utilities	-	-	0.10	-	0.16	-	0.16
Other costs (lease, membrane, etc.)	-	-	15.00	-	5.32	-	10.63
Total			139.38		140.57		149.45

*Cost per metric ton NaOH (100%) and 0.89 metric ton Cl_2; Japanese base December 1976; (300 yen = 1 U.S. $).
**Metal Anode.
***Modified asbestos expandable anode.

These costs demonstrate that chlorine can be manufactured economically using the membrane cell.

Use of the Membrane Cell with Either the Oxygen Cathode or a Catalytic Hydrogen Electrode

To use the membrane cell most effectively, additional approaches have been suggested which, if implemented, would further reduce overall process electrical power requirements.

Power needs for the electrolysis of brines can be reduced by two means:

(a) Reduce the overvoltage required to liberate hydrogen from brine solutions, or

(b) Substitute an oxygen electrode for the hydrogen evolution reaction as the cathodic reaction in electrochemical chlorine production.

Catalytically Active Cathodes: The first approach involves reducing the extra amount of electrical energy required to liberate hydrogen from aqueous solutions. The following reactions are involved:

(1) $H_3O^+ + e^- \rightarrow H_2O + H^1$ (adsorbed on electrode)

(2) $2H^1$ (adsorbed) $\rightarrow H_2$ (gas)

Currently, industry generally uses mild steel as the cathode material, although it is a mediocre catalyst for the above reaction sequence. Its use results in an overvoltage penalty of about 0.2 to 0.4 V. Substitution of better catalytic material than mild steel would greatly reduce the overvoltage values and reduce power consumption.

The problem of hydrogen overvoltage has been previously studied in relation to fuel cell development efforts and a number of promising approaches have been developed. These are:

(1) Use of a catalytic material in the cathode electrode structures. Materials with superior activity include nickel and platinum group metals. Coating of the cathodes with one of these materials should reduce the overvoltage requirements (11)(12).

(2) Development of a hydrogen diffusion electrode (13)(14). As atomic hydrogen is formed (step 1), it diffuses directly into and through a hollow electrode structure. If the electrode material selectively diffuses hydrogen, pure hydrogen gas can then be collected, free of any water vapor. United Technologies Corporation and Leesena Moos Corporation developed such types of electrodes in their fuel cell study efforts. These hollow, selective electrodes consist of activated palladium and palladium alloys. Such electrodes might conceivably eliminate the overvoltage problem and provide a simple means for direct generation of pure hydrogen. Much of the work in this area has already been performed and patented. What is required is a "technology transfer" from the fuel cell area to the chlor-alkali industry. Energy saved by use of this approach may be as high as 10%. However, economics may not favor the use of more expensive cathode materials.

Oxygen Cathode: A second approach to reducing power consumption is the use of an oxygen electrode as the cathode. The electrochemical reaction: $O_2 + 2H_2O + 4e^- \rightarrow 4OH^-$ is used to replace the hydrogen evolution which normally occurs at the cathode in membrane cell electrolysis. An oxygen cathode is generated by introducing oxygen into the cell cathode compartment and fabricating the cathode electrode of a material that is active for the electrochemical reduction of oxygen. The concept of applying the oxygen electrode to the chlor-alkali electrolytic process was developed by Gritzner of Dow Chemical (15)-(17) and resulted in the issuance of three patents. Work described in these patents is laboratory scale; however, this type of effort is being continued not only by Dow, but by other chlorine manufacturing organizations as well.

Actually, the oxygen electrode concept itself is not new. A considerable effort was expended in the late 1950's through mid-1960's to develop oxygen electrodes suitable for use in acid and basic media. Electrolyte systems studied included phosphoric acid, molten carbonates and moderately concentrated solutions of potassium hydroxide. Electrode materials recommended include:

(a) Various transition metal oxides and doped oxides including those of nickel and cobalt;

(b) Various transition metals such as nickel, palladium, platinum and rhodium formed into Teflon-bonded electrode structures;

(c) Carbon electrodes coated with various noble metals (chiefly platinum and palladium);

(d) Silver and gold black-type electrodes;

(e) Electrodes using dispersed high surface area blacks prepared as alloys of palladium with either silver or gold;

(f) A number of perovskite-type double oxides, chiefly those of cobalt and rare earth metals; and

(g) Various redox-type systems where a higher valence state salt of transition or rare earth metal is dissolved in the electrolyte, reduced electrochemically to a lower valence state and then chemically oxidixed in solution to continue the cycle.

With the exception of (g), the above approaches warrant further study to determine their applicability to the chlor-alkali industry. The last approach is not likely to be useful in chlor-alkali production because of the potential contamination of the caustic products by redox salts.

The oxygen cathode concept is of interest for the following reasons:

(a) The theoretical oxygen electrode potential at 25°C is 1.23 volts higher than the hydrogen electrode.

(b) While overvoltage problems will exist for oxygen electrochemical reduction, at least 0.8 volt of the 1.23 volt difference between the hydrogen and oxygen electrodes could be saved, based on data from past fuel cell efforts. As current chlor-alkali cells operate in the 3 to 4 volt range, a savings of 0.8 volt translates into a power consumption savings of 20 to 25%.

The disadvantage to this approach is that hydrogen is no longer coproduced. For plants where hydrogen is used, this approach may not be desirable; however, at most plants, the hydrogen is burned for fuel values and the oxygen cathode is an appropriate approach.

The oxygen cathode is not expected to require modifications in the overall membrane process other than substitution of electrodes and the use of oxygen in the catholyte cell compartment. This means that capital investment costs for using the oxygen cathode should not be significantly different from more conventional membrane cells.

Use of the Membrane Cell to Produce Coproducts Other than Caustic Soda

Since the membrane is permeable only to positive ions, negative ions other than hydroxyl may be used in the catholyte chamber. Based on this principle, at least four modifications of the membrane cell have been developed to produce sodium and potassium carbonates, sodium and potassium phosphates, and organic alcohols as coproducts.

Buttre et al, proposed the use of phosphoric acid as the catholyte liquid (18). As sodium ions pass through the membrane and hydrogen is evolved during cell operation, the phosphoric acid present is converted to sodium phosphate salts. With continuous makeup of phosphoric acid and withdrawal of equal amounts of catholyte liquid, sodium dihydrogen phosphate can be produced as a solid product by subsequent evaporation of the withdrawn catholyte liquor. This modification was developed through the laboratory stage. No pilot plant efforts have been reported.

A similar effort by F.L. Ramp of B.F. Goodrich electrolyzes potassium chloride in place of the normal sodium chloride-based brine (19). Again, phosphoric acid is used in the catholyte section of the membrane cell. Simultaneous migration of potassium ions through the membrane and liberation of hydrogen at the cathode results in generation of potassium dihydrogen phosphate. Because catholyte liquors are continually withdrawn from the system during operation and fresh phosphoric acid is added, the withdrawn liquors need only be evaporated to obtain the phosphate coproduct. Efficiencies of 99% were reported using a Nafion membrane and a dimensionally stable anode in the test cell. According to patent data, this work was successfully carried through the pilot plant development stage. Current use of this approach has not been determined.

A third proposed modification of the membrane cell process allows simultaneous production of either soda ash or potassium carbonate as a coproduct. In both cases, carbon dioxide is continually introduced into the catholyte cell chamber in which a 25 to 30% sodium or potassium hydroxide solution is circulating. Reaction of the carbon dioxide with the caustic soda or caustic potash generates the corresponding carbonates. All other operating conditions remain the same. The carbonates can be recovered from the catholyte liquors by evaporation. All tests of the process were performed using Nafion membranes and dimensionally stable anodes. Current efficiencies of over 97% were reported for both cases, and chloride impurity levels from the carbonates generated were found to be about 800 ppm for the sodium salt and from 50 to 860 ppm for the potassium carbonate. These carbonate coproduct schemes have been successfully carried through the pilot plant evaluation stage. Patents were granted to K.J. O'Leary, C.J. Hora, and D.L. DeRespiris, assigned to Diamond Shamrock Corporation. Comparative cost data for these and other methods of preparing sodium and potassium carbonates have not yet been released. Note that sodium carbonate is currently produced at low cost from natural sources.

A fourth approach to the production of coproducts other than caustic soda has recently been reported in a patent issued to H.B. Johnson of PPG Industries (20). There, the envisioned process involves reduction of tert-butyl hydroperoxide or some other organic peroxide to the corresponding alcohol in the catholyte compartment of the membrane cell. As much as 0.6 V reduction has been claimed for cases where noble metal-coated electrodes were used with such a scheme. Data were also presented showing that the unreacted peroxide, alcohol and caustic coproducts could be separated and recovered in such a system. The overall catholyte process operating for this case is: $ROOH + H_2O + 2e^- \rightarrow 2OH^- + ROH$.

No scale-up data have been reported for this system and comparative cost information on other current methods of preparation of tert-butyl and other alcohols has not yet been released. Also, projections on the availability and costs of large quantities of tert-butyl hydroperoxide have yet to be published.

CHLORINE FROM HYDROGEN CHLORIDE

Shell Hydrogen Chloride Catalytic Oxidation Process

Process History: The first major investigation of chlorine recovery was performed by LeBlanc in the early nineteenth century in connection with a process he developed for alkali production. His process liberated large quantities of hydrogen chloride which could be later used to obtain chlorine. The chemical process was:

$$NaCl + H_2SO_4 \rightarrow NaHSO_4 + HCl$$

$$NaCl + NaHSO_4 \rightarrow Na_2SO_4 + HCl$$

No attempt was made to trap the hydrogen chloride gas. In 1836, Gossage took advantage of this excess gas and developed a means to produce chlorine. He used gas wash towers with manganese dioxide sorbent to obtain chlorine. His process was infeasible, however, except on a laboratory scale, because large quantities of costly manganese were required and yields were poor.

In 1866, Weldon improved this process by devising a method to cut down on manganese waste using reconversion and recycling techniques on the manganese products. His yields were 30% with a chlorine concentration of up to 90%.

Deacon addressed the problem of chlorine production using hydrogen chloride in 1868. His first developments, which are traceable to Oxland, Laurens and Tregomain, generated chlorine by the oxidation of gaseous hydrogen chloride in the presence of an inert porous catalyst. The reaction is:

$$4HCl + O_2 \rightarrow 2Cl_2 + 2H_2O$$

Deacon patented his invention in 1868 (British patent 1403). He continued developing his process for the next 10 years, and in this time he registered over 20 patents.

Various two-stage processes of chlorine formation were developed beginning in 1885. These processes were based on forming a metal chloride by reacting a metal oxide with hydrogen chloride, and then oxidizing the metal chloride with air to liberate chlorine. Oxides of iron, magnesium and nickel were used. Although theoretically these processes looked attractive, operating procedures were inefficient and difficult. One serious drawback was the need for high temperatures (300° to 600°C) to oxidize the metal chlorides. At such high temperatures some metal chlorides (e.g., nickel chloride and ferric chloride) vaporized before oxidation took place. One process using magnesium oxide was commercially developed by Weldon and Pechiney, but it proved unsatisfactory and was soon discarded. Thus, the primary processes established in the late 19th century were the Weldon and Deacon processes.

In 1937, Grosvenor Labs developed an oxidation process that used a four-component contact mass. The contact mass contained ferric chloride, potassium chloride, cadmium chloride, and diatomaceous earth. Chlorine was formed by treating this mass with oxygen at temperatures around 500°C. In 1947, Dow Chemical patented a similar process in which they used ferric chloride, potassium chloride, cupric chloride, and Celite on a moving bed. Based on Dow and Grosvenor patents, Hercules Power Company built a plant to produce 35 tons per day. The plant was shut down, however, because of low yields and high operating and maintenance costs.

An electrolytic hydrogen chloride-based process which appeared promising was developed in 1943 by I.G. Farben. The pilot plant had limited success. A later version of the electrolytic process, using a two-step method, was also investigated by Westvaco. Both of these methods are currently inactive. However, a direct electrolysis method has since been developed by Uhde Corporation and used commercially.

From a historical perspective the four main hydrogen chloride/chlorine formation processes are:

(1) Catalytic oxidation of gaseous hydrogen chloride;

(2) Direct oxidation of hydrogen chloride by an inorganic oxidizing agent;

(3) Two-stage processes involving intermediate formation of a metal chloride from either its oxide or oxychloride and release of the chlorine by treatment with air or oxygen or by heating; and

(4) Electrolysis of hydrogen chloride.

Shell Oil Company researched these four methods and showed that complications existed for each.

Shell found that the catalytic oxidation processes similar to the Weldon type would not be economically or commercially feasible because yields were low. Despite modifications, large losses of chlorine occurred and recovery remained below 50%.

Direct oxidation was abandoned due to extreme corrosion problems. Process profitability required favorable market conditions and specific market locations. Based on technical and economic studies, the two-stage processes were regarded as having no future for commercial development. Shell's conclusion came at a time when considerable work was still being done. On this two-stage process, four patents were granted during this period to: Grosvenor Laboratories (in 1940), Dow Chemical (in 1951), Standard Oil Co. (in 1948), and Socony Vacuum Oil Co. (in 1947) (21)-(24).

The electrolysis process, though technically suitable, showed poor economics for all locations, except those with low electricity costs.

Shell Oil Company concluded that the modified Deacon Process was best for future chlorine development.

Scientific Basis: The Deacon Process produces chlorine by oxidation (using air or pure oxygen) of hydrogen chloride in the presence of a cupric chloride catalyst as follows:

$$4HCl + O_2 \xrightarrow{CuCl_2} 2Cl_2 + 2H_2O \quad (\Delta H < 0 \text{ exothermic})$$

Complete process thermodynamics are available from Arnold and Kobe (1952) (25). They studied the effects of temperature, pressure, dilution with inert gases, and the ratio of reactants on the conversion of hydrogen chloride to chlorine.

They found that greater hydrogen chloride conversions were possible at lower temperatures, higher pressures, and increased air to hydrogen chloride ratios. The use of pure oxygen instead of air slightly increased the conversion percentage (about 2%) and produced a higher chlorine content in the product. There are problems associated with using pure oxygen, including difficulties in heat removal, temperature control, and high costs for oxygen.

The optimum conditions determined by Deacon were within the narrow temperature range of 430° to 475°C. His experiments produced conversion percentages in this temperature range of 60 to 70%. Many problems plagued the process, including:

- High reaction temperature requirements;
- Moderate yields;
- Volatilization of cupric chloride, causing a rapid decline in catalyst activity;
- Difficulty in controlling temperature because the highly exothermic oxidation reaction caused hot spots which aggravated the volatilization; and
- Severe corrosion of equipment at reactive temperatures.

Engineering: The Shell Process operates by passing a gaseous mixture containing hydrogen chloride and oxygen over the catalyst-containing silica bed. It is presented in Figure 1.2.

The main aim of the Shell research was to devise a process (similar to the Deacon Process) that had both a satisfactory reaction rate and high conversion. This required development of a catalyst that had sufficient activity, life, and thermal stability.

Figure 1.2: Flow Diagram for the Shell Hydrogen Chloride Oxidation Process

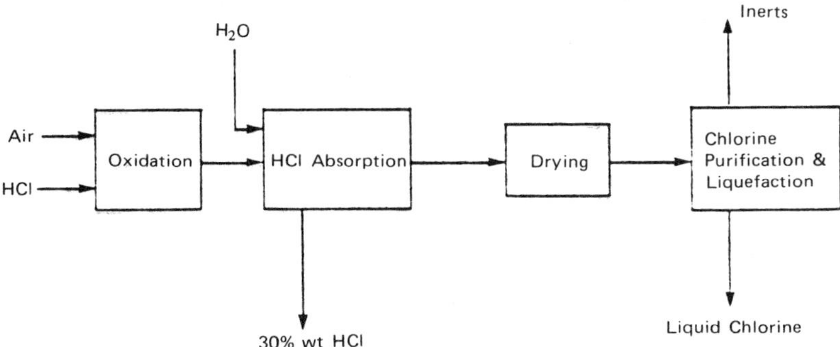

Source: ANL/OEPM-79-1

The most effective catalyst developed by Shell was cupric chloride with small quantities of rare earth metal chlorides such as scandium, yttrium, zirconium, thorium and uranium and one or more alkali metal chlorides. The metal chlorides are used in a molten state to increase their activity. All catalyst material is supported by silica gel. This silica carrier has an optimal surface area of 200 m^2/g with an average pore diameter of 60 Å. Proportional ratios are also important here in determining the optimal conversion. For best results, the rare earth metals to copper ratio should be about 0.15, and the ratio of alkali metals to copper can range from 0.8 to 1.2.

In experiments with actual industrial waste hydrogen chloride, this catalyst material performed satisfactorily. Catalyst activity remained unaffected by acid contaminants. High conversions were attained at both low temperatures and high space velocities.

Preferable process temperatures are 330° to 400°C. In this temperature range, increased pressure does not significantly increase conversion. Because of the low temperatures, copper volatilization is eliminated and corrosion due to heated reactants is greatly reduced.

Sample yields are as follows:

(1) Fixed bed, 350°C: 76.5% conversion

(2) Fluidized bed, 365°C: 77.5% conversion

The use of pure oxygen slightly increases conversion and subsequent purification, but problems of temperature control preclude its use.

Economics: Shell claims that its process is commercially competitive and has a rate of return significant enough to warrant a large capital investment. Shell claims that the process has the highest return and lowest capital cost of any of the processes used by major hydrogen chloride/chlorine producers.

The process, while apparently attractive from an energy and economic standpoint, has never enjoyed large scale commercialization. Currently, the only operating unit using this method is a 19,000 tpy pilot unit at Pernis in the Netherlands.

Kel-Chlor Catalytic Oxidation Process

Process History: The Kel-Chlor process was developed by the M.W. Kellogg Company (now Pullman-Kellogg). Its principal inventor, C.P. van Dijk, filed for a patent in 1975 and was granted the patent on May 31, 1977 (26). This patent was a continuation of a similar patent which was never issued. The Kel-Chlor process is referred to by its developers as an up-to-date Deacon process.

Pullman-Kellogg, in conjunction with Du Pont Corporation, operates a 600 ton per day facility at Corpus Christi, Texas. Pullman-Kellogg has made improvements on this facility to increase production to 1,000 to 1,100 tpd using the existing train.

A process that operates by oxidation of hydrogen chloride using nitrosyl chloride (NOCl) catalyst did not have its start with Kellogg. As early as 1919, Datta described such a process (27). Another process using nitrogen dioxide as a catalyst for the oxidation of hydrogen chloride was studied by Stow in 1962. A third process also designed to use nitrosyl chloride was patented by Jonas Hamlet of New York in 1964 while working for Du Pont (28). Development of all three processes was only on a bench scale and they were dropped due to poor results.

Scientific Basis: The Kel-Chlor method is based on the oxidation of hydrogen chloride. The Kel-Chlor process combines an active homogeneous catalyst with a strong dehydrating agent. An overall representation of the process is:

$$4HCl + O_2 \xrightarrow[NOCl]{N_2O_3} 2Cl_2 + 2H_2O$$

This overall reaction equation can be reduced into the four steps of the total reactive process, including recycle-acid stripping, oxidation, absorption-oxidation, and recycle-acid flashing. After conversion by the above process, chlorine gas can be dried and liquefied.

In the recycle-acid stripping step, gaseous hydrogen chloride contacts hot aqueous sulfuric acid. The countercurrent sulfuric acid stream has a concentration of 80% and contains the water of reaction and a catalyst which were taken up in the absorber. In this step the following reaction occurs:

$$(1) \qquad HNSO_5 + HCl \rightarrow NOCl + H_2SO_4$$

This represents a stripping of the catalyst from the acid. The catalyst is primarily nitrosylsulfuric acid ($HNSO_5$). In the second section of the stripper, the unreacted hydrogen chloride is stripped out by introducing oxygen. The original acid mixture exits the stripper carrying only small amounts of $HNSO_5$ and reaction water. Equation (1) proceeds practically to completion because of the countercurrent flow.

The oxidation occurs in a series of reactions as follows:

(2) $$2NOCl \rightarrow 2NO + Cl_2$$

(3) $$2NO + O_2 \rightarrow 2NO_2$$

(4) $$NO_2 + 2HCl \rightarrow NO + Cl_2 + H_2O$$

The hydrogen chloride gas mixture contains nitrosyl chloride, excess hydrogen chloride, oxygen and water. In the oxidizer, this mixture is heated to increase the rate of nitrosyl chloride decomposition, which is an exothermic reaction. The generated heat must be removed for the desired conversion to take place. This removed heat is then used to generate steam to run the sulfuric acid vacuum flash system and to provide all necessary process heat.

During the absorption-oxidation phase the three reactions (2, 3, 4) continue until complete hydrogen chloride conversion is achieved. Besides the oxidation reactions, the sulfuric acid absorbs all nitrogen oxides. This additional step is represented by:

(5) $$NO + NO_2 + 2H_2SO_4 \rightarrow 2HNSO_5 + H_2O$$

(6) $$NOCl + H_2SO_4 \rightarrow HNSO_5 + HCl$$

(7) $$NO_2 + 2HCl \rightleftharpoons NO + Cl_2 + H_2O$$

The chlorine gas produced is virtually free of hydrogen chloride, oxygen, and nitrogen compounds. The product chlorine gas is cooled and dried with a small stream of cold sulfuric acid.

The final phase of the Kel-Chlor system is known as recycle-acid flashing. In this step the reaction water is removed from the sulfuric acid by means of an adiabatic flash. The flashing mechanism occurs in a reduced pressure environment. The flashed acid stream is recycled to the absorber-oxidizer.

The majority of the recycled acid is hot, but a small percentage is cooled and used to provide a cooling medium for the product gas at the top of the tower.

Engineering: A flow diagram of this four-stage process is shown in Figure 1.3. Hydrogen chloride enters the system under pressure at the bottom of the stripper. The hydrogen chloride reacts with hot nitrosylsulfuric acid coming from the absorption-oxidation vessel, then contacts with oxygen in the oxidizer. The effluent gas mixture from this phase contains chlorine, steam, oxygen, catalyst components and unconverted hydrogen chloride and is sent to the absorber-oxidizer where it contacts first hot, then cool, sulfuric acid. The catalyst components react with the sulfuric acid, water is absorbed, and hydrogen chloride conversion is completed. Stripped hot sulfuric acid is dried by sending it to the flasher. Cooled sulfuric acid is recycled to the absorption-oxidation stage.

Equipment design and construction are important considerations for this process because of the highly corrosive mixture of chemicals which flow through the system.

Figure 1.3: Flow Diagram for the Kel-Chlor Process

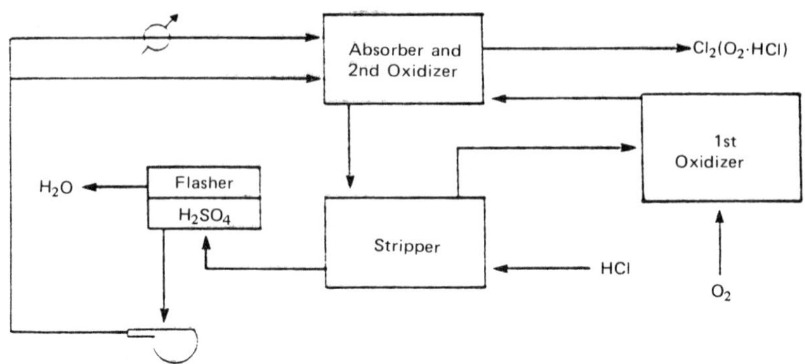

Source: ANL/OEPM-79-1

The two factors that control the extent of this corrosive atmosphere are temperature and pressure. The range of these factors governs the types of materials used, with some ranges allowing glass or Teflon-lined equipment to be used, while others rely on acid-resistant brick or expensive alloys (tantalum), and metals.

The operation of this unit in terms of start-up time, shut-down time, and response to perturbations has been studied in depth by process developers using the following variable conditions:

Pressure (absolute)	1–15 atm
Acid strength	70–85 wt %
Acid circulation temperature	127–182°C
Excess oxygen	-15 to +40%
HCl to oxidizer	0.2–1.0
HCl conversion	80–100%

Upon completion of testing at these conditions only minor catalyst losses occurred, nitrogen impurities in the product were below detectable levels, and yields above 99% were consistently obtained.

It was found that the required quality product could be obtained in a few hours once acid circulation rate, temperature, and unit pressure were stabilized. Key control parameters include oxidizer conversion, catalyst inventory, and acid strength. Expensive combinations of alloys and metals are used to combat corrosion in an effort to reduce downtime.

Economics: The basic economic principle behind the Kel-Chlor process is economy of scale. Capital requirements follow the six-tenths power rule. Table 1.3

gives an estimate of production cost for two levels of Kel-Chlor operation.

Table 1.3: Cost of Chlorine Production via the Kel-Chlor Process (Based on 90% Steam Factor)

	Units per Ton Cl_2	Unit Cost ($)	Dollars per Ton Cl_2	Plant Size: 900 Ton/Day ($20 million)*	Plant Size: 500 Ton/Day ($14.1 million)*
Oxygen	0.2576 ton	10/ton	2.576	–	–
Lime	39 lb	0.01/lb	0.390	–	–
98% H_2SO_4	11 lb	0.01/lb	0.110	–	–
HCl	1.0547 ton	–	–	–	–
Total ($/ton)				3.076	3.076
Power	40 kWh	0.007/kWh	0.280	–	–
Steam	40 lb	0.50/thousand lb	0.020	–	–
Cooling water	9,100 gal	0.02/thousand gal	0.182	–	–
Refrigeration	10 ton-hr	0.05/ton-hr	0.500	–	–
Total				0.982	0.982
Labor, 2 men/shift at $4.00/man-hr				0.237	0.427
General overhead at 100% of labor				0.237	0.427
Maintenance at 5% of investment/yr				3.382	4.292
Taxes and insurance at 3% of investment/yr				2.030	2.576
Depreciation at 10% of investment/yr				6.764	8.584
Total manufacturing cost ($/ton)				16.708	20.364

*Fixed capital investment based on U.S. Gulf Coast plant location and price. Excluded are utilities such as cooling tower, boiler and associated equipment, feed and product storage, acid storage and disposal.

The incremental cost incurred by raising output by 400 tons per day is $12.14 per ton. The rates of return, as a function of chlorine market value, assuming negligible hydrogen chloride costs, are provided below:

Plant Capacity (tons per day)	Cl_2 ($/ton): 45	50	55
	(%)		
500	39	60	50
900	52	60	67

Studies are concentrating on reducing temperature and pressure levels so that cheaper construction materials can be used.

The Kel-Chlor process is cost competitive with current chlorine manufacturing techniques and the power and steam requirements are less for Kel-Chlor. The process is also proven commercially. The major drawback is a limited supply of hydrogen chloride raw material.

Mobay Direct Electrolysis Process

Process History: This particular electrolysis process was invented by Stefon Payer of Germany and developed by the Hoechst-Uhde Corporation (29)(30). The first hydrogen chloride electrolysis plant was built at Farbwerke Hoechst in Leverkusen, West Germany, in 1964 with an output of 43,000 tons per year of chlorine. Since 1970, seven other plants have been built in various countries with combined output of more than 400,000 tons per year. The major U.S. plant of this type is in Baytown, Texas, and is operated by Mobay Chemical Company, a subsidiary of Mobil Oil. The facility, built in 1972, was expanded in 1976. The driving force for using this technology was the desire to recover and reuse waste hydrogen chloride rather than sell dilute hydrochloric acid in an uncertain market.

Scientific Basis: This electrolytic-type process uses direct current to decompose hydrogen chloride. The chemical equation for this mechanism is as follows:

$$2HCl \xrightarrow{\text{electricity}} H_2 + Cl_2$$

The waste hydrogen chloride raw material has inert gas components which are removed by absorption in a dilute hydrochloric acid solution prior to electrolysis. The acid product is fed to the electrolytic cells for chlorine production. The unreacted dilute acid is recycled to the absorption system. Similar to the membrane cell, the inlet acid product concentration should be maintained at 20 to 26% for best electrical efficiency. The dilute acid recycle stream should be maintained above 16% to avoid oxygen evolution in the cell.

The above reaction can be divided into two separate electrode processes:

$$H^+ + e \rightarrow \tfrac{1}{2}H_2 \text{ (cathode)}$$

$$Cl^- - e \rightarrow \tfrac{1}{2}Cl_2 \text{ (anode)}$$

The electrolytic cell uses a polyvinyl chloride diaphragm to separate the electrodes and prevent gas bubble migration. As a result, the chlorine product stream contains only about 0.2% hydrogen by volume and the hydrogen stream contains 1.0% chlorine by volume.

The product gases are saturated with water vapor and hydrogen chloride. Each gas is cleaned separately to remove these impurities. The chlorine is cooled with water to condense the water vapor and dried with sulfuric acid before liquefaction and storage. The hydrogen gas is also cooled, but with dilute hydrogen chloride. The cooled hydrogen gas stream is then scrubbed of chlorine with caustic to form sodium hypochlorite as follows:

$$H_2 + Cl_2 + 2NaOH \rightarrow NaOCl + NaCl + H_2O + H_2$$

Product recovery for direct electrolysis of hydrogen chloride is about 98 to 99% for both chlorine and hydrogen.

Engineering: Figure 1.4 is a flow diagram for the hydrogen chloride electrolytic process (29). Depleted aqueous hydrogen chloride is withdrawn from the catholyte cycle and sent to the absorption unit where it is saturated with fresh gaseous hydrogen chloride produced in a separate chlorination plant.

A 30% acid mixture is sent to the electrolytic cell from the absorption stage. This procedure ensures a continuous supply of fresh acid to replace the decomposed material.

Figure 1.4: Flow Diagram of the Electrolytic Process for Recovery of Chlorine from Hydrogen Chloride Solution

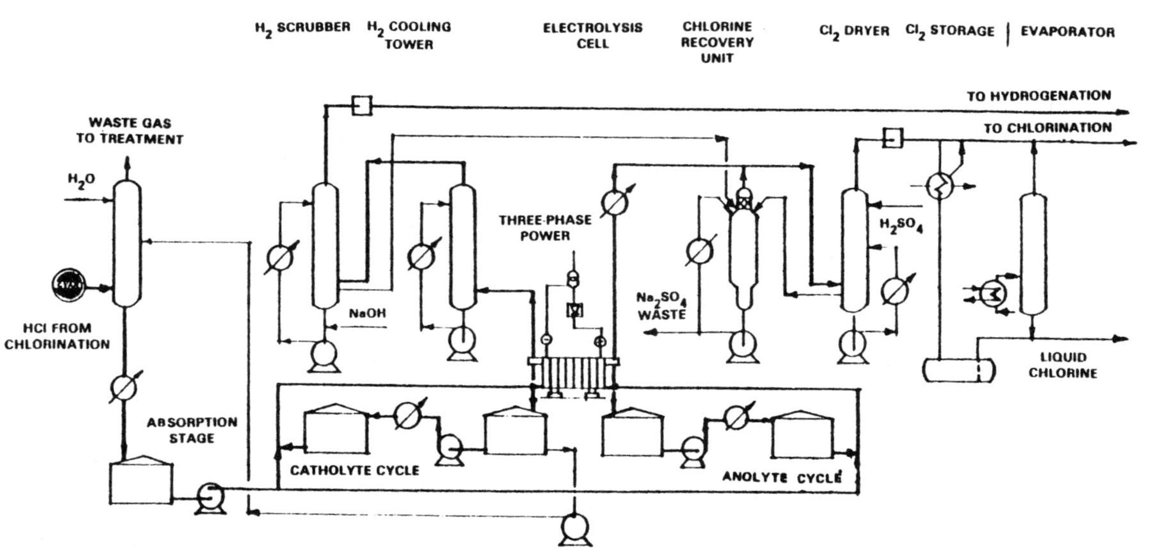

Source: ANL/OEPM-79-1

The electrodes used are vertical graphite plates. To prevent product mixing (i.e., hydrogen and chlorine), these plates are separated by polyvinyl chloride fabric diaphragms which are permeable to liquids but impermeable to gases. The anode, cathode, and diaphragm form a cell and 30 to 36 cells together form the electrolyzer. The end-plates of the electrolyzer act as the monopolar electrodes connected to the dc power supply. All other electrode plates are bipolar. The monopolar end-plates are steel lined with rubber.

A typical electrolyzer is shown in Figure 1.5 (29).

Figure 1.5: Cross Section of an Electrolyzer

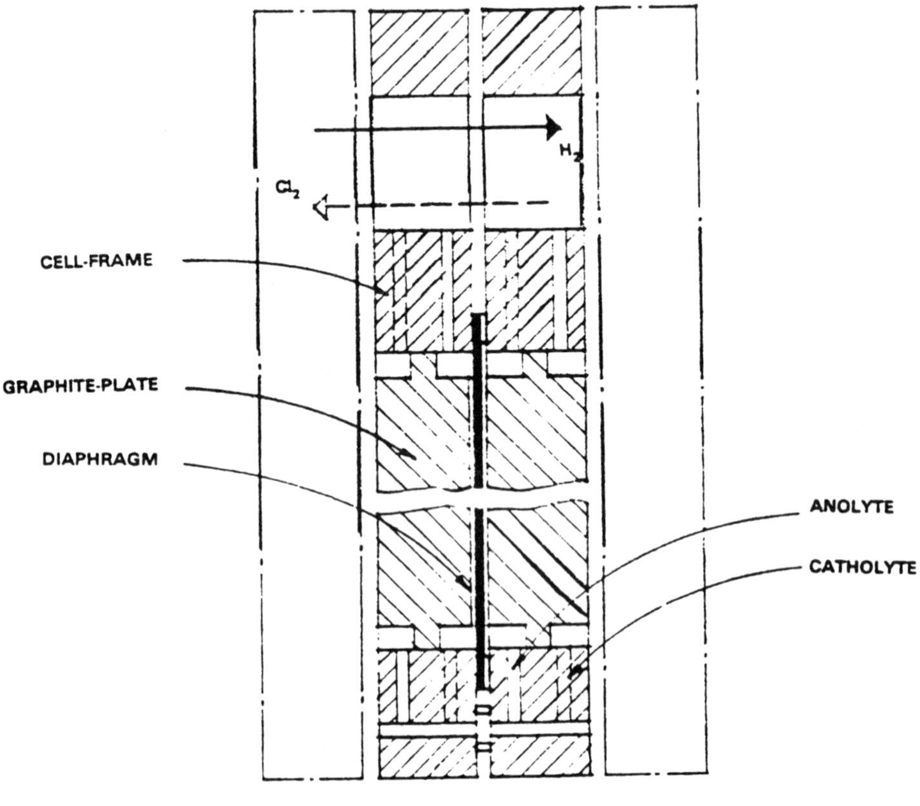

Source: ANL/OEPM-79-1

Inlet channels for fresh hydrogen chloride and individual cell inlet openings are located along the bottom half of the frame. The top of the frame is for product discharge and depleted acid removal.

Electrolysis produces chlorine and hydrogen gases at 75° to 80°C. Each is saturated with water vapor and hydrogen chloride gas at concentrations corresponding to their respective partial pressures. Both gases are cooled and scrubbed. The gaseous chlorine is then dried with sulfuric acid. As appropriate to requirements of the chlorination plant, the chlorine may be compressed and further purified.

The chlorine and hydrogen need no further treatment. Using new diaphragms, chlorine yields of 99.8% have been obtained. Due to deterioration of the diaphragms over time, average yields of 99.5% are expected with diaphragm replacement every 2 to 3 years.

Power consumption in any electrolytic process is of major importance. The maximum electrolyzer load is 12 kA. This is equivalent to a specific current density of 4.8 kA/m^2. With proper operating practices, the current efficiencies of cells can reach 98%. In 1965, commercial electrolyzers of hydrogen chloride used cell voltages at 2.5 V at a cell load of 10 kA. This corresponds to a dc power consumption of 1,720 kWh per kkg of chlorine. Under the same cell load, facilities operate at cell voltages of 2 V with power consumption of 1,405 kWh per kkg of chlorine. These power reductions are a result of improvements in diaphragms, narrower electrode gaps, and the addition of noble metal salts to the catholyte. The salts were added to reduce hydrogen overvoltage.

Besides using the metal salts to increase efficiency, other material criteria must also be met. The hydrogen chloride used should be dry gaseous material to minimize addition of water to the process. Waste hydrochloric acid may be converted to the gaseous form and reintroduced into the electrolyzer by absorption distillation with sulfuric acid.

Other impurities, particularly organics, will severely affect the process. These impurities can cause deposits to form on the electrodes and diaphragms, reducing product quality. Chlorine derivatives of cyclic compounds (benzene or phenols) containing hydrogen chloride have two effects on the electrolysis cycle:

(1) Diaphragm clogging; and

(2) Electrode deterioration due to crystal growth.

Both these effects increase cell voltage. This situation can be ameliorated through adiabatic absorption of the hydrogen chloride followed by adsorption of the chlorine derivatives by activated charcoal. Inorganic impurities, specifically multivalent ions, can reduce current efficiency. Also, metal ions that have a deposition potential more positive than that of hydrogen will plate on the electrodes and thus increase the hydrogen overvoltage.

Economics: Operational costs (1974) are given below for a hydrogen chloride electrolysis plant (29).

	Dollars/Ton of Cl_2
Raw Materials	
HCl	0 (waste product)
NaOH	0.85
Auxiliary materials (diaphragms, etc.)	0.26
Power and Utilities	
Electricity	12.24
Cooling Water	0.41

(continued)

	Dollars/Ton of Cl_2
Capital Expenditure	10.81
Maintenance	1.63
Personnel	1.53
Subtotal	27.72
Credits	
Hydrogenation hydrogen	7.41
Total Production Costs	20.30

The production costs of $20.30 per ton of chlorine is contrasted with the $26 to $27 per ton figure for chlorine by the Kel-Chlor process. Even if the hydrogen could only be used as fuel, the costs are comparable. Kel-Chlor is cost-effective for large facilities because of increased economies of scale.

The Bayer-Hoechst-Uhde process uses about 40% less electricity per ton of chlorine than other commercial electrolytic processes.

OXIDATION OF AMMONIUM CHLORIDE WITH NITROGEN DIOXIDE

Because of the complexity of nitrogen oxide chemistry and its temperature dependence, two processes, both developed by Imperial Chemical Industries, Ltd. (ICI), to convert ammonium chloride to chlorine are presented in this section.

The first process operates at ambient temperature and generates ammonium nitrate and chlorine as coproducts. From the overall process chemistry, it consumes oxygen and nitrogen dioxide in addition to ammonium chloride.

The second process operates at higher temperatures and uses more nitrogen dioxide in the initial oxidation. Nitric acid and chlorine are the final products.

ICI Ammonium Nitrate Process

History: This process was patented in the United Kingdom by Gordon Diprose in 1960 and assigned to Imperial Chemical Industries, Ltd. (31)(32). All work quoted in the patent was performed on the laboratory scale, and no development work or other commercialization has since been reported.

The process is similar to the Vicksburg and the U.S. Department of Agriculture processes which produce chlorine by reaction of nitric acid and/or nitrogen dioxide with potassium chloride. For the ICI process, ammonium chloride is used as the raw material in place of potassium chloride.

Scientific Basis: Solid ammonium chloride containing 1 to 5% water is reacted with nitrogen dioxide on a slowly moving bed under ambient conditions. With agitation, a 97.7% conversion of the ammonium chloride is claimed via the reaction: $NH_4Cl + 2NO_2 \rightarrow NH_4NO_3 + NOCl$. The nitrosyl chloride (NOCl) is then oxidized to nitrogen dioxide and chlorine by already established means. Although the separation of nitrogen dioxide from chlorine is not discussed in detail in the patent, it is known from other processes (i.e., the Vicksburg process) that these materials may be successfully separated by a fractional distillation operation. The separated nitrogen dioxide may then be returned to the initial process step.

Process kinetics have not been developed; however, thermodynamic calculations show that, at room temperature, the process is strongly exothermic, liberating 47.5 kcal/mol of nitrosyl chloride generated in the initial process step. The secondary reaction step (i.e., $2NOCl + O_2 \rightarrow 2NO_2 + Cl_2$) is also exothermic, liberating 7.0 kcal/mol. Process energy needs would be minimal, if use can be made of the considerable volumes of heat released by the initial reaction.

Engineering: The process of producing chlorine from ammonium chloride and nitrogen dioxide consists of the following operations and is shown in Figure 1.6.

(1) Reaction at room temperature of nitrogen dioxide with moist ammonium chloride in an agitated system to produce ammonium nitrate and nitrosyl chloride (NOCl).

(2) Recovery of the ammonium nitrate coproduct (NH_4NO_3) from the reactor and drying it to remove moisture.

(3) Oxidation of the nitrosyl chloride product with oxygen at temperatures below 200°C to yield nitrogen dioxide and chlorine.

(4) Separation of nitrogen dioxide and chlorine in liquid phase by fractional distillation at -20°C.

(5) Compression of the chlorine product and recycle of the nitrogen dioxide to the initial process step.

Figure 1.6: Ammonium Nitrate Process Flowsheet

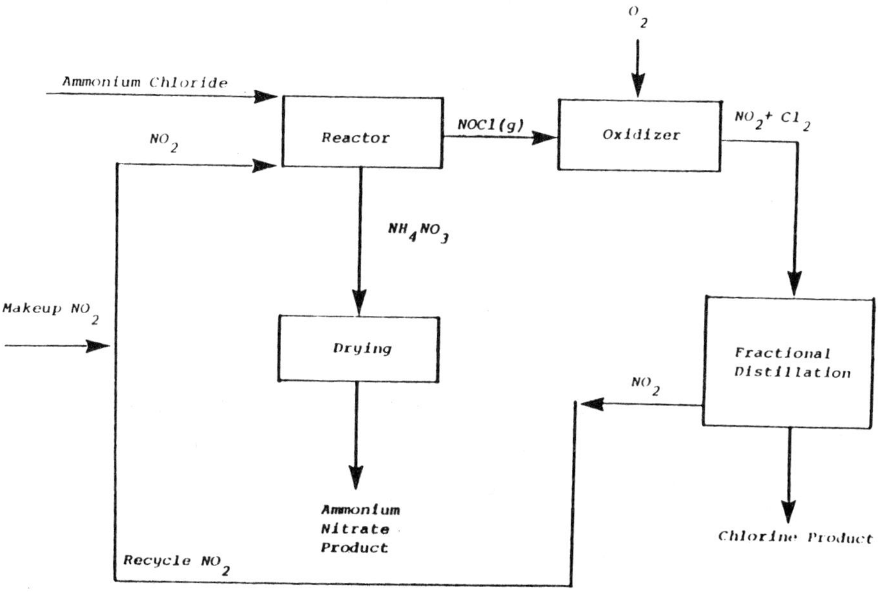

Source: ANL/OEPM-79-1

For this process, corrosion resistant materials such as titanium or stainless steel will be required for all system units in contact with either chlorine or nitrogen dioxide.

Economic Considerations: This process consumes ammonium chloride, nitrogen dioxide and oxygen and produces ammonium nitrate and chlorine. According to 1977 figures, U.S. consumption of ammonium nitrate is 8.3×10^6 tons per year; this product should be marketable in significant quantities. This process, although only developed to the laboratory stage, may be of some promise because of its extremely low energy requirements.

Nitrogen Dioxide/Ammonium Chloride High Temperature Process

Process History: This process was developed as a parallel process to the Vicksburg and Allied processes using ammonium chloride instead of potassium chloride as a raw material. Work in this area was performed by ICI on a laboratory scale in the late 1950's and culminated in the issuance of a number of British, Belgian, and German patents to ICI in the early 1960's (32)(33). No further work was performed on this process after the issuance of the patents.

Scientific Basis: Laboratory scale efforts demonstrate that ammonium chloride would react with gaseous nitrogen dioxide to yield nitrosyl chloride and nitrogen oxide (and water), and that the nitrogen dioxide-chlorine mixtures produced by oxidation of these products could be separated by fractionation. No other laboratory efforts were performed to establish a complete process.

The gaseous nitrogen dioxide reacts in a mobile bed of solid ammonium chloride to yield nitrosyl chloride, according to the reaction:

$$3NO_2 + NH_4Cl \rightarrow NOCl + 2H_2O + 3NO$$

The resulting gaseous products are further oxidized after removal of water, as follows:

$$2NO + O_2 \rightarrow 2NO_2$$

$$2NOCl + O_2 \rightarrow 2NO_2 + Cl_2$$

The chlorine and nitrogen dioxide are then separated and the nitrogen dioxide is recycled to the process. Kinetics of the initial reaction have not been reported. Thermodynamic calculations show the reaction is exothermic at room temperature by 20.2 kcal/mol.

Engineering: The Nitrogen Dioxide/Ammonium Chloride High Temperature Process flowsheet is shown in Figure 1.7. The process is carried out in three steps as follows:

(a) Solid moist ammonium chloride is introduced into a mobile bed reactor where it is contacted with gaseous nitrogen dioxide at about 100°C. The chemical reactions produce nitrogen oxide, nitrosyl chlorine, and water vapor, which exit the reactor.

(b) The gaseous mixture is then stripped with either concentrated sulfuric acid or other solvent to remove water vapor, and the dried gas stream is fed to an oxidizing chamber. The spent solvent, which contains the removed water, is reconcentrated by heating and is recycled.

(c) The dry gaseous products (NO + NOCl) are reacted with oxygen at about 200°C to convert these materials into a mixture of nitrogen dioxide and chlorine. This mixture is fractionated, most of the nitrogen dioxide is recycled, and the chlorine is compressed and sold. The additional nitrogen dioxide created by the process could be converted on-site to nitric acid for sale. For this process, special corrosion resistant equipment will be required.

Work on this process was continued only on a laboratory scale, and only certain process steps were studied (33). No pilot plant efforts were ever conducted. One unresolved question concerns the potential loss of nitrogen oxide or nitrosyl chloride to the sulfuric acid or other solvent used to remove water from the reaction product mixtures. Another question concerns loss of nitrogen dioxide values by conversion to nitrogen or nitrous oxide during ammonium chloride oxidation because of unwanted side reactions; the degree to which such reactions may occur is not known.

This process depends on the use of ammonium chloride as a raw material. However, only 30,000 tons per year of ammonium chloride are generated directly in the United States. As a result, this process cannot have widespread application and is limited to sites close to ammonium chloride production facilities.

This process would be ideal, however, for incorporation into a Solvay process soda ash plant. The Solvay plant produces ammonium chloride during the initial reaction step, which generates sodium bicarbonate and ammonium chloride. Instead of recovering ammonia, as in the normal Solvay process, the ammonium chloride could be converted to chlorine and nitrogen dioxide.

The nitrogen dioxide would then be further processed to nitric acid. While this process increases ammonia usage, sales of additional chlorine and nitric acid products would more than offset the lost ammonia sales revenue.

However, a number of questions involving nitrogen oxide losses and ammonium chloride purity requirements need to be resolved before this process can be fully assessed. The process is also less energy efficient than the low temperature process described in the previous section.

The Solvay soda ash process is in declining use in the United States. Only one plant remains which uses this technology, thus limiting supplies of cheap ammonium chloride as a raw material for this process. However, the capacity of this facility could provide 0.7×10^6 tons per year of ammonium chloride raw material which would generate about 400,000 tons per year of chlorine using this process. Therefore, about 5% of the nation's chlorine supply could be theoretically generated via this route with no reopening of closed Solvay plants or construction of new plants. Based on the above discussion, this process will not be studied further.

Figure 1.7: Process Flowsheet for the Nitrogen Dioxide/Ammonium Chloride Process

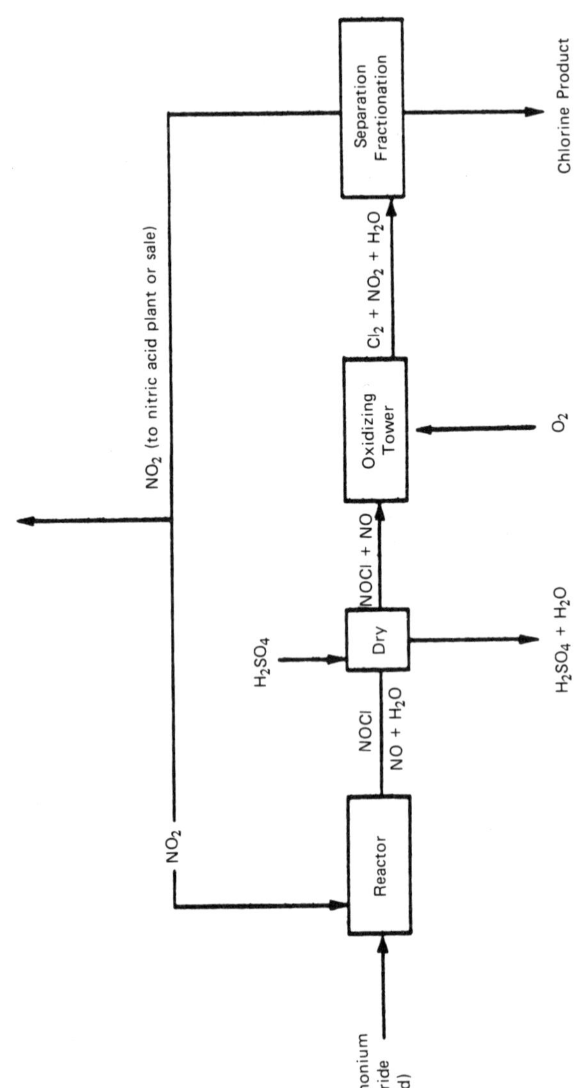

Source: ANL/OEPM-79-1

Comparison of the Two Processes

Energy Use: Intuitively, the process generating ammonium nitrate (ICI Process) will have lower energy requirements because (a) the initial reaction involves no heat input and, in fact, is a considerable generator of heat; (b) there is no need for removal of water from the product gas mixtures, and therefore, reconcentrating spent, dilute sulfuric acid is not necessary.

Raw Material Requirements: For raw material needs, the ammonium nitrate producing process is a net consumer of both oxygen and nitrogen dioxide. The higher temperature process, without side reactions, requires only oxygen. However, at higher temperatures, ammonium nitrate is known to decompose to nitrous oxide and water. Therefore, formation of nitrous oxide via side reactions is possible, which could produce corrosion problems.

Input ammonium chloride quality is less important for the ammonium nitrate process. Most ammonium chloride used will be generated via the Solvay process and will contain small amounts of sodium bicarbonate and sodium chloride. In the ammonium nitrate process, these are converted to sodium nitrate, which do not pose a problem for prospective fertilizer producers who would use the ammonium nitrate coproduct. For the higher temperature process, the presence of sodium salts results in a buildup of sodium nitrate solids in the initial reactor where only gaseous products are desired. This could lead to operational problems. Capital costs are expected to be lower for the ammonium nitrate process since a nitric acid plant is not needed.

PROCESSES WHICH APPEAR THUS FAR TO BE NONCOMPETITIVE

These processes, deemed noncompetitive by the Versar study, are included (1) to complete the overall picture, (2) because the patents reviewed in Chapter 8 indicate that improvements in some of the processes are still being made, and (3) because some have simply not yet progressed beyond the laboratory stage.

β-Alumina Membrane Process (Japanese Molten Salt Process)

The Japanese Molten Salt process is being developed under Professor S. Yoshizawa at Kyoto University (34). All work reported to date has been from laboratory scale experiments to establish process feasibility (34)(35).

In this process, chlorine and caustic soda are produced by electrolysis of molten salt. The reaction is similar to the electrolytic processes:

$$H_2O\ (g) + NaCl\ (NaCl\text{-}ZnCl_2) \rightarrow \tfrac{1}{2}H_2\ (g) + NaOH + \tfrac{1}{2}Cl_2\ (g)$$

A cell similar to the aforementioned cell process is used, except the membrane material is a β-alumina diaphragm. On the anode side of the diaphragm, a 50% sodium chloride/50% zinc chloride molten salt mixture is used; on the cathode side, molten caustic soda containing about 5% water is employed. During operation, chlorine is liberated at the anode and excess sodium ions migrate through the alumina diaphragm to the cathode compartment where hydrogen is liberated from the water present. The chlorine formed at the anode is already water-free and can be compressed and liquefied for use or shipment. The molten caustic from the cathode side is recirculated, with fresh water added and some caustic withdrawn in the molten state, cooled, and recovered as solid product.

This process has several advantages and disadvantages as listed below:

(1) The cell operating voltage (3.18 V) is lower than those reported for the membrane cell (3.5-4.0 V), and hence direct electrical energy requirements are lower. There is an additional energy penalty incurred in this process, however, because of the need for maintaining the cells and circulating molten salts at the 330°C operating temperatures. Part of this thermal energy requirement includes steam generation, since the water must be injected into the molten caustic as steam.

(2) Severe corrosion problems are anticipated with this system caused by both molten caustic and a molten sodium chloride-zinc chloride eutectic at 330°C. Use of these molten salts requires that the electrolytic cells and salt recirculation systems be constructed of special corrosion-resistant materials such as tantalum or titanium. These special materials increase the capital cost of the cells compared to conventional membrane or diaphragm cells.

(3) There is considerable evidence that the β-alumina diaphragms will not stand up under the proposed operating conditions. β-alumina is a sodium aluminate with the approximate formula of $Na_2O \cdot 11Al_2O_3$. This material, like other aluminates, is expected to exhibit some degree of solubility in molten caustic. Hence, membrane lifetimes may be short. Any contamination of the caustic coproduct due to β-alumina membrane dissolution or migration of zinc ions through the membrane can result in serious problems involving product caustic salability.

One further disadvantage of this process, as with the membrane cell, is the stringent raw material purity requirements which necessitate that the salt raw materials be prepurified. The industry standard method of brine purification is removal of impurities by chemical precipitation. If this method is used for the β-alumina membrane process, the purified brine will require evaporation to recover the solid salt used in the process. This entails additional energy requirements. Also, prepurification of the zinc chloride may be required.

It has been reported that the presence of calcium or potassium ions in molten salts in contact with β-alumina accelerates degradation of this proposed diaphragm material (35)(36). The sensitivity to such impurities implies that stringent salt purification methods will be required for this proposed process, which will add considerably to process capital and operating costs.

Since this process has not been demonstrated at the pilot plant scale, the overall importance of corrosive problems on process viability is indeterminate and only an estimate of capital or operating costs is possible. Materials of construction and corresponding operating parameters have yet to be specified. Most important, there is considerable doubt that the β-alumina diaphragm can withstand the operating conditions for a reasonable period of time. Further laboratory efforts in this area would be beneficial to determine whether a viable process can be developed. To become competitive, energy requirements for steam generation and salt purification would have to be considerably reduced.

Downs Cell

The Downs cell process for coproduction of chlorine and sodium metal was developed and patented by J.C. Downs of the E.I. Du Pont de Nemours Company in 1924 (37). The process was commercialized by Du Pont at their Niagara Falls, New York plant in the late 1920's. The process rapidly replaced the older Kastner electrolytic process for sodium manufacture, which involved electrolysis of molten caustic soda. By 1940, all Kastner process plants had been replaced. The Downs cell process has been continually operated by Du Pont since 1940 and there are now five plants producing sodium via this method. These are:

Company	Plant Location	Capacity (millions lb/yr)
Du Pont	Niagara Falls, NY	113
	Memphis, TN	44
Ethyl Corporation	Baton Rouge, LA	90
	Pasadena, TX	60
RMI Corporation	Ashtabula, OH	74
Total		381

The Downs cell process supplies all sodium produced in the United States (37).

In the Downs cell process, prepurified sodium chloride is mixed in a 42/58 ratio with dry calcium chloride ($CaCl_2$). This mixture is then introduced into a molten salt diaphragm-type cell where the temperature is raised 580°C to liquefy the salt eutectic mixture. The molten salt is then electrolyzed as follows:

$$2NaCl\ (melt) \longrightarrow 2Na + Cl_2\uparrow$$

The chlorine is formed at the graphite anodes, collected above the melt, purified, compressed and readied for sale. The sodium coproduct forms at a pipe-type electrode of special design. The molten sodium exits the cell through the pipe-type electrode and is then partially cooled to precipitate impurities, especially the sodium-calcium alloy. This material is then removed from the molten sodium by filtration at 110°C. The sodium is then cooled, solidified and readied for sale. The sodium-calcium sludges removed from the product sodium are converted to the corresponding hydroxides for plant use or disposal.

Although the Downs cell process has been operated on a commercial scale for over 40 years, this process is limited in its ability to generate chlorine by the demand for the coproduct sodium. All United States requirements for this metal are met by the five operating plants. In addition, one major use for sodium, production of tetraethyl lead, is expected to decline sharply as lead additives are removed from gasoline. As a result, the United States market for sodium is expected to decrease in the future, and Downs cell capacity in the United States is unlikely to expand in the near future.

There is another commercial molten salt electrolysis process which produces magnesium and chlorine. This process uses a magnesium chloride feed produced on-site from magnesium hydroxide recovered from seawater and waste hydrochloric acid. The process is used by Dow Chemical Company at one site where sufficient quantities of all raw materials are available. No other chlorine plants have sufficient quantities of hydrochloric acid or magnesium hydroxide to operate this process.

Kellogg Modified Deacon Process

This process for chlorine production using clay materials was developed and expanded by the M.W. Kellogg Company. The company applied for an English patent on August 23, 1960 and the patent was granted on January 15, 1964 (38). Overall, this process involves the continuous oxidation of hydrogen chloride or other halogen acids to produce chlorine, bromine or iodine in a higher yield and in a higher degree of purity than was normally obtained in the late 1950's. Work was only performed on the laboratory scale. Once the patent was obtained, further development ceased. The project is inactive.

The M.W. Kellogg Company process produces chlorine by oxidizing an inorganic halide such as hydrogen chloride in the presence of a metal- or silica-containing oxidation catalyst and a dissimilar material containing a clay desiccant. The clay desiccant must have a reversible water content of at least 0.5% by weight and a crystalline structure stable to at least 1,400°F.

The possible sources of inorganic halides include the following: hydrogen chloride, hydrogen bromide, hydrogen iodide, nitrosyl chloride, nitrosyl bromide, nitrosyl iodide, ammonium chloride and ammonium bromide. The chlorides and bromides are preferred.

The primary oxidizing agents used for these halides include molecular oxygen, air, ozone, nitric acid and nitrogen oxides. Catalysts that can be used for this reaction include: chromium sesquioxide (Cr_2O_3), elemental chromium, potassium chromate (K_2CrO_4), cerium oxide (CeO_2), manganese dioxide (MnO_2), uranium trioxide (UO_3), cupric chloride ($CuCl_2$), a ferric oxide/ferric chloride mixture ($Fe_2O_3/FeCl_3$), a chromium sesquioxide/alumina mixture (Cr_2O_3/Al_2O_3), calcium chloride ($CaCl_2$), and silica gel. For the oxidation of hydrogen chloride with molecular oxygen, chromium sesquioxide is preferred.

The basic conversion reaction for the process is the same as that for other oxidation processes using hydrogen chloride: $4HCl + O_2 \rightarrow 2Cl_2\ (g) + 2H_2O$.

The clay desiccant used in the process absorbs water as it is formed in the reaction zone. This shifts the equilibrium toward the right, increasing yield and purity of the halogen product. The clays are preferably bentonitic in nature. Montmorillonite is the most desirable of these types. They are regenerated after absorbing about 1% by weight of water.

Although production conversions were high (average 93%), three major disadvantages exist:

(1) Low availability of large quantities of hydrogen chloride;

(2) High process temperatures required, i.e., high energy and power costs; and

(3) More energy efficient processes exist using the same basic

raw materials for making chlorine, namely Kel-Chlor, Mobay, and the Shell process. In fact, Kel-Chlor was developed by the same company which invented this process and M.W. Kellogg chose Kel-Chlor for commercial development.

These factors indicate that the Kellogg halide oxidation process for making chlorine is not a viable alternative to existing production methods.

Catalytic Oxidation of Hydrogen Chloride Using Metal Oxide Catalysts

Preparation of Chlorine Using Hydrogen Chloride with a Vanadium Catalyst: This process of chlorine formation results from the oxidation of hydrogen chloride in the presence of a catalyst similar to the Shell process. The original process was developed by Imperial Chemical Industries (ICI) in 1967 (39). Their patent made claims of 63% conversion. It used vanadium pentoxide (V_2O_3) as a catalyst. Also in 1967, BASF patented an improvement to this process (40). Their improvement dealt with the addition of several other catalytic materials (CeO_2, MnO_2, CoO, NiO, etc.) to the vanadium. Once patented on a bench scale, BASF abandoned the project. The status of the process, including improvements, is inactive.

The process operates by thermal, catalytic decomposition of hydrogen chloride. The reaction mechanism for the ICI process is:

$$4HCl\ (g) + O_2 \xrightarrow[heat]{V_2O_5} 2H_2O + 2Cl_2$$

The final chlorine product is obtained by drying the liberated water/chlorine mixture with sulfuric acid. The dilute sulfuric acid is reconcentrated and recycled.

In the BASF method, the catalyst material is specially prepared. Vanadium is added to a melt of potassium pyrosulfate ($K_2S_2O_7$) and sodium bisulfate ($NaHSO_4 \cdot H_2O$) and heated to 375°C for one hour. The product catalyst and catalyst activators then are pulverized and an inert support (silica gel) added. The catalytic bed is formed from this heated (350°C) mixture. Once the bed is formed, the hydrogen chloride/oxygen mixture is in a ratio of 4/1 and at a temperature of 380°C. The hourly output of product gas contains the following percentages of materials: 22.2% chlorine; 44.5% hydrogen chloride; 22.2% oxygen; and 11.1% water. Yields can be increased by using temperatures greater than 380°C.

For the overall reaction of this process, the free energy change was equal to 12.67 kcal/mol Cl_2 at 25°C. This makes the reaction endothermic. The overall operation is similar to the Shell process except that a different catalyst and slightly different operating temperatures are used.

This process is infeasible for commercial chlorine preparation for several reasons:

(1) Other catalytic processes (e.g., Kel-Chlor) already exist for more efficient conversion of hydrogen chloride to chlorine.

(2) Current cost of vanadium ($1.00 per lb) means even small catalyst losses would impose a severe operating cost penalty on the process.

(3) Yields obtained are relatively low, and the product exits the reaction as a dilute constituent of a gas stream. In general, this dilute gas stream will require pressurization to recover the chlorine by liquefaction followed by fractional distillation and drying.

A process yielding a more concentrated product gas stream would be preferred. The process thermodynamics are the same as that discussed for other gas phase hydrogen chloride oxidation processes.

Hydrogen Chloride Oxidation over Chloride/Cerium Oxide Catalysts: This process was investigated on a laboratory scale by personnel of the Farbwerke Hoechst AG and patented by that organization in several West European countries. The data presented here are abstracted from Belgian Patent 658,435 assigned to that organization (41). The patent was granted in 1965, and no other work concerning this process has been reported.

The process involves the oxidation of gaseous hydrogen chloride with oxygen over catalysts of special composition. The catalysts are supported mixtures of cerium dioxide (CeO_2) with either iridium trichloride ($IrCl_3$) or iridium tetrachloride ($IrCl_4$). The ratio of cerium dioxide to iridium salts is specified as being between 10 and 1,000:1. The support material was not specified in the patent. The oxidation reaction is:

$$4HCl + O_2 \xrightarrow[\text{Catalyst}]{IrCl_3\text{-}CeO_2} 2Cl_2 + 2H_2O$$

and is conducted in the low (30° to 250°C) temperature range. Yields vary with catalyst composition, temperature, and retention time. At temperatures below 50°C, yields of 21 to 38% were reported for a 20-minute retention time with low iridium content catalysts. For higher iridium content, yields of up to 83% were found with the same retention times and temperature.

At 70°C, the same low iridium content catalysts gave 71 to 96% yields with 200-minute retention times. Above 150°C yields greater than 80% were reported for low iridium content catalysts at 15-minute retention times. No data relating either to the catalytic reaction or to recovering chlorine from the product gas stream were reported. It can be assumed, however, that recovery processes would be similar to other hydrogen chloride catalytic oxidation processes.

The data presented in the patent were obtained from a bench-scale evaluation of potential catalysts. A full process was never studied. However, the overall process should be similar to others which use catalytic oxidation of hydrogen chloride. A complete description of catalytic oxidation of hydrogen chloride processes is given for the Shell process. What is of particular interest in this process is that the oxidation reaction in the presence of cerium dioxide and iridium trichloride apparently proceeds at a significant rate at temperatures below 250°C. Other gas phase catalytic processes all require temperatures well above 300°C to obtain yields greater than 70%.

Although only laboratory work has been done with this process, the data available indicate that the use of cerium-iridium catalysts might be developed into a viable process for hydrogen chloride oxidation. There already exist two other processes (an electrochemical process developed by Uhde, and the Kel-Chlor process) which are in commercial operation and accomplish the same end of converting waste hydrogen chloride gas to chlorine. The amount of by-product hydrogen chloride available for conversion to chlorine is 600,000 tons per year, and more than half of this material is converted using the other two processes. It would seem that insufficient raw material is available to develop another process in competition with the Uhde and Kel-Chlor processes.

Electrolysis of Hydrochloric Acid with Metal Chloride Catalysts

Three processes have been developed that electrolytically convert hydrochloric acid to chlorine using metal chloride catalysts. The three processes are the Schroeder process (using $NiCl_2$), the Westvaco process (using $CuCl_2$), and the South African process (using $MnCl_2$)(42)-(46). In each case complications prevented development of the process past the pilot plant stage. Since all three processes have similar scientific bases and engineering methods and problems, the Schroeder process is presented in detail as representative of the three.

The Schroeder process was developed by David W. Schroeder of Seattle University. His first patent was filed on August 11, 1958, but he abandoned his efforts before any patent was granted. The existing patent, which was filed on June 13, 1960, was granted on January 28, 1964, and is a continuation of the original process (42). The process (as well as the Westvaco and South African processes) is inactive.

The Schroeder process is cyclic and involves two main steps. The first step is the decomposition by electrolysis of a metal chloride (preferably nickel chloride) into the metal component and chlorine. The second step combines the metal element with hydrochloric acid to reform the metal chloride for recycling. The equations are:

(1) $$NiCl_2 \xrightarrow{\text{electricity}} Ni + Cl_2$$

(2) $$Ni + 2HCl \longrightarrow NiCl_2 + H_2$$

(Overall) $$2HCl \longrightarrow H_2 + Cl_2$$

The reactions are carried out by using two sets of cells. In the first cell, electrolytic deposition occurs, with nickel metal plating out at the cathode and chlorine gas evolving at the anode. In the second cell, the nickel is dissolved in hydrochloric acid with the resultant metal chloride fed back into cell number 1. The use of current in the second cell is optional. A reaction without the use of current proceeds slowly unless nickel chloride or other metal chloride salts are present in the hydrochloric acid solution. By increasing the chloride concentrations, the rate of reaction becomes commercially practical.

Chlorine from Oxidation of Ferric Chloride Salt Complex

Du Pont Process: Of the chlorine production processes that use ferric chloride ($FeCl_3$) for the raw material, the Du Pont process is considered the most efficient and economic.

The principal inventors of this method were Wendell Dunn, Jr. and John Maurer, both assignors to Du Pont Co. of Delaware. The patent for their process was filed on August 3, 1965 and was granted on April 2, 1968.

Prior to this method the primary process considered for iron chloride oxidation was: $3O_2 + 4FeCl_3 \rightarrow 6Cl_2 + 2Fe_2O_3$. This process, however, did not perform satisfactorily in pilot plant operation (47). Alternatives using high temperatures for adequate reaction rates, low temperatures with a salt melt, and high pressures to cause ferric chloride liquefaction all proved inadequate. Each alternative required the operation of complicated equipment and careful control over operating conditions during the process.

The Du Pont patent sought to remedy these problems. One of the main objectives of this patent was to provide improved operation during the rate-limiting step of the ferric chloride oxidation. This process described continuous operation of a commercial plant under a wide latitude of controlled uniform conditions, which would ensure high recovery/increased yields of chlorine and a high degree of purity. However, Du Pont was not able to meet these goals. Development of this process went no farther than pilot plant stage. Current status of the operation is inactive.

The basic step in this method is to flow a molten salt complex of the formula: $XFeCl_4$, where X is an alkali metal, as a thin film over a moving bed of inert particulate material. This molten salt flows concurrently with an oxygen-containing gas.

The salt complex and the oxygen react under the process conditions as follows: $3O_2 + 4NaFeCl_4 \rightarrow 2Fe_2O_3 + 4NaCl + 6Cl_2$. Product gas evolves while the ferric oxide and sodium chloride-coated particulate beds are withdrawn, treated, and recycled.

The major reactant to the process, $NaFeCl_4$, is normally in a liquid state as it enters the reactor. However, a system exists for sodium chloride feed as a solid and ferric chloride as a solid, liquid, or gas. The introduction of chemicals is usually accomplished through the flow stream of the oxygen-containing gas.

Although sodium is the preferred alkali metal to complex with ferric chloride, other alkali metals (potassium, lithium, etc.) may be used as substitutes. These alternates would react in a general equation as follows:

$$3O_2 + 4XFeCl_4 \longrightarrow 2Fe_2O_3 + 4XCl + 6Cl_2$$

where X is the substitute alkali metal. The particular state and concentration of the ferric chloride used is not a critical factor in the preparation of the salt complex. The ferric chloride can be in either a liquid (preferred) or solid state. If solid material is used, the reactor must provide sufficient heat to vaporize the ferric chloride. A liquid final complex reactant is preferred because of its relatively low vapor pressure.

The process is carried out at a temperature range of 500° to 650°C. The main reaction should occur in controlled anhydrous conditions. Within the reactor, pressure levels should be slightly above atmospheric to prevent gas leakage into the bed. Exit pressures may range anywhere from 0.5 to 7 atmospheres. The reactor moving bed rate is dependent on the recycle ratio. Rates from 1.0 to 100 feet per hour are used. This moving bed can operate either concurrently or countercurrently to the reactant flow. However, the oxidizing gas flow must remain concurrent with the reactants. In this particular patent, the optimum gas flow rates defined as the number of pounds of gas fed to the reactor per hour per square foot of reactor cross-sectional area, are from 75 to 325 $lb/hr/ft^2$.

The yields from the experimental runs of this patent were excellent with about 92.5% chlorine recovery. Reaction rates ranged from 0.94 lb/min to 3 lb/min, depending on the relative proportions of the reactant materials involved.

Factors to be considered when deciding on the feasibility of the procedure for commercial chlorine production include the availability of ferric chloride and the high energy cost to sustain the necessary process temperatures. As mentioned previously, ferric chloride is produced at only a small number of plants in the United States. Total ferric chloride production in 1976, if applied to chlorine formation, would amount to about 2% of U.S. chlorine production. Also, high fuel costs pose economic disadvantages for this method of chlorine preparation. On a small scale basis, this process could be feasible for the recovery of chlorine from certain waste ferric chloride. Such an operation would be performed on a localized scale with the probable user of the produced chlorine or source of ferric chloride nearby.

Cyanamid Chlorine Process ($FeCl_3$ 2-step Oxidation): This chlorine preparation process was developed by American Cyanamid Company engineers, John Wikswo and Earl Nelson (48). They applied for a patent on February 26, 1964, and were granted one on June 13, 1967. This particular patent is a continuation of their earlier patent work. Cyanamid has since dropped work on all the patents.

The process involves the production of iron oxide and chlorine by combusting ferric chloride with oxygen. It describes a method to increase the efficiency of the ferric chloride oxidation process in using two combustion zones to generate chlorine. The overall chemical reaction in the two zones is:

$$4FeCl_3 + 3O_2 \longrightarrow 2Fe_2O_3 + 6Cl_2$$

At 25°C, this reaction is endothermic with a change in free energy of 5.3 kcal/mol Cl_2. The yield of chlorine is about one ton of chlorine for every three tons of ferric chloride, or a 50% yield from theoretical.

In the first reaction vessel, the oxygen and ferric chloride react at a temperature between 650° and 1000°C. The product in this first vessel is an equilibrium mixture consisting of iron oxide, chlorine, unreacted ferric chloride and oxygen. The time spent in the first vessel must be sufficient to attain equilibrium. The preferred operating temperature in the first vessel is 700° to 900°C. In the second operating vessel, the temperature is reduced. The reduction must be at least 25°C, but preferably between 50° and 100°C. The temperature drop must be controlled, however, so that conditions in the second vessel do not go below 450°C. This is done so that at least 75% equilibrium is attained. Once the mixture is transferred to the second vessel, it undergoes another equilibrium change at the reduced temperature. The new equilibrium is richer in both iron oxides and chlorine. Less unreacted ferric chloride and oxygen appear in the product of this second combustion zone.

This process was never commercialized due to its high energy requirements and operational problems.

Metaizeau Process

This process was developed by Paul Metaizeau of Belgium in the mid-1960's as a method for recovering chlorine from waste ammonium chloride produced by the Solvay process. Patents were issued for the process and after a few modifications, all were assigned to Solvay et Cie. of Belgium (49)-(52). Following the issuance of the patents, no further work was reported on the process by the Solvay Company although efforts in this area have since been reported by others (53)(54).

The reaction for this process may be written as:

(1) $Fe_2O_3 + 6HCl \longrightarrow 2FeCl_3 + 3H_2O$

(2) $2FeCl_3 + H_2 \longrightarrow 2FeCl_2 + 2HCl$

(3) $4FeCl_2 + 3O_2 \longrightarrow 2Fe_2O_3 + 4Cl_2$

(4) $4HCl + O_2 \longrightarrow 2H_2O + 2Cl_2$

Ammonium chloride is sublimed and dissociated to yield a gaseous mixture of ammonia and hydrogen chloride. This mixture is passed over a bed of supported iron oxide catalyst. The hydrogen chloride reacts with the iron oxide to yield iron chlorides and the separated ammonia can then be recovered for reuse in the Solvay process. After recovery of the ammonia, the iron chlorides formed are reduced with hydrogen to the ferrous state and the hydrogen chloride and ferrous chloride are reacted with oxygen to yield ferric oxide, chlorine and water. The ferric oxide is recycled to the initial process step and the water-chlorine mixture is dried with sulfuric acid to prepare the chlorine product.

The original process is as described above. However, there were several drawbacks which led to a process modification. In the improved version, the reactant mixture circulates through four vessels by means of a moving bed. The first reaction step in the modified process involves reduction of the supported catalyst to ferrous oxide (FeO) by means of hydrogen. This initial reduction is accomplished with the gases and solids moving countercurrently in the first vessel. In the second vessel, into which the reduced catalyst then flows, ammonium chloride vapors are introduced and these react with the reduced catalyst to yield ferrous chloride and free ammonia. The ammonia is received from the second chamber and the solids are then fed to the third and fourth chambers, where the ferrous chloride is oxidized in two steps to chlorine and ferric oxide. The ferric oxide is then recirculated to the initial process step.

Other patents granted for this process involved regeneration of the ferric oxide catalyst, process equipment design modifications and the choice of catalyst support materials (53)(54).

Operating temperatures for the process steps are:

(1) 400° to 500°C for the initial ferric oxide reduction step;

(2) 350° to 480°C for the ammonium chloride reaction step; and

(3) 550°C for the conversion of ferrous chloride to chlorine and ferric oxide.

All reactions are conducted at near atmospheric pressure.

The overall economics of this scheme are unfavorable because:

(1) The process, with its high operating temperatures, requires a considerable energy input. Hydrogen is also required.

(2) The conversion efficiencies for ferrous chloride to ferric oxide are only about 95% and there may be a buildup of unconverted ferrous chloride in the catalytic bed with time.

(3) The process may suffer from operational problems because it is operated at temperatures above the melting point of ferrous chloride (306°C). At higher temperatures small amounts of liquid ferrous chloride and ferric chloride may become entrained with process gas streams.

For the above reasons, the process was apparently abandoned.

Chlorine from Salt and Sulfur Trioxide

This chlorine generation process developed from the commercial production of chlorosulfonic acid ($HOSO_2Cl$), chlorosulfonate salts (e.g., $NaSO_3Cl$), thionyl chloride ($SOCl_2$) and sulfuryl chloride (SO_2Cl_2).

Research efforts can be traced back to 1940, when a patent was issued to Tauch and Iler for a new method of producing sodium chlorosulfonate (55). Other patents issued to Iler of Du Pont dealt with further developments in production of other chlorosulfonate compounds (56). Other patents in this area were issued for other organizations, including American Cyanamid Corporation (57)(58).

In 1942 a patent was issued to A.H. Maude of Hooker Chemical for separation of sulfur dioxide from chlorine (59). Similar patents were received by Hexon and Miller; McAdam, Farrell and Eichelburger; and Bouchard and Iler. During World War II a process evolved to produce chlorine using chlorosulfonate salts as follows:

(1) Preparation of sodium chlorosulfonate from sulfur trioxide and sodium chloride;

(2) Thermal decomposition of this material to sodium sulfate (Na_2SO_4), sulfur dioxide and chlorine; and

(3) Separation of the sulfur dioxide and chlorine with removal of the chlorine as product and oxidation of the separated sulfur dioxide to sulfur trioxide for recycle in the process.

Work continued on this process until 1950 when other processes for chlorine generation became more attractive. The process never developed past the laboratory stage. However, a number of commercial operations evolved from these efforts including:

(1) The manufacture of chlorosulfonic acid and chlorosulfonates;

(2) The use of carbon tetrachloride and sulfur monochloride (S_2Cl_2) for separation of chlorine from sulfur-bearing gases; and

(3) The manufacture of sulfuryl chloride.

The first stage of this chlorine production process is the absorption of sulfur trioxide by salt at relatively low temperatures to form sodium chlorosulfate as shown in Equation (1).

(1) $NaCl + SO_3 \longrightarrow NaSO_3Cl$

Above 225°C sodium chlorosulfonate begins to decompose into pyrosulfate, sodium chloride, sulfur dioxide and chlorine (Equation 2):

(2) $3NaSO_3Cl \xrightarrow{(225°C)} Na_2S_2O_7 + SO_2 + Cl_2 + NaCl$

Reaction kinetics results show that this process is first order with respect to temperature. Laury suggests that thionyl chloride will be produced by the decomposition in a competing reaction as follows (58):

(3) $4NaSO_3Cl \longrightarrow Na_2SO_4 + Na_2S_2O_7 + SOCl_2 + Cl_2$

Furthermore, in the presence of oxygen the thionyl chloride may be oxidized to sulfur dioxide and chlorine:

(4) $2SOCl_2 + O_2 \longrightarrow 2SO_2 + 2Cl_2$

Other side reactions are also possible. One product formed could be pyrosulfuryl chloride. This will occur when the sodium chlorosulfonate is heated in the presence of excess SO_3 (Equation 5):

(5) $2NaSO_3Cl + SO_3 \longrightarrow Na_2SO_4 + S_2O_5Cl_2$

Above 250°C the pyrosulfuryl chloride will decompose into sulfur trioxide, sulfure dioxide and chlorine. Complications in production can arise from these side products.

The main reactions following from the sodium chlorosulfonate decomposition step are:

(6) $2Na_2S_2O_7 + 2NaCl \longrightarrow 3Na_2SO_4 + SO_2\ (g) + Cl_2\ (g)$

(7) $2NaCl + 2SO_3\ (g) \longrightarrow Na_2SO_4 + SO_2\ (g) + Cl_2\ (g)$

(8) $2NaSO_3Cl \longrightarrow Na_2SO_4 + SO_2\ (g) + Cl_2\ (g)$

The adsorption reaction of salt and sulfur trioxide (Equation 7) and the overall yield reaction that produces sulfate and the chlorine/sulfur dioxide gaseous mixture are both exothermic. The intermediate decomposition reactions are endothermic, however. These two facts could be important since they indicate the possibility of technical problems with energy requirements and heat removal during the production process. The high temperature environment (>300°C) is responsible for the exothermic reaction and the energy balance becomes more favorable as the temperature increases. Conversely, at ambient temperatures the overall process is endothermic.

There are two routes, a single-stage and a two-stage process, for obtaining chlorine. The single-stage process is carried out at high temperature and requires that pure sulfur trioxide be used unless the gas product of the reaction will be diluted by oxygen and nitrogen.

The two-stage process isolates the intermediate product sodium chlorosulfonate and uses sulfur trioxide in a contact gas form.

For both processes, there are four main steps as follows:

(1) Manufacture of sulfur trioxide from sulfur or from recovered sulfur dioxide;

(2) Manufacture of sodium chlorosulfonate from salt and sulfur trioxide;

(3) Decomposition of the sodium chlorosulfonate into saltcake and the gaseous products, chlorine and sulfur dioxide; and

(4) Separation of the gaseous chlorine-sulfur dioxide mixture and production of liquid chlorine.

The two-stage process has several drawbacks compared to the single-stage process. First, technical difficulties are reduced with the single-stage because there is no intermediate product separation. Secondly, since these intermediate decomposition reactions are endothermic, heat must be contained within the walls of the reaction chamber. Construction material limitations hamper this process. For these reasons, the single-stage process is more attractive and is discussed further.

The primary equipment used in the single-stage system is:

(1) A unit for the production of oleum which uses either sulfur or pyrites as raw materials;

(2) An oleum desorber made up of Reymersholms sulfur trioxide stills;

(3) Grinders and preheaters for salt preparation;

(4) A series of reactors, such as fluidized bed or rotating drums; and

(5) A means of separating chlorine and sulfur dioxide and producing liquid chlorine.

Maximum conversion efficiency for this process is reached when the salt particles are suspended in the gas. Caking of the reactive materials occurs, however, which decreases the effective contact between gas and salt. High temperatures in excess of 500°C are necessary to prevent caking which is caused by the 300°C adiabatic temperature rise of pure sulfur trioxide. Since the chlorine/sulfur dioxide mixture is extremely active at these high temperatures, reactor walls should be constructed of a highly corrosion-resistant material.

A main study area for this process was the separation of chlorine from the chlorine/sulfur dioxide mixture. Two separation techniques were considered—selective absorption in a suitable solvent and formation of sulfuryl chloride (SO_2Cl_2) from the azeotrope. Selective oxidation was suggested by one researcher but was not actively pursued (60).

Selective absorption uses a suitable chlorinated solvent, such as sulfur monochloride or carbon tetrachloride, to separate the mixture. Study results showed that carbon tetrachloride is the preferred solvent (61).

For the second separation procedure, the chlorine/sulfur dioxide mixture is fractionated to give sulfur dioxide and the azeotrope which is further converted to sulfuryl chloride. Some chlorine is separated in the fractionation process itself. The sulfuryl chloride is then converted to sulfur dioxide and chlorine and the separation cycle is repeated. This process encountered corrosion problems in the fluidized bed reactors during laboratory tests. Discussions with the original researchers revealed that this problem was solved satisfactorily. An economic and environmental problem is the disposal of two tons of salt cake (sodium sulfate contaminated with about 10% salt) and one ton of sulfur dioxide for every ton of chlorine made.

Another process for separating the chlorine/gas mixture was developed after World War II in Oppau, Germany, by I.G. Farben (60). Chlorine was obtained by the oxidation of hydrogen chloride. An important aspect of this process was the mechanism for chlorine gas separation. The dry gas mixture produced in the plant was scrubbed at 8 atm of pressure with sulfur monochloride. Chlorine was absorbed and the oxygen was recycled for use in oxidation. The chlorine product was desorbed under pressure and liquefied.

When determining the economic feasibility of this process, one must consider the production costs versus revenue generated by the sale of coproducts. In the salt-sulfur trioxide process, chlorine, sulfur dioxide and sodium sulfate are potentially marketable. Sulfur dioxide is used in the sulfite pulp industry and sodium sulfate is a raw material for Kraft paper and glass production. It is probable, however, that the sodium sulfate will not be salable because it will contain salt up to 10% by weight.

If chlorine production by this process were more than 700,000 tons per year, the amount of sodium sulfate produced would exceed total U.S. demand. Disposal of excess sodium sulfate would be necessary and might cause environmental as well as economic problems.

Discussions with Du Pont and the original research personnel revealed that this process was abandoned in the late 1940's because of the inability to separate sulfur dioxide and chlorine gases completely and the disposal problems associated with impure sodium sulfate (62). Difficulty in carrying out certain process steps on a large scale, uncertainties about equipment function, corrosion, and operational methodology are further reasons for questioning the efficacy of this process. The relative magnitude of these problems is unknown because no full-scale operations of this type have ever been undertaken by which data could be obtained and studied.

Chlorine from Hydrogen Chloride and Sulfur Trioxide

Holme and Hooker et al, Process: Sulfur trioxide, when used in an oxidation reaction with hydrogen chloride, produces chlorosulfonic acid (HSO_3Cl) from which a final chlorine product can be retrieved.

Work on this decomposition reaction using mercuric salts as catalysts was done by Ruff in the early 1900's (63). He conceptually designed some commercial applications for the manufacture of sulfuryl chloride (SO_2Cl_2). Decomposition of the acid without using a catalyst was performed by Mellor, but his work was not carried through extensively. Sanger and Reige also worked on the noncatalytic decomposition reaction to obtain a sulfuryl chloride product. The kinetics of this particular reaction have been studied in depth by Payne, Pearce, and Fernelius (64).

A process for using higher temperatures to prevent pyrosulfuryl chloride formation was also set forth by Farbenind (65). An expansion of these pyrosulfuryl control methods was later set forth by Edwards in the early 1940's (66). In the 1940's Kallal concentrated once again on the use of mercuric salts as catalysts (67). Kallal worked extensively on decomposition rates of chlorosulfonic acid using varying catalyst concentrations.

Once the basic processes have been established, the major effort was to improve process efficiency, especially chlorine separation procedures. Johnstone in 1942 did important studies in this area along with Bouchard who focused on absorption processes for separation (68)(69). Carbon tetrachloride was the preferred absorber. These absorption processes and ongoing improvements were continually patented until the late 1950's. Two of the more prominent patents were by Holme in 1956 and Hooker, Geeroog, and Maude (70).

All the research work was laboratory scale. No extensive test facilities were developed. Further work on all the patents received has been abandoned.

The initial step of this process is the formation of chlorosulfonic acid. This is accomplished as follows:

$$HCl\ (g) + SO_3\ (g) \xrightarrow[70-100^\circ C]{} HSO_3Cl\ (l)$$

Commercially chlorosulfonic acid is produced from contact plant gas and dry hydrogen chloride. The chlorosulfonic acid product is then decomposed by using mercuric chloride with the whole system under pressure. This decomposition reaction is:

$$2HSO_3Cl \xrightarrow[HgCl_2]{160^\circ C} H_2SO_4\ (l) + SO_2Cl_2\ (g)$$

With pressure maintained between 3 and 11 atm the reaction proceeds at a continuous and faster rate. The change in free energy at the temperature of reaction is $\Delta F = 6190 - 16.27(T)$. At 110°C the reaction is spontaneous. Note that the sulfuric acid produced is sufficiently pure for sale as a by-product.

The next step involves the decomposition of the sulfuryl chloride to sulfur dioxide and chlorine. This is accomplished using activated charcoal as a catalyst and high temperature as follows:

$$SO_2Cl_2 \xrightarrow[\text{catalyst}]{\text{heat}} SO_2 + Cl_2\ (g)$$

At the temperature of reaction the free energy change is $\Delta F = 7760 - 22.67(T)$. At room temperature, decomposition does not occur, but at temperatures above 70°C, the reaction theoretically becomes spontaneous.

The mixture of sulfur dioxide and chlorine is then passed through a solvent absorber using carbon tetrachloride or sulfur monochloride. In this solvent, sulfur dioxide is relatively insoluble, while chlorine is soluble.

The next two steps in this process are recycling procedures. The first deals with the recycling of sulfur trioxide to the initial process step. The sulfur dioxide produced in step 3 is combined with oxygen to make the sulfur trioxide:

$$2SO_2 + O_2 \longrightarrow 2SO_3$$

In the second operation the absorbing solvent is recycled and the chlorine product is recovered:

$$Cl_2 \text{ in } CCl_4{}^* \xrightarrow{\text{heat}} \underset{\text{(product)}}{Cl_2\uparrow} + CCl_4{}^*$$

*S_2Cl_2 may be used as the absorbing solvent.

The overall free energy change of the total reaction is about 115 kcal/mol at 25°C. The overall process reaction is:

$$2SO_3 + 2HCl \longrightarrow H_2SO_4 + Cl_2 + SO_2$$

The overall free energy change is exothermic. Process electricity necessary for the production of one kkg of chlorine is 305 kWh.

The operation can be described as follows:

(1) Manufacture of the chlorosulfonic acid using sulfur trioxide contact gas and hydrogen chloride. This is carried out by passing both gases to the base of a packed tower. Cool chlorosulfonic acid is circulated to the tower such that it dissolves the sulfur trioxide and exposes it to the hydrogen chloride for reaction. Circulation must be maintained so that the excess heat of reaction can be removed. A temperature of around 100°C is required.

(2) The catalytic (mercuric chloride) decomposition of the chlorosulfonic acid then occurs. Sulfuryl chloride and sulfuric acid are generated at temperatures from 150° to 300°C. This reaction is slow and is the controlling process in continuous operation. With pressure maintained between 3 and 11 atm the reaction proceeds at a continuous and faster rate. The only difficulty is separation of the process catalyst (mercuric chloride). The use of these higher pressures requires high capital expenditures

for purchasing both the pressure generating equipment and pressure resistant systems.

(3) The catalytic (activated charcoal) decomposition of sulfuryl chloride into sulfur dioxide and chlorine occurs at a temperature above 200°C.

(4) Separation of the sulfur dioxide/chlorine mixture by selective adsorption onto carbon tetrachloride or sulfur monochloride.

(5) Recycling of the carbon tetrachloride to step (4) and the removal of the chlorine from the carbon tetrachloride.

Some process problems did arise with the bench testing of this method. Unsatisfactory yields were obtained due to clogging in the tower. Combined with this problem was that of balancing the feed rate and heat input rate. Further testing is needed to improve process efficiency.

Corrosion problems must also be addressed. In the absence of leaks and moist air, steel and iron casings are protected; however, surfaces of copper, nickel, and silver alloys are eroded slowly. The acids present are strong enough to damage cement and brick, so all substances of this type used should be acidproof. Leakages from these towers could have serious environmental effects, especially from mercury salt catalysts and sulfur dioxide.

This hydrogen chloride/sulfur trioxide process has some advantages which make it attractive. The raw materials needed are easily obtained, although hydrogen chloride supply may be limited, and the process reactions are easily controlled. Its disadvantages are having to separate sulfur dioxide and chlorine, and remove mercuric chloride from the sulfuric acid product, but if these problems can be solved, then large-scale quantities may be generated.

Kuhlmann Chlorine Production Method: This chlorine production process uses hydrogen chloride, oxygen, sulfur dioxide and a catalyst and is considered an improvement over the Holmes and Hooker et al process. The principal developer was the Kuhlmann Corporation of Paris, France, which received a patent on June 6, 1968 (71). Development work on this method was carried out only on a laboratory scale and further process refinement has ceased.

The chemistry of the Kuhlmann method is an improvement on a similar process developed by the Holmes and Hooker, et al, which utilized sulfur trioxide as an oxidizing agent for converting hydrogen chloride into chlorine. The following equations summarize this mechanism:

$$(1) \quad HCl\ (g) + SO_3\ (g) \xrightarrow{70^\circ-100^\circ C} HSO_3Cl\ (l)$$

$$(2) \quad 2HSO_3Cl \xrightarrow[HgCl_2]{160^\circ C} SO_2Cl_2\ (g) + H_2SO_4\ (l)$$

$$(3) \quad SO_2Cl_2\ (l) \xrightarrow[\text{activated charcoal}]{} SO_2\ (g) + Cl_2\ (g)$$

Pragmatically, these reactions must be carried out separately because reactions (1) and (3) are rapid, while reaction (2) is slow. As such, chlorosulfonic acid must react under pressure in the liquid phase to attain sufficient speed. This condition is detrimental since continuous operation of the process is difficult and separation of the catalyst, sulfuric acid and sulfuryl chloride is a problem.

Kuhlmann's improvement involves oxidizing the hydrogen chloride in a single step to produce liquid sulfuric acid and gaseous chlorine. This reaction is represented by the following equation:

$$SO_2\ (g) + O_2\ (g) + 2HCl\ (g) \xrightarrow[\text{catalyst}]{} H_2SO_4\ (l) + Cl_2\ (g)$$

The catalyst in this case is an inert support material combined with 3 to 15% vanadium compounds and 5 to 20% alkali metal/alkaline earth metal oxides. The resultant products are gaseous chlorine, liquid sulfuric acid, sulfur trioxide, chlorosulfonic acid and a small quantity of unreacted sulfur dioxide. Further treatment of this mixture involves condensing the chlorosulfonic and sulfuric acids, absorbing the sulfur trioxide in sulfuric acid and separating out the desired chlorine.

The conceptual method of operation for this process involves a two-stage process: heating, then cooling. In the heating state an equilibrium is reached between the reactants (i.e., SO_2, O_2 and HCl) and the products (i.e., H_2SO_4 and Cl_2).

The temperature at atmospheric pressure must be between 400° and 600°C. These high temperature ranges must be maintained; however, the pressure level may vary from below atmospheric up to 10 atm.

The obtained equilibrium, however, does not permit a complete conversion of hydrogen chloride to chlorine. This requires a cooling stage in which the reaction of the hydrogen chloride and sulfur trioxide is completed. In the cooling stage, chlorosulfonic acid is formed from the hydrogen chloride and the sulfur trioxide. The reaction temperatures present in this zone are between 100° and 120°C.

Under the above operating conditions, the contact times between the gaseous starting mixture and the catalytic mass should range between 1 and 20 seconds. Optimal contact time is between 3 and 10 seconds. Catalytic contact can be made on either a fixed or fluidized bed, which promotes continuous process operation.

The advantages of this process are that 95% yields are attainable and continuous operation is easily achieved. The major problems are: (1) the process is based on hydrogen chloride, so raw material supplies are limited; (2) materials involved are of a highly corrosive nature; and (3) high process temperatures are required.

Oxidation of Hydrogen Chloride with Hydrogen Peroxide

In this process hydrogen chloride is reacted with hydrogen peroxide and calcium chloride. The principal developers of the process are Hans Klebe, Alfred Meffert, and Albert Logenfeld (72). The process patent was assigned to the Deutsche Gold- and Silver-Scheideanstalt vormals Roessler of Frankfurt, Germany. The original patent was filed on July 5, 1972, and granted on April 23, 1974. The patent is an improvement on earlier work performed by Klebe in 1970. The main refinement to the original patent work is the addition of dissolved calcium chloride to the reactant mixture to improve yields. The complete process operation was carried out on only a bench scale and commercial feasibility was not explored. Process development is inactive.

The basic equation for this technique is:

$$H_2O_2 + 2HCl \longrightarrow 2H_2O + Cl_2$$

As an additive to this equation, calcium chloride is used. This is added in the amount of 100 to 500 grams of calcium chloride per liter of reaction solution. The aqueous hydrogen chloride solution concentration ranges from 20 to 40% by weight with a preferred range of 25 to 35%. The aqueous hydrogen peroxide solution has a preferred concentration range from 40 to 70%, with a maximum value of 90%. During solution preparation hydrogen peroxide is usually added in the ratio of 0.8 to 0.9 mol for every 2 mols of hydrogen chloride. The temperatures at which the various components react depend on the particular procedural mechanism used.

The reaction can occur either at the temperature provided for the absorption agent of 27° to 40°C or at the temperature provided by the oxidation reaction activation temperature, which is below 60°C. Operating pressures of 0.2 to 0.95 atm enhance chlorine liberation. Chlorine can also be liberated by using a rinsing gas that strips the chlorine from the reaction chamber. A combination of low pressure and a rinsing gas often is used.

The process can be carried out by batch or continuous operations. With a continuous operation, water must be constantly removed to maintain adequate hydrochloric acid concentrations. Water formed during reaction can be withdrawn in the form of dilute hydrochloric acid. Hydrogen chloride gas is added to maintain hydrochloric acid concentration.

The overall process can be accomplished using any of three separate subprocesses which differ in how the hydrogen chloride is extracted from the hydrogen chloride-containing gas.

The first method uses a multistep countercurrent wash with hydrochloric acid to separate the hydrogen chloride, steam, and air mixture and remove heat prior to reaction with hydrogen peroxide. The cooled product flows concurrently to the hydrogen peroxide through a packed tower to form the chlorine gas. One advantage of this technique is that chlorine can be obtained from gas mixtures having a low concentration of hydrogen chloride.

The second technique uses the same multistep wash, except that the resultant solution is heated until enough hydrogen chloride is removed to form an azeotrope. The desorbed hydrogen chloride is then contacted countercurrently with concentrated hydrochloric acid and hydrogen peroxide. Chlorine is formed in the reaction tower and removed. Excess water is discharged in the form of azeotropic hydrochloric acid.

The third alternative subprocess does not separate the gaseous mixture. The hydrogen chloride is introduced immediately into the aqueous medium and reacts with the hydrogen peroxide. The product chlorine gas must be reconcentrated, however, because it is mixed with accompanying gases of the hydrogen chloride mixture.

Regardless of the procedure used, the yields are good, in excess of a 90% conversion.

An advantage of this process is that no special equipment is needed to perform the operation. Standard reactions such as the desorption of hydrogen chloride from hydrochloric acid to form the azeotrope can be accomplished by existing commercial processes.

Although this process has low energy requirements and high yields, it is ruled out as a viable means of commercial chlorine production because of the high cost and low availability of hydrogen peroxide. Hydrogen peroxide producers have no incentive to increase production without an increase in price. Precise economic data were not available in the patent.

Chlorine from Alkali Metal Chlorides and Nitrogen Oxide Compounds

Vicksburg Process: The concept of producing alkali metal nitrates and a chlorine coproduct from alkali metal chloride and nitric acid and/or nitrogen dioxide is quite old. As early as the 1930's, Allied Chemical worked on a process for coproduction of sodium nitrate and chlorine. A facility was built before World War II at Hopewell, Virginia, and was operated until the mid-1960's when it was shut down due to a lack of demand for sodium nitrate. Also in the 1930's, the U.S. Department of Agriculture developed a process for potassium nitrate (KNO_3) production based on the reaction of nitrogen dioxide with potassium chloride. This process, however, was never commercialized (73).

In the late 1950's interest developed in a process based on the reaction of nitric acid with potassium chloride to yield a potassium nitrate by-product usable for fertilizer applications. At least three organizations performed independent work in this area. In 1962 a patent was granted to J.L. Chadwick of Delhi-Taylor Oil for this process; however, this patent was not commercialized due to corrosion problems (74). Almost at the same time another patent was granted to H.A. Beekhuis (unassigned) for a similar process (75)(76). This work also saw no commercial application.

In 1961, the Southwest Potash Division of American Metal Climax Corporation (AMAX) began work on a process for converting potassium chloride to potassium nitrate and chlorine. After pilot plant evaluation, a plant was opened in 1965 at Vicksburg, Mississippi to produce 63,000 tpy of potassium nitrate and 22,000 tpy of chlorine by-product (77). This facility was operated by AMAX for several years, then was sold to Vicksburg Chemical Company in 1973 (78). Vicksburg Chemical Company has continued to operate the facility and has made a number of minor modifications, mostly in construction materials to reduce corrosion problems. No other facilities using this process have been constructed or are planned.

In this process, 64% nitric acid is first reacted with potassium chloride according to the equation:

$$3KCl + 4HNO_3 \longrightarrow 3KNO_3 + NOCl\uparrow + Cl_2\uparrow + 2H_2O$$

This initial reaction is generally conducted below 10°C to avoid corrosion problems involved in handling hot nitric acid-chloride based mixtures. From this reaction both gaseous nitrosyl chloride and chlorine are liberated.

Next, the liberated gases are passed through a gas column reactor where they are contacted with nitric acid vapors at elevated temperatures to oxidize the nitrosyl chloride present according to the equations:

(1) $NOCl + 2HNO_3 \longrightarrow NO_2Cl + 2NO_2 + H_2O$

(2) $H_2O + 3NO_2 \longrightarrow 2HNO_3 + NO$

(3) $NO_2Cl + NO \longrightarrow NO_2 + NOCl$

(4) $2NO_2Cl \rightleftharpoons 2NO_2 + Cl_2$

For these reactions, nitric acid in excess of 70% is required.

The product mixture of nitric acid vapors, nitric oxide, nitrogen dioxide, nitrosyl chloride and chlorine is separated by fractional distillation. The chlorine fraction is compressed, liquefied and prepared for shipment as product. The other separated materials (NO, NO_2, NOCl and HNO_3 vapors) are fed to a nitrogen dioxide fractionating column, where nitrogen dioxide and nitric acid vapors are separated from the other components. The nitric oxide and nitrosyl chloride are collected and recycled to the second process. The gas column reactor and the nitric acid and nitrogen dioxide are then reconverted to 65% nitric acid for recycle back to the initial reaction step.

Further processing is needed for the initial reaction step solutions which, after release of nitrosyl chloride and chlorine, still contain the potassium nitrate coproduct and unreacted 65% nitric acid. These solutions are stripped of excess water by heating and are then cooled and fed to a series of vacuum crystallizers, where 81% nitric acid is removed as vapor. This acid gas is then recondensed, converted back to 65% nitric acid and recycled to the initial process step. The potassium nitrate recovered in this evaporative process is further separated from residual nitric acid by centrifugation and is then dried, melted, prilled, cooled and marketed. Nitric acid separated during centrifugation is recovered and recycled.

Some key aspects of this process are:

(1) A shift in the composition of the nitric acid-water azeotrope from about 70% nitric acid to over 80% in the presence of high concentrations of potassium nitrate. It has been found that only the nitrates of the alkali metals heavier than sodium shift the nitric acid azeotrope to higher concentrates. The nitrates of the lighter alkali metals (i.e., lithium and sodium) have the reverse effect of shifting the azeotrope to lower nitric acid contents, which was no problem with the old Allied sodium nitrate process.

(2) The oxidation of nitrosyl chloride (NOCl) to chlorine and nitrogen dioxide with nitric acid in the 70 to 80% range at temperatures and pressures which are practical from an operational viewpoint.

There are two facilities producing potassium nitrate in the United States. These two plants satisfy the entire domestic demand. Of these two, the Vicksburg

Chemical plant is by far the larger (78). The second facility, operated by Mallinckrodt Chemical Company, produces only a small amount of reagent grade material.

While this process is technically viable, limited markets for the potassium nitrate coproduct make it unlikely that additional plants of this type will be constructed in the foreseeable future.

Potassium Nitrate Process: The process described below is a forerunner of the present-day Vicksburg process. The process itself was never carried past bench scale and the production of chlorine was not a consideration. The process originated with Donald Reed and K.G. Clark (73) in the mid-1920's. They were working with the Bureau of Chemistry and Soils of the U.S. Department of Agriculture. The main thrust of their work was in the chemical fertilizer industry. They developed a process for the production of potassium nitrate using nitrogen oxides and potassium chloride as follows:

$$2NO_2 + KCl \longrightarrow KNO_3 + NOCl$$

No further work was carried out.

Based on the Vicksburg process, however, the technology is known for converting nitrosyl chloride to chlorine. Fairly good reaction rates were obtained in this process, but the conversions were poor. Reactions were conducted generally under ambient conditions.

Aside from work on the initial process reaction, no other efforts were expended on this process. However, the thermodynamics and engineering considerations for the reactions of nitrogen dioxide with potassium chloride are similar to those for the Vicksburg process.

In conclusion, this process was a predecessor effort to the now commercialized Vicksburg process. Since the Vicksburg process supplies over 70% of U.S. potassium nitrate production, a market does not exist for significant additional amounts of this material.

Allied Chemical Company Salt and Nitric Acid Process

In 1847, Dunlop Company patented a process using sodium chloride and nitric acid to produce chlorine. Many modifications were made to this basic process until the Allied Chemical Company developed a commercially feasible scheme in the mid-1930's. The first and only commercially successful plant to use this technology was built in Hopewell, Virginia, in 1936. The process continued to be modified and improved until the early 1950's. The last major change converted the by-product nitrosyl chloride to chlorine and sodium nitrate. By the 1960's, however, the Hopewell plant was shut down because of the reduced market for the sodium nitrate coproduct. This market had been reduced by the trend away from sodium nitrate-based fertilizers (79).

The first phase of Allied's method is known as the salt process with nitrosyl chloride neutralization. It converts salt, nitric acid, and sodium carbonate to chlorine and sodium nitrate. The overall chemical reaction including salt digestion, chlorine recovery and nitrosyl chloride oxidation is:

$$6NaCl + 12HNO_3 + 2Na_2CO_3 \longrightarrow 3Cl_2 + 10NaNO_3 + 2CO_2 + 2NO + 6H_2O$$

Note that for each pound of chlorine produced, four pounds of sodium nitrate are produced.

There is a tendency in the digestion reaction to form nitrosyl chloride (NOCl), reducing the chlorine yield. To increase yields, Allied oxidized the nitrosyl chloride as follows:

$$2NOCl + O_2 \longrightarrow N_2O_4 + Cl_2$$

The nitrosyl chloride is first vaporized and oxidized with oxygen. This mixture is then cooled, condensed, and distilled. Nitrogen tetroxide is recoverable as a liquid product that can be stored and sold or recycled to make nitric acid.

A major operating difficulty for this process is excessive corrosion. The equipment used for this process must be corrosion resistant and fluid-tight. Construction materials, such as metal alloys and corrosion resistant cement, are needed.

This process is not industrially feasible because of the lack of a viable market for the tremendous quantities of sodium nitrate generated. If none of the sodium nitrate can be sold, disposal costs would be prohibitive from both an economic and environmental standpoint. Combining these problems with those of the corrosion potential, this process appears infeasible for the large scale commercial production of chlorine.

French Salt-Kaolin Process

In 1886, A. Goyeu found that heating a mixture of salt and kaolin with oxygen to a temperature of 1000°C led to the formation of elemental chlorine. Little further work was done with this reaction for over sixty years. In 1948, Briner and Roth performed a more extensive kinetic study of this reaction and corroborated the original findings (80). Further verification was soon reported by Dodson in 1957 (81). In 1966, after further laboratory studies, F. Trombe, M. Foex, P. Courty and G. Hulbt obtained a patent on a proposed process using this scheme (82)(83). No further efforts in this area have been reported since.

According to the patent of Trombe, et al, this process involves the following steps:

(1) Kaolin and salt are precalcined to remove water at 800°C. For the kaolin the following dehydration reaction occurs:

$$2SiO_2 \cdot Al_2O_3 \cdot 2H_2O \longrightarrow 2H_2O + 2SiO_2 \cdot Al_2O_3$$

(2) The hot salt and kaolin are mixed at a temperature of 850°C to create a bed of hot dehydrated kaolin particles covered by a film of molten salt.

(3) Air is passed through the bed forming the coproducts sodium aluminosilicate and chlorine, as follows:

$$2NaCl + 2SiO_2 \cdot Al_2O_3 + \tfrac{1}{2}O_2 \longrightarrow Na_2O \cdot 2SiO_2 \cdot Al_2O_3 + Cl_2$$

(4) The product chlorine exits the process with unreacted oxygen and is recovered by selective absorption in a solvent, such as carbon tetrachloride.

(5) The chlorine is recovered from the carbon tetrachloride by heating.

No specific discussion is made of further processing of the sodium aluminosilicate coproduct, although recovery of this coproduct may be required to develop a viable process. Aside from the Trombe, et al patent, no other recent work has been reported in this area.

The detailed economics of this process depend critically on whether or not a marketable sodium aluminate can be recovered as a coproduct. If a marketable coproduct cannot be recovered, the process must bear the costs of land disposal of about 4 tons of sodium aluminosilicate per ton of chlorine generated. If one assumes a land disposal cost of approximately $10 per ton of material, then a $40 per ton of chlorine waste disposal penalty would result for the process if no marketable coproduct is produced. This cost is sufficiently high to make the process uneconomical.

Du Pont High Temperature Silica and Salt Process

This high temperature method for preparing chlorine was invented by Wallace Ward of E.I. Du Pont de Nemours and Company. The process patent was granted on August 30, 1960. After the patent had been secured the project was terminated. The experience on this process was only laboratory work. No further experimentation has been reported.

The reaction mechanism for this process involves the combination of sodium chloride ($NaCl$) and silica (SiO_2), with silicon tetrafluoride (SiF_4) acting as a catalyst to form chlorine. The reaction occurs in the presence of oxygen (O_2) and intense heat. Besides chlorine, the products are sodium silicate (Na_2SiO_3), an alkali metal silicate, and unreacted oxygen. Separation by condensation of the catalytic silicon tetrafluoride allows catalyst recycle. The reaction is as follows:

$$2NaCl + SiO_2 + SiF_4 \xrightarrow[1000^\circ-1500^\circ C]{\frac{1}{2}O_2} Na_2SiO_3 + Cl_2 + SiF_4$$

The process thermodynamics show the reaction is endothermic with a ΔF for the reaction equal to 35.0 kcal/mol at 25°C. Process temperatures necessary for the conversion range from 1000° to 1500°C. The major discovery from this process is that silicon tetrafluoride is a highly active catalyst for chlorine preparation by thermal methods. Process kinetics were not provided in the patent (84).

The sodium silicate by-product formed by the high temperature reaction is a viscous liquid. This liquid is retained in the molten sodium chloride pool and due to density differences and the insolubility of the sodium chloride and the sodium silicate in each other, easy separation of the by-product is achieved.

One advantage of this process is its use of relatively inexpensive raw materials and the formation of a valuable by-product. In addition to the preferred sodium chloride as a raw material, lithium and potassium chloride are possible alternatives.

The silica used in the process may also be in both a pure or a combined form. Small particle sizes are usually used to provide the largest surface areas upon which reaction can occur.

Upon completion of a sample laboratory experiment, the yield from this process was found to be between 0 and 48% chlorine with the average concentration of chlorine at 28%. The reaction period was 220 minutes.

Three factors: (1) low process yield, (2) slow process rates, and (3) high temperature and energy requirements, make this method highly unsuitable for use.

Israeli Process for Chlorine from Magnesium Chloride

The Israeli process for chlorine production was originally designed to use the highly concentrated magnesium chloride brines present in the Dead Sea. This process was developed and patented by J. Kiperman and never developed past the laboratory stage (85).

The process as described in the Israeli patent uses a diaphragm-type cell modified by introduction of a membrane prepared from asbestos, cloth and plastic. During electrolysis, the anode section contains a sodium chloride brine and the catholyte is a solution of magnesium chloride. As electrolysis proceeds, chlorine is liberated at the anode and magnesium hydroxide, formed at the cathode, deposits as an insoluble precipitate in the cathode section. This material is periodically recovered by an unspecified method, washed and dried. Thermodynamic requirements for this process are the same as for the diaphragm cell.

This process is merely a modification of the diaphragm cell to enable a magnesium chloride brine to be used. Current densities used are low. A potential problem is continuous recovery of magnesium hydroxide from the cell during operation on a large scale. Also, the availability of high magnesium chloride content brines is limited.

Hydrogen-Chloride-Based High Temperature Arc Process

In this process chlorine is produced by a high temperature reaction of hydrogen chloride. The principal developer of this method was Irwin B. Margiloff of New York. His patent work was performed for Halcon International, Inc., of Delaware. The original patent claim was filed on September 17, 1962 and granted on June 7, 1966 (86). This process was not developed past the laboratory stage.

For the high temperature arc process, heat provided by a plasma generator is used to break down hydrogen chloride into chlorine and water. The plasma generator cracks water by means of an electrical arc. The resulting fragments (or plasma) are then allowed to recombine, thereby generating the high process temperatures. The temperature range for this reaction is 2200°C to about 4000°C.

While the reaction is progressing, oxygen is introduced into the system to combine with the hydrogen from the hydrogen chloride to form water. The oxygen can have impurities, although any contaminants will increase power costs. Air may be substituted for oxygen. Besides using water for the plasmas, other suitable substances are argon and nitrogen.

To optimize conversion, pressures range from 1.5 to 7 atm but yields decrease outside this range.

The products are cooled or quenched upon reaction. Water is preferable but other fluids, including hydrochloric acid, may be used. Quenching should be performed rapidly across the temperature interval between the reaction temperature and 300°C. Beyond this temperature normal cooling is performed. The chlorine product is dried and liquefied.

The process yield is about 45% at atmospheric pressure. Increased pressure, hydrogen chloride recycling, and oxygen purification all increased yields somewhat; however, they remained below 75%.

This process is not acceptable for commercial chlorine production because of low yields and extremely high temperature requirements. Inherent with high temperature requirements are high energy costs and high material costs. This process, though simple, is not economic.

REFERENCES

(1) Grot, W.G., U.S. Patent 3,718,627, February, 1973.
(2) Iammartino, N.R., *Chemical Engineering,* 83(13), 86 (1976).
(3) Beck, T.R., *Final Report on Improvements in Energy Efficiency of Industrial Electrochemical Processes,* ANL/OEPM-77-2, Argonne National Laboratory, January 1977.
(4) Berzins, T., "Electrochemical Characterization of Nafion Perfluorosulfuric Membranes in Chlor-Alkali Cells," paper presented at Electrochemical Society Meeting, Atlanta, Georgia, October 1977.
(5) Seko, M., "The Asahi Chemical Membrane Chlor-Alkali Process," paper presented at 20th Chlorine Plants Managers' Symposium, Chlorine Institute, New Orleans, Louisiana, February 1977.
(6) Shirasaki, K., *Chemical Economics Engineering Review,* 7(1), 33 (1975).
(7) Burkhardt, S.F., "Radioactive Tracer Measurement of Sodium Transport Efficiency in Membrane Cells," paper presented at Electrochemical Society Meeting, Atlanta, Georgia, October 1977.
(8) Gierke, T.D., "Ionic Clustering in Nafion Perfluorosulfonic Acid Membranes and Its Relationship to Hydroxyl Rejection and Chlor-Alkali Current Efficiency," paper presented at Electrochemical Society Meeting, Atlanta, Georgia, October 1977.
(9) Hora, C.J., and Maloney, D.E., "Nafion Membranes Structured for High Efficiency Chlor-Alkali Cells," paper presented at Electrochemical Society Meeting, Atlanta, Georgia, October 1977.
(10) Herman, E., et al, "New Design of Mercury Cell for Brine Electrolysis," paper presented at Electrochemical Society Meeting, Seattle, Washington, May 1978.
(11) Jaksic, M.M., Lacnyivac, C.M., Atanasoski, R., and Adzic, R., "Synergistic Electrocatalytic Effects for Transition Metals for Hydrogen Evaluation: Reaction on Titanium and Graphite Cathodes," paper presented at Electrochemical Society Meeting, Seattle, Washington, May 1978.
(12) Kozoil, K.R., Rathjen, H.C., Wink, E.F., and Fisher, A.W.,"Latest Developments and Possibilities with Coated Titanium Electrodes," paper presented at Electrochemical Society Meeting, Seattle, Washington, May 1978.
(13) Adams, A.M., Bacon, F.T., and Watson, R.G., "The High Pressure H_2-O_2 Cell," *Fuel Cells,* W. Mitchell, editor, Academic Press, New York, pp 130-192 (1963).
(14) Oswin, H.G., and Chodosh, S.M., *Advances in Chemistry Series,* 47, 61 (1965).
(15) Gritzner, G., U.S. Patent 3,926,769, December 1975.
(16) Gritzner, G., U.S. Patent 4,035,254, July 1977.
(17) Gritzner, G., U.S. Patent 4,035,255, July 1977.

(18) Buttre, J.L., and Pierrot, F., U.S. Patent 3,763,005, October 1973.
(19) Ramp, F.L., U.S. Patent 3,974,047, August 1976.
(20) Johnson, H.B., U.S. Patent 4,101,394, July 1978.
(21) W.M. Grosvenor Laboratories, Inc., U.S. Patent 2,206,339, 1937-1940.
(22) Dow Chemical Co., U.S. Patent 2,577,808, 1947-1951.
(23) Standard Oil Development Co., U.S. Patent 2,436,870, 1942-1948.
(24) Socony Vacuum Oil Co., U.S. Patent 2,418,930, 1943-1947.
(25) Arnold, C.W., and Kobe, K.A., "Thermodynamics of the Deacon Process," *Chemical Engineering Progress,* 48, 6, p 293 (1952).
(26) Van Dijk, C.P., U.S. Patent 4,037,000, May 1977.
(27) Frischer, H., Canadian Patent 550,813, December 1957.
(28) Hamlet, J., U.S. Patent 3,152,866, October 1964.
(29) Payer, S., *Chemical Engineering Technology,* 47, 142 (1975).
(30) Payer, S., *Hydrocarbon Processing,* 53(11), 147 (1974).
(31) Diprose, G., British Patent 844,789, August 1960.
(32) Diprose, G., Belgian Patent 569,885, January 1959.
(33) Imperial Chemical Industries, Ltd., German Patent 1,077,194, March 1960.
(34) Yoshizawa, S., *Chemical Economics and Engineering Review,* 8(6), 34 (1976).
(35) Yasui, I., and Doremus, R.H., *Journal of the Electrochemical Society,* 125(7), 1007 (1978).
(36) Yasui, I., and Doremus, R.H., *Journal of the American Ceramic Society,* 60, 296 (1977).
(37) Lemke, C.H., *Kirk-Othmer's Encyclopedia of Chemical Technology,* John Wiley and Sons, New York, Volume 18, pp 432-457 (1969).
(38) M.W. Kellogg Company, British Patent 946,830, 1960-64.
(39) Imperial Chemical Industries, British Patent 1,033,300, June 1967.
(40) Nonnenmacher, H., and Andrussow, K., German Patent 1,240,830, May 1967.
(41) Farbwerke Hoechst AG, Belgian Patent 658,435, 1965.
(42) Schroeder, D.W., U.S. Patent 3,119,757, January 1974.
(43) Schroeder, D.W., *Industrial and Engineering Chemistry, Process Design,* 1(2), 141 (1962).
(44) Koninklijke, N.V., *Nederlandische Zortindustrie,* South African Patent 66-4595, Sept. 1966.
(45) Mathieson Chemical Corporation, British Patent 747,772, April 1956.
(46) Gorden, J., *Chemical Engineering,* 60(10), 188 (1953).
(47) Dunn, W.E., Jr., U.S. Patent 3,887,694, June 1975.
(48) Wikswo, J.P., and Nelson, E.W., U.S. Patent 3,325,252, June 1967.
(49) Metaizeau, P., U.S. Patent 3,332,742, July 25, 1967.
(50) Metaizeau, P., U.S. Patent 3,542,557, September 19, 1967.
(51) Metaizeau, P., U.S. Patent 3,383,177, May 14, 1968.
(52) Metaizeau, P., U.S. Patent 3,384,456, May 21, 1968.
(53) Botton, R., and Steinmetz, A., U.S. Patent 3,627,471, December 1971.
(54) Produits Chimiques Pechiney-Saint-Gobain, British Patent 1,098,361, January 1968.
(55) Tauch, E.J., and Iler, R.K., U.S. Patent 2,218,729, October 1940.
(56) Iler, R.K., U.S. Patent 2,219,103, 1941.
(57) Laury, N.A., U.S. Patent 2,254,014, August 1941.
(58) Laury, N.A., U.S. Patent 2,415,358, 1947.
(59) Maude, A.H., et al, U.S. Patent 2,276,079, March 1942.
(60) Johnstone, H.F., "Chlorine Production, Nonelectrolytic Processes," *Chemical Engineering Progress,* 44, 657 (1948).
(61) Langdon, W.M., *Industrial Engineering Chemistry,* anal. ed., 17, 801 (1945).
(62) Personal communication with R.K. Iler, September 6, 1978.
(63) Ruff, O., *Proceedings of the Prussian Academy of Sciences (Ber.)* 34, 3509 (1901).
(64) Payne, J.H., Jr., Fernelius, W.C., and Pierce, D.W., paper presented at A.C.S. Meeting, New York, September 1947.
(65) I.G. Farbenind, German Patent 543,758, May 1929.
(66) Edwards, W.A.M., et al, B.I.O.S. Final Report No. 243.
(67) Kallal, R.J., B.S. Thesis, University of Illinois, 1943.

(68) Johnstone, H.F., "Chlorine Production, Nonelectrolytic Processes," *Chemical Engineering Progress,* 44, 657 (1948).
(69) Bouchard, F.J., U.S. Patent 2,393,229, January 1946.
(70) Maude, A.H., et al, U.S. Patent 2,276,079, March 1942.
(71) Kuhlmann Corporation, British Patent 1,115,971, June 1968.
(72) Klebe, H., Meffert, A., and Longenfeld, A., German Patent 1,963,946, 1974.
(73) Reed, D.L., and Clark, K.G., *Industrial and Engineering Chemistry,* 29(3), 333, 1937.
(74) Chadwick, J.L., U.S. Patent 3,062,616, November 1962.
(75) Beekhuis, H.A., U.S. Patent 3,062,619, November 1962.
(76) Fogler, M.F., *Chlorine, Its Manufacture, Properties and Uses,* J.S. Sconce, editor, R.E. Kreiger Publishing Co., Huntington, New York, Chapter 8, pp 235-249 (1972).
(77) Spealman, M.L., *Chemical Engineering,* 62(11), 198 (1965).
(78) Personal communications from Fred Ahlers, November 1, 1978.
(79) Fogler, M.F., *Chlorine, Its Manufacture, Properties and Uses,* J.S. Sconce, editor, R.E. Kreiger Publishing Co., Huntington, New York pp 235-249 (1972).
(80) Briner, E., and Roth P., *Helvetica Chimica Acta,* 31, 1352 (1948).
(81) Dodson, V.H., "A Study of the Reaction Between Oxygen and Mixtures of Kaolinite and Certain Metal Chlorides at Elevated Temperatures," *The Ohio Journal of Science,* 57(1):29 (January 1957).
(82) Trombe, F., Foex, M., and Courty, P., *Compt Rand,* Ser. C, 262, 910, 1966.
(83) Trombe, F., Foex, M., and Courty, P., French Patent 1,478,418, 1967.
(84) Ward, W., U.S. Patent 2,950,957, August 1960.
(85) Kiperman, J., Israeli Patent 12,195, August 1960.
(86) Margiloff, I.B., U.S. Patent 3,254,958, June 1966.

DIAPHRAGM CELLS

IMPROVED ASBESTOS DIAPHRAGMS

In the electrolysis of aqueous sodium chloride solutions or other brines to produce chlorine and caustic, one of the principal types of equipment used has a porous asbestos diaphragm separating the anode and cathode chambers. The diaphragm can be formed directly on the side of the cathode facing the anode chamber by vacuum deposition of asbestos and binders by techniques similar to those used in paper making. The deposited diaphragm is normally heated to fuse the binder.

The diaphragm must be porous enough to permit the flow of brine from the anode chamber into the cathode chamber under a small hydrostatic head of pressure, but it should also inhibit the diffusion of hydroxyl ions from the cathode chamber back into the anode chamber. The flow of the brine from the anode chamber to the cathode chamber aids in minimizing diffusion from the cathode chamber back into the anode chamber. Also, excessive leakage of hydrogen or chlorine gases through the diaphragm could contaminate the products being produced and require costly purification or even produce hazardous mixtures of the two gases.

Although the nature of asbestos is not completely understood, it has been theorized that hydroxyl ion diffusion is inhibited by negative charges and a concentration of hydroxyl ions in the hydrated magnesium silicate at the surface of the asbestos. These features, combined with the chemical resistance of asbestos, make it a desirable component of chlor-alkali cell diaphragms.

Chlor-alkali cell diaphragms made only or mainly of asbestos, however, have a short life. The cathode chamber has a highly basic pH, such as 11 to 14, while the anode chamber has an acid pH, such as 3 to 5. Combined with the flow of brine through the diaphragm, these factors cause erosion and chemical and dimensional changes in an asbestos diaphragm, requiring replacement of the diaphragm when the cell becomes too inefficient.

Asbestos diaphragms have been improved by using various binders and modifiers. Fluorocarbon resins, such as polytetrafluoroethylene (PTFE), and copolymers of tetrafluoroethylene and hexafluoropropylene, known as fluorinated ethylene-propylene (FEP), are effective as binders due in part to their chemical inertness. Such polymers can be provided as an aqueous codispersion with asbestos from which the diaphragm is deposited. Fibers of such resins can also be used in the dispersions.

Upon heating to fuse the fluorocarbon resin, the binder adheres to the asbestos in places, generally without completely coating the asbestos. Leaving much of the surface of the asbestos exposed is desirable since asbestos is hydrophilic, that is, it wets readily, aiding the brine in flowing through the diaphragm, and it is thought that its surface characteristics can inhibit the back diffusion of hydroxyl ions.

In addition to such fluorocarbon resins which are hydrophobic, fluoropolymer resins containing hydrophilic functional groups such as carboxylic, sulfonic and phosphonic groups can be used as asbestos diaphragm modifiers. They can completely coat the asbestos, substituting their own functional groups for the surface charge and hydrophilic characteristics of the asbestos which then functions as a stable filler. Such resins can react with the asbestos rather than merely sticking to it.

Each of the developments of the prior art is less than ideal. The fluoropolymer resins with functional groups are generally more expensive than fluorocarbon resins without the functional groups.

Diaphragms with exposed asbestos remain subject to attack. Also, magnesium compound tends to be dissolved from the asbestos fibers themselves at the acid (anode) side of the diaphragm and be deposited as magnesium hydroxide on the basic (cathode) side of the diaphragm. This causes restrictions in the size of pores through the diaphragm and can clog the pores to the point where the diaphragm is no longer useful. Alternatively, fine particle size magnesium hydroxide can be washed all the way through the diaphragm, leaving a silicate surface. Excessive flow rates and voltages can result.

This section discusses the most recent developments in alleviating the problems of asbestos diaphragms.

Addition of Magnesium Compounds

J.C. Fang; U.S. Patent 4,173,526; November 6, 1979; assigned to E.I. DuPont de Nemours and Company provides a diaphragm for a chlor-alkali cell comprising asbestos and at least one magnesium compound selected from magnesium oxide, hydroxide, carbonate, oxyhalide and hydroxyhalide (where the halide is at least one of fluorine and chlorine), the magnesium compound being present in an amount of about 5 to 50%, preferably 35 to 45% by weight based on the asbestos plus the magnesium compound. Also provided is a process for treating a chlor-alkali cell diaphragm which comprises passing through the diaphragm a slurry containing the magnesium compound and depositing such magnesium compound within the diaphragm.

While the mechanism is not fully understood, it is thought that the magnesium compounds added to an asbestos diaphragm, either initially or after some operation of the chlor-alkali cell, rejuvenates or improves the operating characteristics of the asbestos. By stabilizing a desirable level of magnesium hydroxide throughout the cross section of the diaphragm, any tendency of the magnesium hydroxide to leach out of the asbestos at the acid side of the diaphragm is counteracted. The hydrophilicity and surface characteristics of the diaphragm become more constant and reliable. Thus, the diaphragm can be used longer before replacement is required.

Generally, such diaphragms permit the use of lower voltage and give higher current efficiency than those of the prior art. Also, the treatment can be used to rejuvenate used diaphragms, especially ones that have begun to pass fluids too readily.

Comparative Example: A slurry of 3,000 ml water, 290 g NaCl, 290 g NaOH, 40.0 g 4D 12 asbestos (Johns-Manville Co.) and fluoropolymer binder was sparged for 1 hour. A wet diaphragm was prepared by pouring the asbestos-polymer slurry into a 5 ℓ stainless steel beaker, placing a cathode screen in the beaker and connecting to a vacuum set-up. Vacuum was applied at 5 cm for 1 minute, then 1 cm for 2 minutes, then 15 cm for 1 minute and finally 20 cm for 1.5 minutes.

The cathode was removed from the beaker and allowed to dry under vacuum for 20 minutes. Holes were repaired with wet asbestos-polymer slurry from the beaker, but with no compaction.

The wet diaphragm is allowed to dry overnight in air at 25°C without baking. The weight of diaphragm was 12.4 g. The diaphragm was placed in the cell and run for 7 days at 8.3 A (182 A/dm^2) at 95°C with a head of 24 to 55 cm to produce 2.1 to 2.26 N caustic at a flow rate of 125 to 146 ml/hr. The initial voltage was 3.68, the final voltage was 3.38, and the current efficiency was 94 to 98%.

Example: Using techniques basically the same as those of the comparative example, but adding magnesium oxide USP grade 90 to the slurry in an amount of about 20% of the weight of the asbestos plus magnesium oxide, it was found that operating voltages of only about 3.10 could be used. This is a substantial improvement over the voltages of the comparative example.

Incorporating Polyelectrolytes Insoluble in Aqueous Alkali Metal Halide Solutions

L. Degueldre and E. Nicolas; U.S. Patent 4,125,450; November 14, 1978; assigned to Solvay & Cie, Belgium have found that the stability of the thickness of diaphragms based on inorganic fibers can be largely ensured while avoiding the disadvantages of the prior art.

Pervious diaphragms for cells for the electrolysis of aqueous solutions of alkali metal halides are provided comprising inorganic fibers and a polymer which is selected from polyelectrolytes insoluble in aqueous solutions of alkali metal halides.

The term "polyelectrolytes" denotes all polymeric substances which comprise monomer units containing ionizable groups, following the generally accepted definition (*Encyclopedia of Polymer Science and Technology,* Vol. 10, p. 781, John Wiley and Sons).

In each of the following examples, an asbestos diaphragm was made directly on a cathode consisting of a disc of 120 cm^2 surface area made of a steel lattice. The cathode, covered with the diaphragm, was then set up vertically in a laboratory-type electrolytic cell, facing an anode made up of a succession of vertical titanium vanes carrying an electrocatalytic coating consisting of a mixture of ruthenium oxide and titanium dioxide. The distance between the cathode and the vanes of the anode was adjusted to 5 mm.

In the cell made up in this manner, a brine saturated with sodium chloride was electrolyzed at 85°C at an anodic current density of 2 kA/m^2 and a hydrostatic pressure on the diaphragm equal to a 30 cm head of electrolyte.

Example 1: *Comparative* – 17.5 g of chrysotile asbestos fibers were dispersed in 0.9 ℓ of an aqueous solution of sodium chloride and sodium hydroxide containing about 170 g/ℓ of NaCl and 120 g/ℓ of NaOH coming from a diaphragm cell in which a sodium chloride brine was being electrolyzed. The suspension thus obtained was then filtered through the cathode lattice of the laboratory cell by applying suction corresponding to 200 mm of mercury.

The recovered filtrate was filtered a second time through the cathode lattice covered by the diaphragm under a suction of 200 mm of mercury. The diaphragm was then dried at ambient temperature, applying beneath the cathode lattice successively a suction of 200 mm of mercury for 15 minutes, then a suction of 400 mm of mercury for 30 minutes. The cathode furnished with the diaphragm was then set up in the laboratory cell, where an electrolysis test was carried out under the conditions stated above.

After 20 days' electrolysis, a voltage of 3.56 V was recorded at the cell terminals and the permeability of the diaphragm was measured as $K = 0.118\ h^{-1}$.

Example 2: A diaphragm of chrysotile asbestos was formed on the foraminate cathode of the cell, using the method described in Example 1. The diaphragm was then treated on the cathode with 0.5 ℓ of a solution of polyvinyl alcohol in water of concentration of 40 g/ℓ, the polyvinyl alcohol being Polyviol W25/140 (Wacker-Chemie GmbH). The diaphragm was then dried at 90°C for 16 hours. The cathode with the diaphragm was then set up in the cell, the anode-cathode distance being adjusted to 5 mm.

In the cell, the diaphragm was treated with a brine saturated with sodium chloride, while proceeding to electrolyze the brine under the conditions stated above. At the end of a period of 20 days' electrolysis, the electrolyzing voltage, measured at the cell terminals was 3.18 V and the permeability of the diaphragm has risen 0.114 h^{-1}.

Polymeric Fluorocarbon Binders

Improved asbestos diaphragms for use in electrolytic chlor-alkali cells are prepared by *R.N. Beaver and C.W. Becker; U.S. Patents 4,093,533; June 6, 1978 and 4,142,951; March 6, 1979; both assigned to The Dow Chemical Company* by using polymeric fluorocarbons as binder for mixtures of chrysotile asbestos and crocidolite asbestos. These diaphragms are particularly effective in permitting operation of the cells at low pH, thereby extending the life of graphite anodes.

The chrysotile fibers and the crocidolite fibers are preferably about ¼" or more in length and the fiber bundles, as normally mined, have been refined to open up the bundles. Commercially available refined asbestos is suitable.

The fluorocarbon polymers may be solid, particulate polymers or copolymers of tetrafluoroethylene, trifluoroethylene, vinylidene fluoride, vinyl fluoride, monochlorotrifluoroethylene or dichlorodifluoroethylene or may be fluorinated ethylene/propylene copolymer commonly known as FEP. Also, a copolymer of ethylene/chlorotrifluoroethylene (Halar) may be used. Preferably, the fluorocarbon polymer is polyvinylidene fluoride, fluorinated ethylene/propylene copolymer, or polytetrafluoroethylene. Most preferably, the fluorocarbon polymer is polyvinylidene fluoride.

Example 1: A diaphragm was prepared for use in a test cell as follows. A dispersion of polyvinylidene fluoride (Kynar) powder was prepared by mixing in a Waring blender 80 g Kynar, 250 ml H_2O and 8 ml of a nonionic surfactant (alkylaryl polyether alcohol plus about 20% isopropanol).

Crocidolite asbestos [Type 713 (North American Asbestos Company)] was mixed at 0.5 lb/gal of water and vigorously agitated for about 5 minutes in a Cowles dissolver.

Chrysotile asbestos [Plastibest (Johns-Manville)] was mixed at 0.5 lb/gal of water and vigorously agitated for about 5 minutes in a Cowles dissolver.

Equal portions of the asbestos dispersions were mixed together and diluted with water to give a slurry containing 10 g/ℓ of asbestos with equal parts of chrysotile and crocidolite.

A volume of the slurry, sufficient to give 21 g of asbestos, was thoroughly blended with a portion of the Kynar slurry, sufficient to give 3.15 g of Kynar. The resulting slurry was substantially uniformly deposited onto a 13 gauge, 33 square inch perforated steel plate cathode by vacuum filtration. Thus, the deposited material was in an amount of about 0.73 g/in^2. The so-coated cathode was placed in a 180°C oven for 3 hours to effect bonding.

After being cooled, the diaphragm was subjected to a stream of water and it was found that the fibers remained adhered in place and none washed off. Nonbonded diaphragms are easily washed off by a stream of water.

The diaphragm-covered cathode was installed in a small laboratory test chlorine cell used to evaluate diaphragm integrity and operability. After 4 days of operation at a pH in the range of 1.0 to 1.5, the cell was found to be performing excellently.

For comparison purposes, another diaphragm was prepared the same way, except that no binder was used; this nonbonded diaphragm had to be removed from service after 18 hours of operation because of disintegration which caused some fibers to wash off the cathode, stop circulation and cause hot spots which resulted in the cell boiling and high voltage drop across the cell.

Example 2: In a manner substantially as shown in Example 1, polytetrafluoroethylene (Teflon) powder is used as a bonding agent for a 50:50 mixture of crocidolite and chrysotile. There are commercially available Teflon dispersions which are suitable for use directly in this process.

In this example, micron size Teflon, available in a spray can, is used. Also in this example, TiO_2 is also mixed into the asbestos to aid in the wetting (since Teflon resists wetting) and to impart greater permeability. The bonding is effected by placing the vacuum-deposited diaphragm in a 300° to 400°C oven for about 4 to 8 minutes under a nitrogen atmosphere.

The diaphragm-covered cathode is placed in the chlor-alkali test cell and prewetted with methanol and then flushed with water, thereby decreasing the hydrophobicity of the diaphragm. The test cell is operated at a pH in the range of about 1.0 to 1.5 pH and the diaphragm is found to resist disintegration at this low pH and has substantially longer life and greater operability than non-bonded asbestos diaphragms.

K. Motani, S. Matuura and R. Kataoka; U.S. Patent 4,070,257; January 24, 1978; assigned to Electrode Corporation disclose an electrolytic process using a diaphragm of a size of at least 1 cm in length in at least two directions, the diaphragm comprising a diaphragm base composed of fibers of fibrous structure containing inorganic fibers and a binder containing a fluorine-containing resin, the fluorine-containing resin being present in an amount of 0.5 to 60% by weight based on the weight of the diaphragm.

Among the inorganic fibers, asbestos fibers, rock wool and cloths and papers prepared therefrom are normally used with advantage, since they are inexpensive and easily available.

The fluorine-containing resins to be used as the binder include polymers of compounds which contain at least one fluorine atom in their molecules, as well as an ethylenic double bond, copolymers thereof, copolymers of at least one of such compounds with other compounds having an ethylenic double bond and fluorinated rubbery or resinous polymers of polyolefins, polyvinyl halides, etc.

The fluorine content of the resins is not particularly limited, but normally those containing at least 10% by weight, preferably at least 20% by weight, of fluorine are used. More specific examples of such resins include known fluorine-containing resinous polymers and copolymers, such as polytetrafluoroethylene, tetrafluoroethylene-hexafluoropropylene copolymer, polychlorotrifluoroethylene, polyvinyl fluoride, polyvinylidene fluoride, etc.

Selection of a specific fluorine-containing resin should be effected in each individual case, according to the intended utility of the diaphragm to be produced. For example, polytetrafluoroethylene, tetrafluoroethylene-hexafluoropropylene copolymer and polychlorotrifluoroethylene exhibit excellent chemical resistance properties, i.e., acid resistance, alkali resistance and oxidation (chlorine) resistance, etc.

Example 1: 6 g of white asbestos (No. 6) were immersed in 200 cc of water containing 0.3% of nonionic surfactant (Solgen 90) and thoroughly dispersed under violent stirring. 0.6 cc of aqueous dispersion of polytetrafluoroethylene at the concentration of 60 wt % (Polflon D-1), which was diluted to one-tenth with water, was added to the asbestos slurry under continuous stirring.

Approximately 10 minutes thereafter, 50 cc of acetone were poured into the system, followed by approximately 10 minutes stirring. The resulting slurry was made into paper in accordance with JIS P8209. (The fiber slurry was thoroughly stirred in a cylindrical vessel with wire gauze attached on the bottom, with perforated plate stirrer. The draining cock at the lower part of the vessel was then fully opened and the contents were packed under reduced pressure to form wet paper on the wire gauze. The paper was couched, pressed and dried.)

After drying, the paper was pressed under a pressure of approximately 150 kg/cm^2 and allowed to stand in a 400°C electric oven for 10 minutes. A paperlike membrane of 0.37 mm in average thickness was obtained thusly, which had an average tensile strength of approximately 75 kg/cm^2 and permeability to saturated brine of 0.30 ml/hr/cm^2/cm H_2O.

The membrane showed no weight change after more than 50 weeks' immersion in anolyte and catholyte for diaphragm electrolysis of sodium chloride and had a permeability to saturated brine of 0.30 ml/hr/cm^2/cm H_2O.

The properties of diaphragms of varied polytetrafluoroethylene content prepared similarly to the above, with the exception that the polytetrafluoroethylene concentration in Polyflon D-1 was varied for each run, were measured similarly to the above, with the results as given in the following table.

Run No.	Polytetrafluoroethylene Content (wt %)	Thickness (mm)	Permeability (ml/hr/cm^2/cm H_2O	Tensile Strength (kg/cm^2)
1	1	1.0	0.32	8*
2	9	0.37	0.30	75
3	20	0.40	0.14	89
4	30	0.39	0.04	112
5	56	0.50	0.003	102*

*The tensile strength in this run was measured under wet conditions.

Control 1 – The paper-making procedures of Example 1 were repeated, except that no Polyflon D-1 was used. Thus, the paperlike membrane of 0.39 mm in average thickness which was obtained had a low mechanical strength, i.e., average tensile strength of approximately 4 kg/cm^2 and very quickly disintegrated and broke in aqueous liquid.

Example 2: The paper-making procedures of Example 1 were repeated, except that Polyflon D-1 was replaced by 0.8 cc of tetrafluoroethylene-hexafluoropropylene copolymer in aqueous dispersion, 50 wt % concentration (Neoflon ND-1) which was diluted to one-tenth with water.

A paperlike membrane of 0.38 mm in average thickness was thus obtained which contained approximately 9 wt % of the copolymer of tetrafluoroethylene-hexafluoropropylene. The membrane was then heated for 7 minutes at 295°C. The resulting diaphragm had an average tensile strength of 76 kg/cm^2 and a permeability to saturated brine of 0.30 ml/hr/cm^2/cm H_2O.

By way of testing the chemical resistance of the diaphragm, it was immersed in the anolyte and catholyte for sodium chloride electrolysis for more than 50 weeks in a manner similar to Example 1. The diaphragm showed no change in weight or permeability, thus demonstrating its excellent performance.

Through the procedures similar to the above, except that polyvinylidene fluoride was used as the fluorine-containing resin, a 0.45 mm thick diaphragm was obtained. Using measurements similar to the above, the diaphragm was confirmed to have a tensile strength of 72 kg/cm^2 and a permeability of 0.28 $ml/hr/cm^2/cm\ H_2O$.

Example 3: A 100 cm^2 piece of asbestos paper [No. 2,500 (Asahi Sekimen Co.)], 0.5 mm thick, was hung in hot water with its one end held with a pincher and boiled to remove the starch binder. Then, from both surfaces of the paper, several thin layers were peeled off to control its thickness and the paper was hung and immersed in approximately 500 ml of Polyflon D-1, which was controlled to 5° Bé (5.7 wt %).

After 5 hours' immersion the paper was withdrawn, drained of water, dried in a drying oven and pressed for 5 minutes under a pressure of approximately 150 kg/cm^2. Finally, the membrane was heated in an electric oven at 400°C for approximately 10 minutes so that the polytetrafluoroethylene in the resinous binder was sintered.

The resulting paperlike membrane contained approximately 31 wt % of polytetrafluoroethylene, and an average thickness of 0.36 mm, average tensile strength of 53 kg/cm^2 and a permeability to saturated brine of 0.03 $ml/hr/cm^2/cm\ H_2O$.

Example 4: The paperlike membrane obtained in Example 3 was immersed in methanol for 10 minutes and the impregnated alcohol was exchanged with water. The resulting membrane was used in the diaphragm process for sodium chloride electrolysis under the conditions shown in the table below, with the results given in the same table.

Conditions	
Brine composition	
NaCl	316 g/ℓ
Na_2CO_3	0.5 g/ℓ
Ca + Mg	No more than 10 ppm
Brine pH	3.50
Electrolytic cell	Filter press-type electrolytic cell with metallic electrode (Ir)
Current density	30 A/dm^2 (cell voltage, 4.5 V)
Results	
Catholyte composition	
NaOH	145 g/ℓ
$NaClO_3$	0.2 g/ℓ
Na_2CO_3	Trace
NaCl	183 g/ℓ
Current efficiency on cathode base	96%
Liquid temperature	
Anolyte	91°C
Catholyte	95°C

This diaphragm has been used for more than one year continuously with no trouble whatsoever. When the current density was varied as in the following table, the resulting NaOH concentration and current efficiency were as also given in the same table. As clearly demonstrated by the results, better electrolysis performance was obtained under higher current density.

Run No.	Current Density (A/dm^2)	NaOH Concentration (g/ℓ)	Current Efficiency (%)
1	10	110	82
2	15	139	94
3	20	145	96

Control 2 – The electrolysis of Example 4 was effected similarly, except that the conventional asbestos paper (Neoplene binder) was used as the diaphragm. Satisfactory results were obtained at low current density (~10 A/dm^2), but the diaphragm was damaged after approximately 8 hours' use and could no longer be used. When the current density was raised to 20 to 30 A/dm^2, the diaphragm was damaged within a very short time.

Diaphragm Containing Thermoplastic Resin

C.R. Dilmore, E.V. Hoover and A.B. Kriss; U.S. Patent 4,186,065; January 29, 1980; assigned to PPG Industries, Inc. disclose a method of preparing a resin-containing asbestos diaphragm. The diaphragm is prepared by depositing asbestos fibers and resin from an aqueous slurry onto a liquid permeable cathode and subsequently heating the deposited asbestos fibers and resin to cause the resin to bond the asbestos fibers together. As disclosed, air flow is maintained through the diaphragm until the diaphragm is substantially free of entrained water.

The heated air is maintained at a temperature below the boiling temperature of entrained water so as to avoid boiling the entrained water. Thereafter, the temperature of the deposited diaphragm is heated to cause the resin to bind the asbestos fibers together.

The polymeric material used is not critical as long as the material used is a thermoplastic and has some chemical resistance to nascent chlorine when used in combination with asbestos. Suitable resins may be hydrocarbons, halogenated hydrocarbons, halocarbons or copolymers thereof.

Example: A series of tests were conducted to determine the effect of drawing a vacuum during drying on a series of diaphragms drawn from a slurry of asbestos and Halar (Allied Chemical) alternating ethylene-chlorotrifluoroethylene resin in aqueous sodium hydroxide-sodium chloride solution.

The slurries contained 1.6 to 1.9 wt % solids in an aqueous solution of 115 to 135 g/ℓ of sodium hydroxide and 175 to 200 g/ℓ of sodium chloride. The solids were Chlorobestos asbestos (Johns-Manville), Halar 5004 alternating ethylene-chlorotrifluoroethylene polymer powder and Merpol SE surfactant (DuPont). The concentration of the ethylene-chlorotrifluoroethylene ranged from 3.3 to 9.0 wt % and the concentration of the surfactant was 1 wt %, based on the weight of the ethylene-chlorotrifluoroethylene.

The diaphragms were deposited on the cathodes by placing an individual cathode unit into a tank of the slurry, drawing a vacuum on the cathode and drawing the slurry through the foraminous surfaces of the cathode.

The diaphragm deposition was accomplished by filling the drawing tank with 2,000 gal of the slurry and submerging the cathode in the slurry so as to provide at least 2" of slurry above the highest foraminous areas of the cathode.

Drawing is started by drawing a vacuum of 1.5" of mercury within the cathode. After 3 minutes, the vacuum is increased to 2.5" of mercury and maintained at 2.5" of mercury until a film of fibers is present on the foraminous surfaces of the cathode. Thereafter, the vacuum is increased to 15" of mercury for 1 minute and then to over 25" of mercury.

The cathodes are maintained in the slurry under a vacuum of over 25" of mercury for about 10 minutes so as to deposit from about 0.3 to 0.4 lb of asbestos per square foot of foraminous cathode area. Thereafter, the cathodes are removed from the slurry tank and, still under a vacuum of over 25" of mercury, allowed to dry at a temperature of about 65° to 75°F for about 20 minutes.

The diaphragms were then dried according to one of two drying cycles. In one drying cycle no vacuum was applied to the cathode during drying. The drying cycle without vacuum was carried out as follows:

(1) The cathode assembly with the deposited diaphragm was placed in an oven and the air feed to the oven was heated from ambient temperature to 200°F over a period of 2 hours;

(2) The oven air temperature was then increased from 200° to 220°F over 1 hour and maintained at 220°F for 5 hours;

(3) The oven air temperature was then increased from 220° to 480°F over 13 hours at the rate of 20°F/hr and from 480° to 532°F in 1 hour. The oven air temperature was then maintained at 532°F until the cathode and diaphragm attained a temperature of 530°F and then maintained at that temperature for 1½ hours;

(4) The cathode and diaphragm were then cooled to 464°F, the melting point of Halar 5004 alternating ethylene-chlorotrifluoroethylene, over a period of 20 minutes;

(5) The cathode and diaphragm were then cooled naturally to ambient temperature. The resulting diaphragms were blistered; and

(6) After heating, the cathode units were then reinserted in a slurry of 1.5 to 1.9 wt % Chlorobestos 25 asbestos in aqueous cell liquor to deposit a second coat containing from 0.03 to 0.45 lb of asbestos per square foot atop the diaphragm.

In the second drying cycle, the drying was carried out while applying a vacuum to the cathode assembly. The alternative drying cycle with vacuum was carried out as follows:

(1) The cathode assembly with a deposited diaphragm was placed in an oven. A vacuum of 5" of mercury was applied to the cathode and diaphragm. The oven air temperature was heated from ambient to 210°F over a period of 2 hours and maintained at 210°F for 2 hours;

(2) The oven air temperature was then increased to 220°F and maintained at 220°F until the relative humidity of the air in the vacuum

line was below about 1%. This took about 6 hours;

(3) The oven air temperature was then increased from 220° to 400°F at the rate of 20°F/hr for 9 hours. When a temperature of 400°F was attained, the vacuum pump was turned off;

(4) The oven air temperature was then increased at the rate of 20°F/hr for 4 hours from 400° to 480°F and 480° to 532°F in 1 hour;

(5) The oven air temperature was maintained at 532°F until the diaphragm reached 530°F (~½ hour). The diaphragm was then maintained at 530°F for about 1½ hours;

(6) The diaphragm was then cooled to 464°F in 20 minutes and thereafter allowed to cool naturally to ambient temperature. The resulting diaphragms appeared to be free of blisters; and

(7) After heating, the cathode units were then reinserted in a slurry of 1.5 to 1.9 wt % Chlorobestos 25 in aqueous cell liquor to deposit a second coat containing from 0.03 to 0.045 lb of asbestos per square foot.

The electrolytic cells were then assembled by placing the cathode units atop the anode-equipped base members so that the anode fingers extended upward between the cathode fingers. The cell top was then placed atop the cathode unit.

The cells were started up by filling the anolyte compartment with water up to a level above the top of the cathode fingers. Thereafter, saturated brine was fed to the anolyte compartment and dilute brine was recovered from the catholyte compartment for 1½ hours prior to start-up. Current was then passed through the cell.

The results were statistically analyzed: at the 99% level of significance, the vacuum-treated, resin-containing asbestos diaphragm had a lower voltage than both nonresin-containing asbestos diaphragms and resin-containing asbestos diaphragms dried at ambient pressure without forced convection of air through the diaphragm.

Impregnation with Divinylbenzene and Vinylpyridine Copolymer

O. de Nora, L. Giuffre and G. Modica; U.S. Patent 4,186,076; January 29, 1980; assigned to Oronzio de Nora Impianti Elettrochimici, SpA, Italy describe porous and electrolyte permeable composite diaphragms for electrolysis cells comprising a chemically inert, fibrous porous matrix or substrate impregnated with a copolymer of divinylbenzene and at least one member of the group consisting of 2-vinylpyridine and 4-vinylpyridine. Preferably, the copolymer is directly formed on the inert fibrous substrate in the absence of a solvent.

The advantages of the composite diaphragms of the process are their great resistance to mechanical abrasion caused by brine flow and by chlorine bubbles. Their dimensional stability allows thinner diaphragms and smaller interelectrodic distances with a considerable saving of electric energy. Moreover, because the distance between the anode side of the diaphragm and the active surface may be accurately preset and is thereafter maintained, a better faraday efficiency, together with a reduction of chlorate concentration in the catholyte, is obtained even when operating with a higher caustic soda concentration in the catholyte.

Preferably, the inert fibrous material is asbestos paper or asbestos mat, such as that produced by pulling a slurry of asbestos fibers through a foraminous metal electrode, but other fibrous inert materials may be used such as polyester fibers in the form of woven or unwoven felt or cloth and woven or unwoven carbon fiber felts.

The ratio of divinylbenzene to the vinylpyridine is 1:16 to 1:1, preferably between 1:9 and 2:3. The composite diaphragm preferably contains 5 to 25% by weight of the final diaphragm of the copolymer, most preferably 8 to 16% by weight.

Example: 38 g of methylcellulose and 80 mg of antifoam surfactant were dissolved in 7,800 ml of water and 52 g of asbestos fibers (3T) and 104 g of asbestos fibers (4T) (Quebec Producers' Association Quebec Screen Test Classification) were added to the solution, which had a viscosity of 35 cp at 22.8°C. The slurry thus obtained was thoroughly stirred for one-half hour by compressed air injection. A steel cathode of 220 x 330 mm provided with a chamber where an outlet was connected to a suction system was immersed in the slurry and then an increasing vacuum up to 690 mm Hg was applied thereto.

5.5 ℓ of slurry were drawn onto and through the cathodic structure resulting in the formation of a mat of 14 g/dm^2 of asbestos fibers. Several coated cathodes were prepared by this procedure.

The coated cathodic structures were then dried at 100°C for 3 hours and were then soaked at room temperature in benzene solutions of 4-vinylpyridine and divinylbenzene at different concentrations and with different molar ratios of the monomers, as indicated in the table.

Run	Percent Weight Increase Due to Copolymer	Molar Ratio of 4VP/DVB	Cell Voltage (V)	Percent Cathodic Current Efficiency	State of Diaphragm at End of Run
1	5	9:1	3.2	92	slight swelling
2	9	9:1	3.25	92	very slight swelling
3	10	9:1	3.28	92.5	very slight swelling
4	10	6:1	3.3	94	no swelling
5	10	4:1	3.35	92.5	no swelling
6	20	9:1	3.7	93	no swelling

The solutions also contained dibenzoyl peroxide as polymerization initiator at 1% molar with respect to the mols of the monomers. After evaporating the solvent under vacuum at room temperature, the exposed surfaces of the supporting cathodes were washed with benzene to avoid coating of the cathodic surfaces with the copolymer.

The cathode assemblies were then heated in a closed reactor designed to reduce to the minimum the space for the gas, at 80°C, for 3 hours. After cooling, they were washed with benzene to remove possible unreacted monomers. The diaphragms, after drying were weighed to determine the increase in weight of different amounts of 4-vinylpyridine/divinylbenzene copolymer as shown in the table.

The diaphragms were then tested in pilot diaphragm cells for the electrolysis of sodium chloride, operating at the following conditions: anode, titanium mesh activated with mixed oxides of titanium and ruthenium; brine feed, 310 g/ℓ of sodium chloride; current density, 2,000 A/m^2; and temperature, 80°C.

Caustic concentration in the catholyte was constantly kept at 160 g/ℓ by suitably varying the anolyte heads on the diaphragms. After 1,000 hours of continuous operation, the results were obtained, as a function of the copolymeric load on the diaphragms and the molar ratio of monomers, as illustrated in the table.

The table also shows that copolymer weight increases from 5 to 20% of the total weight of the diaphragm are effective in rendering it dimensionally stable, with elevated cathodic current efficiencies and not exceedingly high cell voltages.

Treatment with a Silicate Containing Magnesium

Asbestos diaphragms in commercial cells for the electrolysis of brines are treated by *I.V. Kadija and H.M. Patel; U.S. Patent 4,169,774; October 2, 1979; assigned to Olin Corporation* with a dispersion of a silicate containing magnesium. Deposition of the dispersion within the diaphragm results in a decrease in hydrogen concentration in chlorine gas produced as well as an increase in current efficiency. Preferred as magnesium-containing silicates are magnesium silicate, sepiolite, meerschaum, palygorskite, attapulgite and antigorite, with sepiolite and meerschaum being more preferred.

Particles of sepiolite, for example, are admixed with solutions of sodium chloride to form a dispersion. Suitable concentrations of sepiolite include those in the range of from about 1 to 1,000, preferably from about 50 to 300 g/ℓ of sodium chloride brine. Upon admixing the sepiolite particles with the brine, the sepiolite is dispersed throughout the brine solution. The dispersion is accomplished without the need of a dispersing agent. This dispersion is then fed to the anode compartment of the diaphragm cell.

Any suitable amount of the dispersion may be added to the diaphragm cell. For example, amounts of dispersion added to the anolyte include those of from about 0.1 to 10% by volume of anolyte brine.

When dispersed in the brine, the sepiolite particles are hydrated and swell to become gel-like. Upon swelling, the specific gravity of the gel-like particles approaches the specific gravity of the sodium chloride brine solution. As the brine contacts and passes through the porous diaphragm, these hydrated particles are readily deposited on and throughout the porous asbestos diaphragm.

When deposited within the diaphragm, the gel-like particles of the magnesium-containing silicate blend with the gel layer formed within the asbestos diaphragm. This deposition results in the renewal and reinforcement of the diaphragm and thus in the prevention or reduction of hydrogen molecules or hydroxide ions entering from the cathode compartment.

Hydrogen concentration in the chlorine gas removed from the anode compartment is lowered substantially, the anode current efficiency with respect to chlorine production is increased and chlorate formation is reduced.

Example: A commercial chlorine cell for the electrolysis of sodium chloride (315 g/ℓ) used a porous asbestos diaphragm modified by the incorporation of a polymer of fluorinated hydrocarbon. Measurement of the chlorine gas from the anode compartment showed hydrogen was present in an amount of 3.5% by volume, and the catholyte cell liquor produced had a sodium hydroxide concentration of 68 g/ℓ. Power consumption per ton of chlorine at 130 kA was found to be 2,715 kW hours.

A dispersion was prepared by admixing 50 lb of sepiolite in 30 gal of alkaline sodium chloride brine. The sepiolite, having particle sizes in the range of 0.1 to 5 ml, had an analysis indicating oxides of the following elements were present as percent by weight: Si, 79.1; Mg, 9.3; K, 4.8; Ca, 4.8; Al, 1.4 and Fe, 1.4. The sepiolite was dispersed in the brine using a stirrer. To the diaphragm cell were added 3 gal of the dispersion in a 10-minute period. The addition was repeated hourly for 4 hours until 12 gal of the dispersion had been added to the cell.

The following day, 4 additional batches of the dispersion were added to the cell; each of the 5 gal batches were added in a 10-minute period. Within 48 hours, the hydrogen content of the chlorine gas had been reduced to 0.1%.

Catholyte liquor containing 122 g/ℓ of sodium hydroxide was being produced with the power consumption of the cell at 2,595 kW hours per ton of chlorine produced. After a period of 2 weeks, the hydrogen content had increased to 1%. 10 gal of the dispersion were added. Within 48 hours, the hydrogen level had been reduced to 0.1%. 2 weeks later, 10 gal of the dispersion were added to the cell. The next day, the cell was opened and less than 1 lb of the dispersion was found on the bottom of the cell, indicating that essentially all of the dispersion had been deposited on the asbestos diaphragm.

This example shows the effective reduction of the hydrogen level in chlorine gas produced in a cell treated by the method. Further, the example shows improved cell operation resulting in a reduction of the power consumption from 2,715 to 2,595 kW hours while increasing the sodium hydroxide concentrate in the cell liquor.

Deposition of New Diaphragm Material

T.C. Jeffery; U.S. Patent 4,174,266; November 13, 1979; assigned to PPG Industries, Inc. discloses an electrolysis cell and method of operation in which metal anodes (preferably titanium) provided with an electrically-conducting electrocatalytic coating in an anode compartment face diaphragm-covered cathodes in a cathode compartment, in which the anodes are spaced from an imperforate valve metal separating partition by a separating wall behind which the anolyte can recirculate downward; the anodic gases rising in the anode compartment gently circulate the anolyte and the gases are discharged into a brine box above the anode compartment near the center thereof and the anolyte recirculates downward near at least one end of the anode compartment.

Also described is a method of operation which provides circulation from front to back of the anode compartment and from center to sides of the anode compartment and means for adding new diaphragm material to the circulating anolyte to deposit the new diaphragm material on the diaphragms and methods of determining when and how much new diaphragm material to add.

As shown in Figure 2.1, the anode support bars **8b** provide a wall behind the hollow interior of the anode fingers **5**, whereby a portion of the electrolyte which is carried upward in the interelectrodic gap **5c** and in the interior of the hollow anode fingers **5** by the rising anodic gas bubbles (chlorine) is recirculated downward in the spaces **5k** behind the wall of bars **8b** to the bottom of the anode compartments, thus providing up and down circulation from the front to the back of each anode compartment.

Figure 2.1: Anode End of a Bipolar Cell

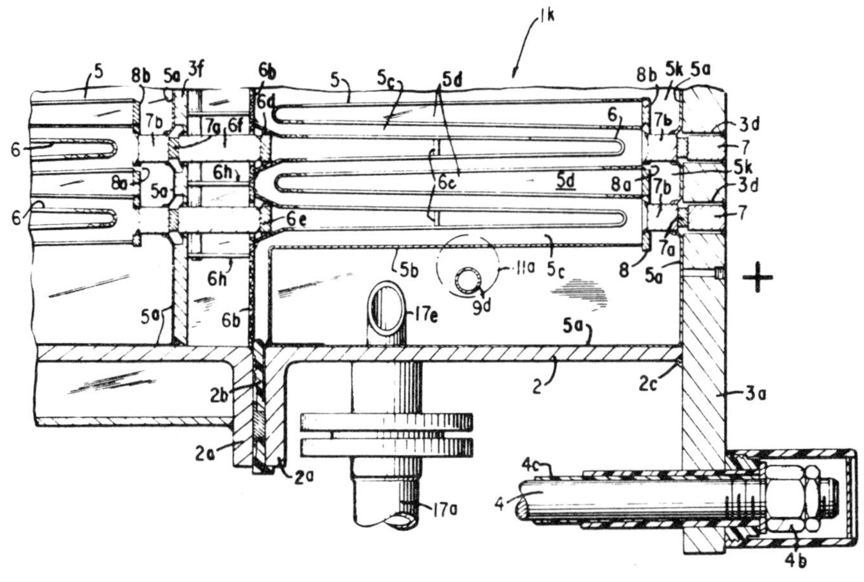

Source: U.S. Patent 4,174,266

The open spaces in the wall permit some equalizing flow of electrolyte between the back space **5k** and the front portion of the anode compartment, but do not interfere with the up and down recirculation of the anolyte.

The higher current density at which these cells operate produces a large volume of bubbles in the front portion of the anode compartment, which causes a vigorous upward flow of the electrolyte contained in this portion of the anode compartment and induces a downward movement of the electrolyte within the back space **5k**.

The volume of the gas release space inside the anode fingers and in the electrodic gap between the anodes and the diaphragm-covered cathodes is more than twice the volume in the electrodic gap, so that in spite of the greater volume of gas bubbles, the flow of the electrolyte upward along the diaphragms is gentle so the new diaphragm can be deposited on the more porous portions of the diaphragms and old diaphragm material on the diaphragms is not disturbed.

Example: An 11-cell bipolar electrolyzer, having titanium-clad steel anolyte chambers and ruthenium dioxide-titanium dioxide-coated titanium anodes interleaved between steel cathodes, was operated for a period of 974 days with the periodic addition of asbestos. The cell, of course, may be so operated for much longer periods of time.

The diaphragms were 3T/4T asbestos diaphragms, deposited on the steel cathode fingers, backing screens, etc., and ranged in weight from 0.35 to 0.42 lb of asbestos per square foot.

The diaphragms were prepared by mixing a 1.5 to 1.8 wt % slurry of 1 part 3T-12 asbestos and 2 parts 4T-12 asbestos in a cell liquor solution containing 11 wt % sodium hydroxide and 19 wt % sodium chloride. The asbestos content was maintained at 1.5 to 1.8 wt % (db) by the periodic addition of asbestos to the slurry.

The diaphragms were deposited by immersing the cathode section of each bipolar unit in the asbestos slurry and drawing the slurry through the screen cathodes.

After assembly of the cell units, operation of the 11-unit cell was commenced and continued in this example for 29 days without the addition of diaphragm material. Asbestos addition in 1 cell of the electrolyzer was commenced at the 29th day and asbestos was added at later times in each cell. The cell operating parameters are shown in the table below. The first addition was made to cell No. 10.

Asbestos Addition to Individual Cells of an 11-Cell Bipolar Electrolyzer

Cell No.	1	2	3	4	5
Initial diaphragm (lb/ft^2)	0.36	0.40	0.35	0.40	0.39
Days to first addition	175	140	107	96	120
No. of Additions	23	46	47	57	57

The number of additions of asbestos varied from 23 to 61 additions and the pounds per square foot per addition varied from 0.0047 to 0.0067 pounds per square foot and from 0.001 to 0.008 pounds per square foot per week.

The sodium chlorate content of the anolyte liquor varied from 0.03 to 0.19 wt % (anhydrous basis) during the first 966 days of operation, but never attained a higher value than about 0.19 wt % (anhydrous basis) during the first 966 days of operation.

When the sodium chlorate content of the anolyte liquor of a particular cell exceeded 0.2 wt % (anhydrous basis) of the anolyte liquor of the cell, and preferably when it exceeded 0.15 wt % of the anolyte liquor of the cell, asbestos addition to the cell was begun and continued until the chlorate content of the anolyte liquor in the cell was brought within the desired limit, and periodic additions were made thereafter as needed.

By continued monitoring of each cell unit and continued addition of asbestos to the cells as indicated by the sodium chlorate content of the cell liquor from each cell, the average sodium chlorate content of the cell liquor from the 11-unit cell was kept at about 0.1% or below for a period of 966 days, with high current efficiency throughout this period, at which time this test was terminated.

Similar monitoring of the hydrogen concentration in the chlorine gas will give similar information and will provide a check on the efficiency of the sodium chlorate monitoring.

By applying similar monitoring to cells equipped with membrane diaphragms and the addition of membrane material to the circulating cell liquor as necessary, the performance of membrane diaphragms may be continued at an approximately steady efficiency and high current efficiency.

ASBESTOS-FREE DIAPHRAGMS

The making of diaphragms for brine-electrolysis cells from asbestos has been widely practiced throughout the world for many decades. The techniques for making diaphragms of this kind which yield satisfactory performance characteristics (such as tolerably low cell voltage at a current density sufficiently high, a desirably low chlorate content in the caustic product, a satisfactory current efficiency and good service life) are well known to those skilled in the art.

Since the brine-electrolysis industry has adopted dimensionally stable anodes, it is necessary for the diaphragm material to give a service life on the order of several hundred days if it is not to become a limiting factor with respect to how long a cell can be operated between renewals.

Asbestos meets these requirements, but most of the materials which have heretofore been tried as a replacement for asbestos have failed in some respect. Either the performance characteristics are poor or they are adequate, but can be maintained only for a relatively short service life, such as one month or less.

The desirability of finding a material to replace asbestos has become increasingly apparent. The mining and handling of asbestos presents a health hazard to the workers dealing with it and this health hazard can be overcome only by adopting measures to protect the involved personnel which add very considerably to the cost of producing and using the asbestos. Not only from the standpoint of the hazard to the personnel involved, but also from the consideration that the spent asbestos diaphragms must be disposed of (and this creates a pollution problem), the widespread use of asbestos is becoming increasingly regarded as intolerable.

The problems to be solved in arriving at an adequate substitute technology are formidable. In the first place, it is not easy to obtain a synthetic substance in a physical form that will approximate the performance of fibers of asbestos. Most of the techniques known have produced fibers that are relatively too coarse, such as tens or dozens of microns in diameter or similar dimension, where what is needed in order to obtain the permeability desired in the product diaphragm is a fiber much finer, on the order of $1 \times 4\ \mu$ in cross section or less.

The environment in which the synthetic fibrous material must operate is a hostile one. On one side of the diaphragm there is a hot caustic solution with a temperature of about 90°C and a pH of 14 or greater. On the other side of the diaphragm is the brine solution, which is also hot, but may be, on the contrary, acidic, with a pH of about 2 to 4.

During operation, there is a considerable evolution of gas taking place on both sides of the diaphragm, so that the solutions in contact with the diaphragm are also turbulent. It is not simple to find materials of the strength and chemical inertness required to suit them for use in such a hostile environment.

The materials which seem most promising in terms of strength and chemical inertness are fluorinated polymers, but they exhibit the concomitant drawback that they are relatively hydrophobic. In contrast, asbestos may be characterized as being hydrophilic. The difficult wettability of the fluorinated polymers is troublesome in that it is difficult to start and maintain a proper flow of liquid through the diaphragm if the diaphragm is difficult to wet. If the diaphragm dewets before (or after) the cell is started, reasonable flow cannot be established through the diaphragm, and the cell is not practically workable.

During operation, partial or total dewetting has a similar bad effect. Accordingly, even if a material of suitable chemical resistance and physical strength is found and produced in a sufficiently divided physical form, other problems indicated above must be solved before a technology to replace the existing practice of making diaphragms from asbestos will be available. Some attempts to produce asbestos-free diaphragms with the desired properties are described in this section.

Fluorinated Polymer with Modified Surface Plies

According to *E.N. Balko, S.D. Argade and J.E. Shrewsburg; U.S. Patent 4,183,793; January 15, 1980; assigned to BASF Wyandotte Corporation*, diaphragms composed in major or important part of the fibers of synthetic material and being substantially or totally free of any content of asbestos, while yet exhibiting not only satisfactory performance characteristics, but also good service life, can be produced by a method which involves:

(a) Taking an appropriate fluorinated polymer;

(b) Putting it in the form of very fine fibers by a method involving dissolving it in a solvent such as tetrahydrofuran which is miscible with water, although the polymer is not, and leading the polymer-solvent mixture through a nozzle under conditions of high shear into a body of water to cause the polymer to be formed into fibers of very small dimension, such as about 0.01 to 40 μ;

(c) Making a slurry of the polymer fiber solution in water with the aid of a surfactant; and then

(d) Using the slurry so produced to deposit a diaphragm upon a cathode of a diaphragm-type electrolytic cell for the electrolysis of brine.

When this is done, and the cell is placed into service, there develops through a period of approximately 2 weeks, a pair of surface plies on the cell-deposited diaphragm which are separable from the main body of the deposited diaphragm, and they exhibit, when tested, a lower molecular weight when determined by the intrinsic viscosity method (70,000 to 150,000 versus 180,000 to 250,000 for the main body of the polymer). Moreover, the burst strength of the diaphragm changes, going from an initial value of, perhaps, 5 to 7 lb/in^2 to an increased value of 20 to 25 lb/in^2, and as a result of the development of such surface plies, the service life of the diaphragms is accordingly increased from a value initially on the order of 30 days or less to a higher value, such as 200 days or more. The tenacious character of the modified surface plies imparts a substantial erosion resistance to the fiber web.

This development constitutes a substantial and significant advance, making it possible to replace existing asbestos-diaphragm technology with an alternative technology in which the use of asbestos is very greatly diminished, if not eliminated entirely. Thus, while continuing to obtain satisfactory performance characteristics such as high caustic concentration and low chlorate levels in the weak-cell-liquor product and at the same time maintaining adequate service life, there is produced a diaphragm which also has capabilities which an asbestos diaphragm does not. It will withstand an acid wash, using, for example, 1:1 water:hydrochloric acid, even if such wash is continued beyond the time that the impurities that it was intended to remove have been caused to disappear, and the diaphragm will, in some cases, make it feasible to produce a caustic soda product which is of higher concentration than would, other things being equal, be obtained.

With respect to the chemical content of the fibers to be used, there is selected a composition based upon a copolymer of, on the average, 24 molecular units of chlorotrifluoroethylene and 1 molecular unit of vinylidene fluoride. Such material is commercially available as Aclon 2100 (Allied Chemical Company). Also suitable is the homopolymer of chlorotrifluoroethylene [Kel-F 81 (3M Company)].

Example: A cell was operated having a diaphragm made in accordance with the process; the composition of the diaphragm was Aclon 2100 polymer. The average cross-sectional dimensions of the fibers used to form the diaphragm were 1 x 4 μ with a length of 0.25 to 0.5 mm. Such fibers were suspended in water, to the extent of 12.7 g/ℓ (dry weight of fiber used), along with 4 g/ℓ of dioctyl sodium sulfosuccinate and 2 g/ℓ of a fluorine-containing surfactant, namely, Fluorad FC-170.

Fiber dispersion and slurry agitation were performed with the use of a propeller-type mechanical agitator driven by a Lightnin mixer. A two-layered web was formed by drawing two successive volumes of slurry through the cathode screen at a ratio of 8.3 ml of slurry per square centimeter of screen area per layer according to the following schedule:

2 minutes at 25 mm of mercury difference from atmospheric pressure;

3 minutes further at 50 mm of mercury difference in pressure; and

2 minutes further at 100 mm of mercury difference in pressure.

The second layer was then applied:

3 minutes at 50 mm of mercury difference from atmospheric pressure;

8 minutes further at 100 mm of mercury difference in pressure; and

2 minutes further at 150 mm of mercury difference in pressure.

The full vacuum of 615 mm of mercury was then applied for 20 minutes. There was obtained a diaphragm having a gross thickness of 2.7 mm and having a permeability coefficient 1.7 x 10^{-9} cm^2. After being dried at 110°C for 16 hours, such diaphragm was installed in a cell with a 6.4 mm electrode gap. The anode was of the DSA type; the cathode was mild steel.

The following performance data were measured at a current density of 160 mA per square centimeter:

Day of Operation	Cell Temp. (°C)	Cell Voltage	Sodium Hydroxide Concentration (g/ℓ)	Sodium Chlorate Concentration (g/ℓ)
13	70	3.31	116	<0.1
40	73	3.18	124	0.10
63	78	3.21	120	0.12
105	62	-	120	0.1

Comparative Example: For comparative purposes, diaphragms have been prepared from fiber of the same dimensions as those of the Aclon 2100 fiber, but made from the 1:1 copolymer of chlorotrifluoroethylene and ethylene. This material is Halar 5004 (Allied Chemical Company). In operation as a chloralkali cell separator, the Halar polymer does not form the surface plies which confer the desirable properties on diaphragms of Aclon 2100 and Kel-F 81 fluoropolymers.

One such diaphragm, a two-layered web, was prepared by essentially the same procedure described in the example. The diaphragm was installed in a chloralkali cell and operated at 160 mA/cm^2, 80° to 85°C, and at a 6.4 mm electrode spacing. After 7 days of operation, the diaphragm had failed completely. Inspection revealed that the electrolyte turbulence within the cell had so severely eroded the deposited Halar web that no diaphragm remained on most of the cathode screen.

Molecular weight determinations were made on the remaining polymer from several failed Halar diaphragms. The molecular weight determination was made by gel permeation chromatography in ortho-dichlorobenzene at 160°C. There was little, if any, polymer degradation. Diaphragm failure was due to hydraulic effects.

Tetrafluoroethylene with Microstructure of Nodes Interconnected by Fibrils

K.T. McAloon; U.S. Patent 4,089,758; May 16, 1978; assigned to Imperial Chemical Industries, Limited, England, describes an electrochemical cell having an anode and a cathode separated by a diaphragm wherein the diaphragm comprises a porous polymeric material containing units derived from tetrafluoroethylene, the material having a microstructure characterized by nodes interconnected by fibrils. The porous polymeric material comprising the diaphragm is as described in U.K. Patent 1,355,373 (corresponding to South African Patent 713,287).

The porous polymeric material is prepared by a process which comprises forming a shaped article of a tetrafluoroethylene polymer by extruding a paste of the polymer, expanding the shaped article by stretching it in one or more directions, heating the polymer while in its stretched condition to a temperature above the melting point of the polymer, and maintaining the resultant porous article in its stretched condition while cooling.

The porosity that is produced by expansion is retained, for there is little or no coalescence or shrinking on releasing the cooled final article. The optimum heat-treating temperature is in the range of 350° to 370°C and the heating periods may range from about 5 seconds to 1 hour. The stretching is effected biaxially.

The porosity of the sintered sheet may be varied by introducing slight modifications into the manufacturing process; in particular, an increase in stretch ratio gives rise to a product of high porosity. In addition, the temperature of heat treatment of the product is another important parameter as it is possible to enhance the extensibility of the tetrafluoroethylene polymer if the product is heat-treated to 327°C or greater.

Since the porosity of the diaphragm can be varied by altering the processing conditions, diaphragms of different brine permeabilities can be obtained so that the porosity and, therefore, permeability of the diaphragm may be chosen according to diaphragm cell size and shape in order to gain efficient alkali halide conversion.

Example 1: A 12.6 cm x 9.6 cm x 1 mm piece of porous polytetrafluoroethylene [GORE-TEX, Grade L10231 sheet (W.L. Gore and Associates, Inc.) in accordance with the process described in British Patent 1,355,373] was successively treated with a 10% w/w aqueous solution of sodium hydroxide at ambient temperature for 2 hours, a 10% w/w aqueous solution of hydrochloric acid at ambient temperature for 2 hours, and a 10% w/w aqueous solution of sodium dihydrogen phosphate at the boiling point of the solution (~100°C) for 1 hour.

The polytetrafluoroethylene sheet was mounted in a vertical diaphragm cell for the electrolysis of sodium chloride. The cell was fitted with a mild steel mesh cathode and had an anode/cathode gap of 9 mm. Brine was passed through the cell at a rate of 245 ml/hr from a head 9.5 cm high. This corresponded to a permeability of 0.215 ml/hr. Applying current at 2kA/m^2 gave rise to a voltage of 4.03 V. The cell operated at a current efficiency of 95.2% corresponding to a salt conversion of 51%.

Example 2: A 12.6 cm x 9.6 cm x 1 mm piece of porous polytetrafluoroethylene, as above, was successively treated with a 10% w/w aqueous solution of sodium hydroxide at ambient temperature for 2 hours, a 10% w/w aqueous solution of hydrochloric acid at ambient temperature for 2 hours, a 10% w/w aqueous solution of sodium dihydrogen phosphate at the boiling point of the solution (~100°C) for 1 hour, and finally immersed in a constantly agitated 10% w/w suspension of titanium dioxide (0.2 μ average particle size) in isopropyl alcohol for 5 hours.

The polytetrafluoroethylene sheet impregnated with titanium dioxide was removed, washed with isopropyl alcohol to remove excess solid and then mounted in a vertical diaphragm cell for the electrolysis of sodium chloride.

The cell was fitted with a mild steel mesh cathode and had an anode/cathode gap of 9 mm. Brine was passed through the cell at a rate of 315 ml/hr from a head 12.0 cm high. This corresponded to a permeability of 0.218 ml/hr. Applying current at 2 kA/m^2 gave rise to a voltage of 3.26 V. The cell operated at a current efficiency of 95.9% corresponding to a salt conversion of 48.5%.

Hydrophilic Fluorocarbon Resin in Hydrophobic Matrix

R.B. Simmons; U.S. Patent 4,170,539; October 9, 1979; assigned to PPG Industries, Inc. describes a diaphragm having a porous, hydrophobic fluorocarbon matrix, an intermediate layer or film of a hydrophilic fluorocarbon resin on the surfaces of the matrix, and a hydrous oxide of zirconium contained in the void volumes of the matrix. The layer of the hydrous oxide of zirconium may also contain MgO. The hydrophilic fluorocarbon resin is a perfluorinated hydrocarbon having pendant acid groups chosen from the group consisting of $-COOH$, $-SO_3H$ and derivatives thereof.

Example: A diaphragm was prepared by saturating a microporous poly(tetrafluoroethylene matrix with a zirconium oxychloride solution and, thereafter, contacting the matrix with NH_3 vapor.

The matrix was a 25 ml thick poly(tetrafluoroethylene) microporous matrix (Porex P1000) having pores 10 μm in diameter and approximately 80% void volume. The matrix was treated with a 6.5% solution of Nafion 601 polymer (DuPont), a perfluorinated polymer having pendant sulfonic acid groups in ethanol. The polymer was applied to the mat by laying the mat on a flat glass plate and brushing the solution onto the mat until the mat was saturated. The saturated mat was dried in 27°C air for 35 minutes until it appeared dry, followed by heating to 100°C for 30 minutes to drive off any residual solvent and to anneal the coating.

The mat contained 3.39 g of resin per square foot, i.e., 10.7 wt % resin, basis resin plus dry matrix. The mat was then contacted with a solution of zirconium oxychloride, $ZrOCl_2$.

The zirconium oxychloride solution was prepared by adding $ZrOCl_2 \cdot 4H_2O$ 99% assay (PCR, Inc.) to water to obtain a 41 wt % solution of $ZrOCl_2 \cdot 4H_2O$ and then diluting the solution further by adding 9 parts distilled water.

The microporous matrix was then saturated with the zirconium oxychloride solution by submerging the matrix in the solution, drawing a vacuum on the submerged matrix to evacuate the air from the porous matrix, and releasing the vacuum to allow the solution to penetrate and fill the air-evacuated mat.

The drawing and releasing of the vacuum was repeated until there was no further uptake of solution. The matrix was then contacted with NH_3 vapor for 42 hours to hydrolyze the chloride and then stored in distilled water.

The mat was tested, thereafter, as a diaphragm in a laboratory diaphragm cell, With an 0.16" (4.1 mm) anode to cathode gap, a ruthenium dioxide-coated titanium mesh anode and a perforated steel plate cathode, the head was 3" to 6", the average cell voltage was 3.02 to 3.07 V at a current density of 190 A/ft^2, and the cathode current efficiency was 93%.

Thermoplastic Fibers Requiring No Bonding

Diaphragms for electrolytic cells are prepared by *A.S. Patil and E.Y. Weissman; U.S. Patents 4,125,451; November 14, 1978 and 4,154,666; May 15, 1979; both assigned to BASF Wyandotte Corporation* by depositing onto a cathode screen discrete thermoplastic fibers. The fibers are highly branched, and when deposited, form an entanglement or network which does not require bonding or cementing.

Suitable thermoplastic fibers include polyolefins, polycarbonates, polyesters, polyamides, and the like, as well as mixtures thereof. A particularly preferred class of thermoplastic fibers is the fluorinated hydrocarbon, and in particular, fluorinated polyalkylenes.

As is known to those skilled in the art, fluorinated hydrocarbon fibers, per se, are difficult to disperse in an aqueous medium, thereby rendering such fibers difficult to deposit on a cathode screen or support. To alleviate this situation, the disclosure also includes an improved method of dispersing fluorinated hydrocarbon fibers. It has been found that if the fibers are dispersed in an aqueous-acetone medium, and in the presence of a surfactant, to form a slurry, the problems of dispersing the fibers are overcome.

The highly branched fibers can be produced in accordance with the process described in Belgian Patent 795,724 or any other process which produces highly branched fibers. In a preferred method, fibers are produced by a process in which the polymer is dissolved in a suitable solvent such as tetrahydrofuran and then the polymer solution is led through a nozzle under conditions of high energy into an aqueous media in which the solvent is soluble, but the polymer is not.

Example: Into a 1:1 water-acetone medium containing 0.1% by weight of a fluorocarbon surfactant (Fluorad FC-126) were added 6% by weight of polyvinylidene fluoride fibers. The fibers were produced by the process described in Belgian Patent 795,724. The fibers were mixed and dispersed in the medium to form a slurry.

While maintaining the slurry in a state of agitation, a cathode screen mounted in a vacuum box was submerged in the slurry. A partial vacuum of 1" Hg was applied to the box for 3 minutes. The vacuum was then increased to 3" Hg and was applied to the box for 3 minutes. While still maintaining the slurry in a state of agitation, a full vacuum was then applied to the box for 5 minutes.

The so deposited diaphragm on the cathode screen was then dried in an oven for 2 hours at 100°C. The diaphragm was then mounted in a test chlor-alkali cell and subjected to brine electrolysis. The cell with the diaphragm mounted therein produced 98 g/ℓ of caustic at 81% current efficiency.

Support Fabric Impregnated with Nonfibrilic Silica

A diaphragm developed by *I.V. Kadija; U.S. Patent 4,184,939; January 22, 1980; assigned to Olin Corporation* for use in the electrolysis of alkali metal chloride brines in electrolytic diaphragm cells is comprised of a support fabric impregnated with a nonfibrilic active component containing silica where the porous diaphragm has a permeability to alkali metal chloride brines of from about 100 to 300 $ml/min/m^2$ of diaphragm at a head level difference in the cell of from about 0.1" to 20" of the alkali metal chloride brines. The active component containing silica is used in concentrations of from about 10 to 75 mg/cm^2 of support fabric.

Suitable silica-containing materials include sand, colloidal silica, alkali metal silicates, alkaline earth metal silicates, aluminum silicates, as well as minerals such as sepiolites, meerschaums, attapulgites, montmorillonites and bentonites. Support fabrics include, for example, felt fabrics produced from thermoplastics such as polyolefins or polyarylene sulfides.

The diaphragms are physically and chemically stable, can be easily installed in an electrolytic cell, have increased operational life and are produced from inexpensive materials.

Figure 2.2 illustrates a diaphragm suitable for covering a cathode. Diaphragm **1**, comprised of fabric, has end portions **10** attached, for example, by sewing, to diaphragm body **12**. Diaphragm body **12** is a hollow rectangle which is mounted on a cathode (not shown) so that it surrounds the cathode on all sides. End portions **10** have openings **14** which permit end portions **10** to be attached to the the cell walls (not shown).

Figure 2.2: Porous Diaphragm

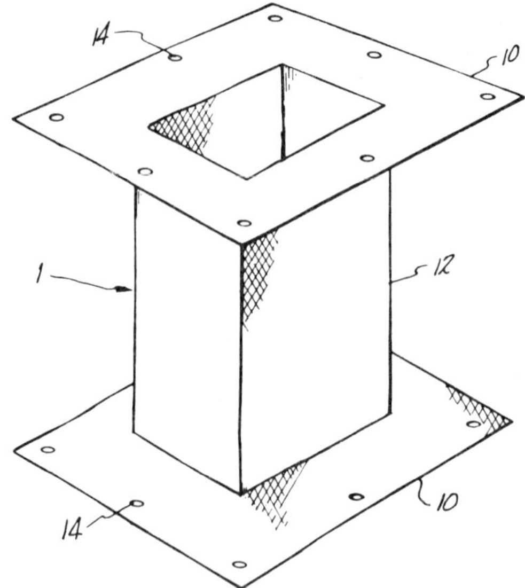

Source: U.S. Patent 4,184,939

Example 1: Sepiolite, having particle sizes in the range between 44 and <1 μ, was added to sodium chloride brine having a concentration of 295 to 305 g/ℓ of NaCl. The sepiolite was dispersed in the brine using a blender until the brine contained about 5% by volume of sepiolite. Analysis of the sepiolite indicated oxides of the following elements were present as percent by weight: Si, 79.1; Mg, 9.3; K, 4.8; Ca, 4.8; Al, 1.4; and Fe, 1.4.

A section of polytetrafluoroethylene felt, 0.048" thick, in the form shown in Figure 2.2, was washed in a caustic soda solution containing 15 to 20% NaOH and at a temperature of 30°C for about 24 hours to remove residues and improve wettability. The felt was then fitted on a steel mesh cathode. The felt had an air permeability in the range of from about 20 to 70 $ft^3/min/ft^2$.

The felt-covered cathode was immersed in the brine containing sepiolite and a vacuum applied to impregnate the felt with the dispersion until a vacuum of 23" to 27" was reached. The vacuum was shut off and the procedure repeated three times.

The impregnated, felt-covered cathode was installed in an electrolytic cell using a ruthenium oxide-coated titanium mesh anode and sodium chloride brine at a pH of 12, a concentration of 300±5 g of NaCl per liter and a temperature of 90°C. Current was passed through the brine at a density of 2.0 kA/m^2 of anode surface. The initial brine head level was 0.5" to 1" greater in the anode compartment than in the cathode compartment. The permeability of the impregnated diaphragm was found to be in the range of from about 200 to 250 ml/m^2 of diaphragm by measuring the rate of catholyte liquor produced.

After about 6 days of cell operation, the premixed dispersion of sepiolite in brine was added to the anolyte. The amount added corresponded to about 3% of the volume of the anolyte compartment of the cell, the addition being made without interruption of the electrolysis process. After a period of 6 weeks, the cell voltage began to increase rapidly and current efficiency was reduced.

While maintaining the cell in operation, a 5% HCl solution was fed to the anolyte compartment and the catholyte liquor was diluted with cold water. Cell performance after treatment of the anolyte and the catholyte was restored to that found earlier. The catholyte liquor produced had a sodium chloride concentration in the range of 130 to 170 g/ℓ.

Example 2: The procedure of Example 1 was duplicated using a polypropylene felt having a thickness of 0.18 of an inch. After 1 week of cell operation, a mixture of colloidal silica and magnesium chloride in a 10% aqueous solution was prepared. The mixture, containing a weight ratio of silica to $MgCl_2$ of 85:15, was added to the anolyte in an amount corresponding to about 3% of the volume of the anolyte compartment. The cell was operated for a period of about 3 weeks at a cell voltage of 3.00 to 3.10 V and produced catholyte liquor containing 122 to 142 g/ℓ of NaOH at a cathode current efficiency of 86 to 92%.

Support Material Impregnated with Sand and Binding Agent

A diaphragm developed by *I.V. Kadija and K.E. Woodard, Jr.; U.S. Patent 4,168,221; September 18, 1979; assigned to Olin Corporation* for use in the electrolysis of alkali metal chloride solutions in electrolytic diaphragm cells is comprised of a mixture of an electrically nonconducting, nonswelling support material impregnated with a mixture of sand and a synthetic thermoplastic polymeric binding agent.

The diaphragm may include a lubricant, a wetting agent or an additive such as alumina, inorganic phosphates, lithium salts, lime, magnesia or inorganic magnesium salts. The diaphragms have increased chemical and dimensional stability, a long operational life and are nonpolluting.

Example: Sand (99% SiO_2), having a particle size smaller than 100 mesh, was added to a tumbler along with polyphenylene sulfide resin [Ryton-PPS V-1 (Phillips Petroleum Company)] particles smaller than 200 mesh and graphite having a particle size of $<$100 mesh.

The components were blended for about 2 hours to provide a mixture containing (by volume) 50% sand, 40% resin and 10% graphite. The mixture was poured into a mold and heated to a temperature of 330°C. Pressure was then applied (12 kg/cm^2) and the mixture allowed to cool down under pressure. Into an electrolytic cell containing brine having a sodium chloride concentration of 315 to 320 g/ℓ, the porous-shaped diaphragm (3.5" x 5.5") was placed adjacent to the cathode.

Electrolysis of the brine was conducted at a current density of 2 kA/m^2 for a period of 20 days to produce Cl_2 gas and sodium hydroxide at a concentration of 115 to 170 g/ℓ at an average power concentration in the range of 2,250 to 2,700 kW hours per ton of Cl_2. During the period of operation, no evidence of plugging was found.

Support Fabric Coated with Electroconductive Metal

I.V. Kadija; U.S. Patent 4,165,271; August 21, 1979; assigned to Olin Corporation has developed additional diaphragms comprising a support fabric of a thermoplastic material which is nonconducting and stable to the gases and solutions found in the cell.

The support fabric is impregnated with an active component containing silica, for example, sand, colloidal silica, alkali metal silicates, alkaline earth metal silicates, aluminum silicate, as well as related mineral products. The impregnated diaphragm has suitable separation properties, however, the electrical resistance is higher than desired. To reduce this resistance, the support fabric is coated on at least one side with an electroconductive metal.

Any suitable electroconductive metal may be used which is stable to the cell environment and does not interact with other cell components. For example, nickel, nickel alloys, gold, gold alloys, platinum group metals, alloys of platinum group metals, and mixtures thereof, are suitable electroconductive metals.

Example: A section of polytetrafluoroethylene felt having a thickness of 0.068" was sprayed on one side with a silver metallizing paint. The silver paint was applied in a manner which provided a noncontinuous coating on the felt and which minimized penetration of the paint into the felt. Electrical conductivity of the painted fabric was determined by contacting the painted surface with two nickel-plated, needlelike electrodes, each having a contact surface of 1 mm^2.

The electrodes, each connected to an ohmmeter, were pressed against the painted side at a pressure of 1 kg/m^2. A distance of 1 cm separated the two electrodes. Silver spraying was discontinued when the resistance was below about 0.1 ohm.

After drying, the painted felt was immersed in an electroplating bath containing an aqueous nickel solution containing: nickel sulfate, 300 g/ℓ; nickel chloride, 60 g/ℓ; boric acid, 6 g/ℓ; sodium molybdate, 0.3 g/ℓ; and vanadyl sulfate, 0.4 g/ℓ.

A current of 0.02 kA/m^2 was passed through the solution for a period of about 4 hours, then the current was increased to 0.1 kA/m^2 for an additional 2 hours. Electroplating was completed using a current of 0.4 to 0.6 kA/m^2 for about 2 hours.

After removal from the plating bath, the felt, coated with a nickel-molybdenum-vanadium alloy, was rinsed in tap water and then washed with a 20% solution of caustic soda. The felt was fitted on a louvered steel mesh cathode with the coated side in contact with the cathode surface.

The felt-covered cathode was immersed in a sodium chloride brine (295 to 305 g/ℓ of NaCl) having dispersed therein about 5% by volume of sepiolite. Analysis of the sepiolite indicated oxides of the following elements were present as percent by weight: Si, 79.1; Mg, 9.3; K, 4.8; Ca, 4.8; Al, 1.4; and Fe, 1.4.

A vacuum was applied to impregnate the felt with the dispersion until a vacuum of 23" to 27" was reached. The vacuum was shut off and the procedure repeated three times.

The impregnated, felt-covered cathode was installed in an electrolytic cell using a ruthenium oxide-coated titanium mesh anode and sodium chloride brine at a pH of 12, a concentration of 315 to 320 g of NaCl per liter and a temperature of 90°C. Current was passed through the brine at a density of 2.0 kA/m^2 of anode surface.

The initial brine head level was 0.5" to 1" greater in the anode compartment than in the cathode compartment. The permeability of the impregnated diaphragm was found to be in the range of from about 200 to 250 ml/m^2 of diaphragm by measuring the rate of catholyte liquor produced. The cell was operated for 6 weeks to produce a catholyte liquor having a concentration of 131 to 188 g/ℓ of NaOH at a cathode current efficiency of 87 to 94%. Cell voltage was in the range of 3.1 to 3.2 V. The catholyte liquid produced had a sodium chloride concentration in the range of 130 to 170 g/ℓ.

Comparative Example: A polytetrafluoroethylene felt having a thickness of 0.068" was impregnated with sepiolite using the procedure described above. The felt, however, had not been previously coated on one side with the Ni alloy. The impregnated felt was then fitted to a louvered steel mesh cathode and electrolysis of sodium chloride conducted in the same cell and using identical conditions and brine concentration. A catholyte liquor having a concentration equivalent to that in the preceding example was obtained at current efficiencies of 87 to 94%; however, the cell voltage was in the range of 3.2 to 3.4 V.

Using the diaphragm of the preceding example having a noncontinuous metal coating results in a substantial decrease in cell voltage over the use of an uncoated diaphragm.

Degassing Procedure

In the process developed by *D.A. Kramer, S.D. Argade and E.N. Balko; U.S. Patent 4,193,861; March 18, 1980; assigned to BASF Wyandotte Corporation* initial cell voltages are reduced by decreasing the resistance of the diaphragm through a degassing procedure prior to or at installation thereof. This degassing procedure involves subjecting the diaphragm to subatmospheric pressure while contacting the diaphragm with electrolyte, the electrolyte being an aqueous saline solution having a surface-active agent therein in an amount sufficient to reduce the surface tension below the critical surface tension for wetting the fibers, and increasing the pressure to atmospheric or cell-working pressure to force electrolyte solution into the interstices of the diaphragm.

The chemical content of one of the preferred fibers to be utilized is a composition based upon a copolymer of, on the average, 24 molecular units of chlorotrifluoroethylene and 1 molecular unit of vinylidene fluoride. Such material is commercially available as Aclon 2000 (Allied Chemical Co.). Another preferred fiber is made from the homopolymer of chlorotrifluoroethylene [Kel-F 81 (3M Company)].

Example: A diaphragm was made and processed according to the method and tested to determine the change in electrical resistance as compared to a diaphragm prepared in accordance with the prior art. The composition of the diaphragm was Aclon 2000 polymer. The average cross-sectional dimensions of the fibers used to form the diaphragm were 1 x 4 μ, with a length of 0.25 to 0.5 mm. Such fibers were suspended in water, to the extent of 12.7 g/ℓ (dry weight of fiber used), along with 4 g/ℓ of dioctyl sodium sulfosuccinate and 2 g/ℓ of a fluorine-containing surfactant, namely, Fluorad FC-170 (3M Company).

Fiber dispersion and slurry agitation were performed with the use of a propellor-type mechanical agitator driven by a Lightnin mixer. A two-layered web was formed by drawing two successive volumes of slurry through a cathode screen at a ratio of 8.3 ml of slurry per square centimeter of screen area per layer according to the following schedule:

2 minutes at 25 mm of mercury difference from atmospheric pressure;

3 minutes further at 50 mm of mercury difference in pressure; and

2 minutes further at 100 mm of mercury difference in pressure.

The second layer was then applied:

3 minutes at 50 mm of mercury difference from atmospheric pressure;

8 minutes further at 100 mm of mercury difference in pressure; and

2 minutes further at 150 mm of mercury difference in pressure.

The full vacuum of 615 mm of mercury was then applied for 20 minutes. There was obtained a diaphragm having a gross thickness of 2.7 mm and having a permeability coefficient of 1.7 x 10^{-9} cm^2. After being dried at 110°C for 16 hours, one of such diaphragms was checked for its resistance factor. Another of such diaphragms was processed further in accordance with the process.

The second diaphragm was treated according to the process by immersing the diaphragm in a container having an electrolyte solution therein. The electrolyte solution contained brine at a concentration of 300 g/ℓ of solution and a surfactant in a concentration of 1 g/ℓ of solution. The surfactant used was Purafac RA-40 (BASF Wyandotte Corporation).

The immersed diaphragm was then subjected to reduced pressure by evacuating means which brought the atmosphere over the electrolyte to about the vapor pressure thereof. This pressure was held for 10 minutes and during this time, entrapped air expanded and left the diaphragm. The pressure was then returned to atmospheric pressure with the diaphragm retained in immersed position in electrolyte, and this forced liquid into the diaphragm pores.

The wet diaphragm was then checked for electrical resistance. The resistance factor determined in the test is defined as the ratio of the diaphragm resistance when flooded with electrolyte to that of an identical volume of the same electrolyte. The diaphragm which was not subjected to the treatment had a resistance factor of 51.1 and diaphragm which was treated had a resistance factor of 4.3.

CELL DESIGN

Non-Melt-Processable Polymer Diaphragm Connected by Melt-Processable Polymer

C. Vallance and P.J. Davies; U.S. Patent 4,153,530; May 8, 1979; assigned to Imperial Chemical Industries Limited, England, provide an electrolytic diaphragm cell for the production of halogen, hydrogen and an alkali metal hydroxide solution by electrolysis of an aqueous alkali metal halide solution, which cell comprises a plurality of vertical anodes vertically providing at least one side of the cell, a cathode box providing at least the opposite facing side of the cell and providing a cathode between adjacent anodes, and a hydraulically-permeable diaphragm between adjacent anodes and cathodes.

The diaphragm comprises a sheet of a porous non-melt-processable fluorine-containing polymer connected to upper and lower slotted supports of a melt-processable, fluorine-containing polymer by means of strips of a melt-processable, fluorine-containing polymer fused to the upper and lower edges of the diaphragm. The supports are located in the cell so that the slots in the upper and lower supports are in vertical alignment with one another and the anodes extend into the space defined by the upper and lower supports and the diaphragms.

The non-melt-processable, fluorine-containing polymer comprising the diaphragm may be of polyvinylidene fluoride, for example, but the preferred polymer is polytetrafluoroethylene.

The diaphragm **6** shown in Figures 2.3a and 2.3b comprises four window frame sheets **1**. It is formed by joining pairs of strips **3** to give overlapping joints at **7**, for example, by hot-pressing to give welded joints or by the application of a suitable cement (e.g., low molecular weight, low-melting point polytetrafluoroethylene). The diaphragm **6** thus obtained has continuous strips **4** of a melt-processable, fluorine-containing polymer along its upper and lower edges respectively, and strips **3** at each end. When in position in a cell, diaphragm **6** adopts the shape as shown in Figure 2.3b.

Figure 2.3: Diaphragm and Cell

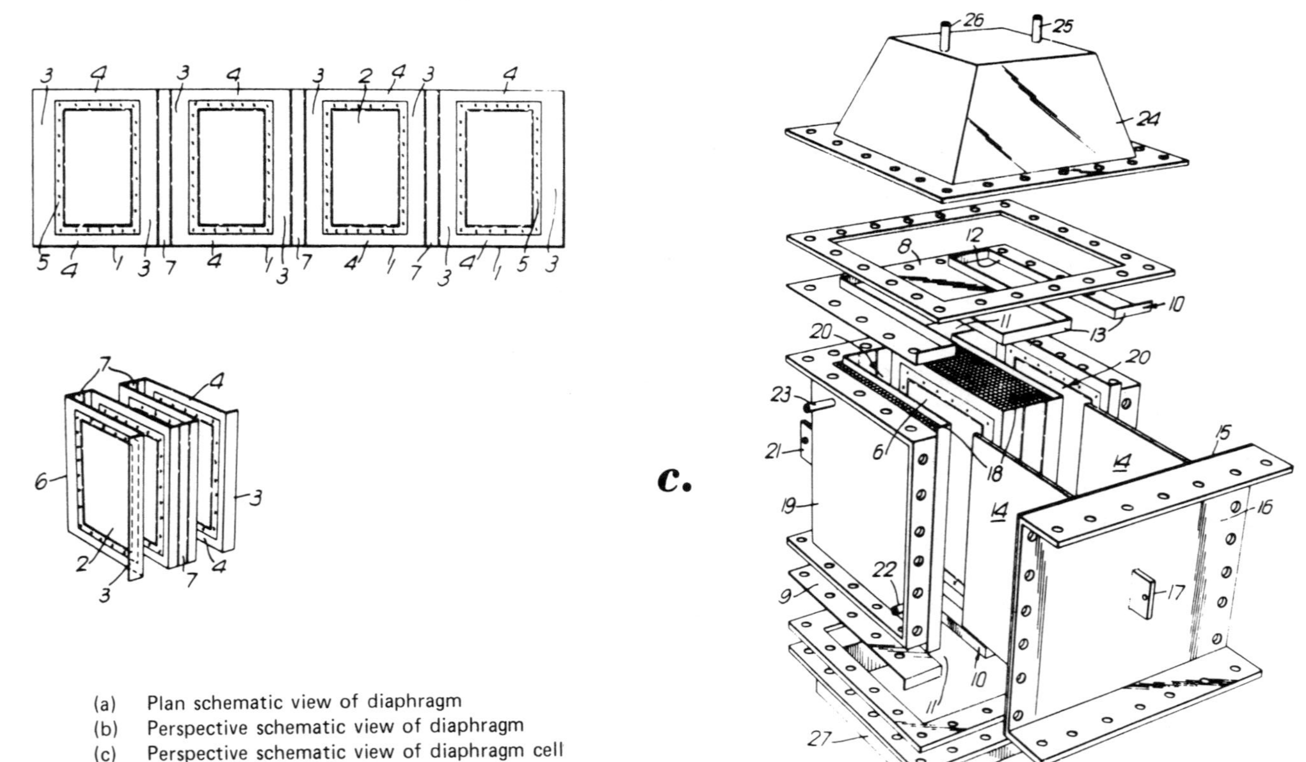

(a) Plan schematic view of diaphragm
(b) Perspective schematic view of diaphragm
(c) Perspective schematic view of diaphragm cell

Source: U.S. Patent 4,153,530

Example: A diaphragm cell of the type shown in Figure 2.3c was provided with three sets of titanium bladed anode plates **14** (blades–6 mm depth, 4 mm apart), coated with a mixture of ruthenium oxide and titanium dioxide, and mounted on a titanium base plate **15**. The anode plates **14** were fitted into the vertically disposed slots **20** of a cathode box **19** provided with mild steel mesh cathodes **18** (2 mm diameter mesh, 2 x 2 mm opening).

The cell was provided with a continuous sheet **6** of polytetrafluoroethylene which was in contact with the cathodes **18**. The diaphragm was fabricated by joining together four window frame sheets **1** (Figure 2.3a) by hot-pressing overlapping strips of a fluorinated ethylene-propylene copolymer fused at or near the edges of the starch-filled polytetrafluoroethylene sheets (2 mm thick).

The diaphragm **6** was, in turn, attached to the upper and lower supports **9, 10** made of a fluorinated ethylene-propylene copolymer by hot-pressing strips of fluorinated ethylene-propylene (previously fused to the upper and lower edges of the diaphragm) to the supports. The anode-cathode gap was 6 mm. The starch was extracted from the diaphragm electrolytically in situ in the cell at a current density of 2 kA/m^2 anode surface.

The cell was fed with sodium chloride brine (300 g/ℓ NaCl) at a rate of 5 ℓ/hr, and the cell was operated at a current density of 2 kA/m^2. The cell operating voltage was 3.2 V. The chlorine produced contained 97.5% by weight of Cl_2 and 2.5% by weight of O_2. The sodium hydroxide produced contained 10% by weight of NaOH. The cell operated at a current efficiency of 96%.

Non-Melt-Processable Diaphragm in Form of Endless Belt

C. Vallance and P.J. Davies; U.S. Patent 4,156,639; May 29, 1979; assigned to Imperial Chemical Industries Limited, England, describe an electrolytic diaphragm cell for the production of halogen, hydrogen and an alkali metal hydroxide solution by electrolysis of an aqueous alkali metal halide solution.

The cell comprises a plurality of anodes vertically mounted on the base of the cell, a cathode box providing cathodes between adjacent anodes, and a hydraulically permeable diaphragm between adjacent anodes and cathodes comprising one or more sheets of a porous non-melt-processable, fluorine-containing polymer joined into the form of an endless belt by a strip or strips of melt-processable, fluorine-containing polymer fused into the sheet or sheets at or near juxtaposed edges of the sheet or sheets, the diaphragms being connected to upper and lower slotted supports of a melt-processable, fluorine-containing polymer by means of strips of a melt-processable, fluorine-containing polymer bonded to the supports at or near the slots therein and fused to the upper and lower edges of the diaphragm.

Referring to Figure 2.4a, the window frame sheet **1** comprises a rectangular sheet **2** of a non-melt-processable, fluorine-containing polymer, for example, polytetrafluoroethylene, which is either porous or contains a removable filler (e.g., starch) which is subsequently removed to provide the desired porosity.

The sheet **2** is provided with strips **3, 4** of a melt-processable, fluorine-containing polymer, for example, a fluorinated ethylene-propylene copolymer, which have been fused into the sheet **1**, e.g., by hot-pressing, to give overlapping joints **5**.

Figure 2.4: Diaphragm and Cell

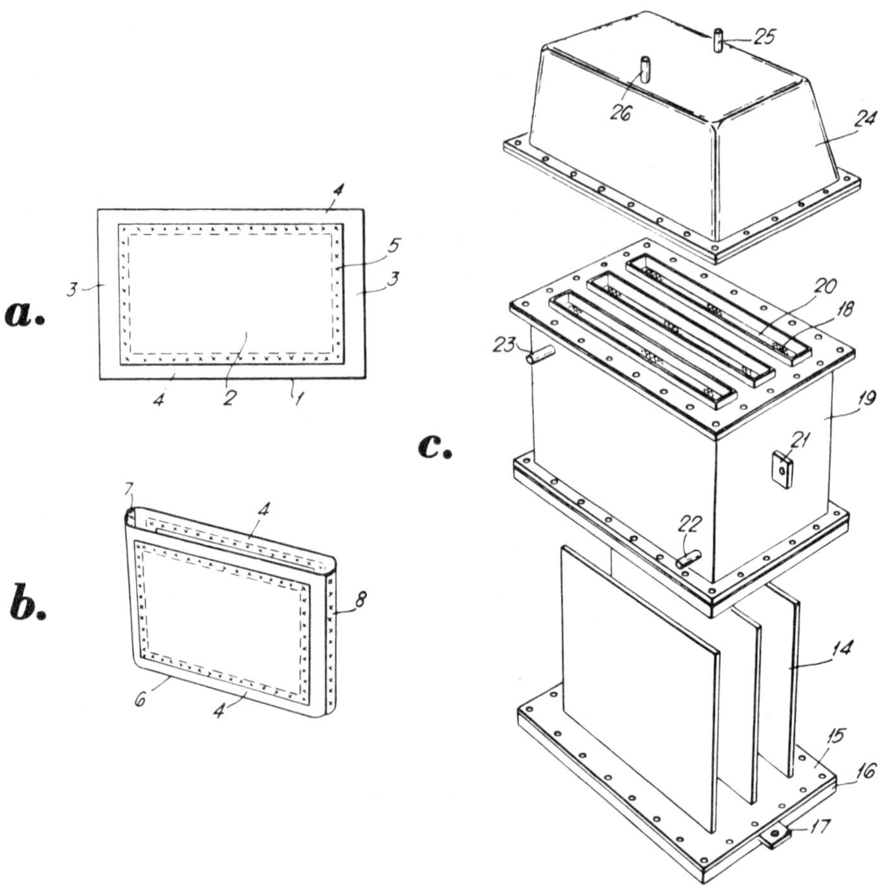

(a) Plan schematic view of window frame sheet
(b) Perspective schematic view of diaphragm
(c) Perspective expanded view of diaphragm cell

Source: U.S. Patent 4,156,639

The endless-belt diaphragm **6** shown in Figure 2.4b comprises two window frame sheets **1**. It is formed by joining two pairs of strips **3** to give overlapping joints at **7, 8**, for example, by hot-pressing to give welded joints or by the application of a suitable cement (e.g., a low molecular weight, low melting point polytetrafluoroethylene). The endless-belt diaphragm thus obtained has strips **4** of a melt-processable, fluorine-containing polymer along its upper and lower edges.

Example: A diaphragm cell of the type shown in Figure 2.4c was provided with three sets of titanium flat anode plates **14** (each 2.5 mm thick), coated with a mixture of ruthenium oxide and titanium dioxide, and mounted on a titanium base plate **15**. The anode plates **14** were fitted into the openings **20** of a cathode box **19** provided with mild steel mesh cathodes **18** (2 mm diameter mesh, 2 x 2 mm opening). The cell was provided with an endless-loop diaphragm **6** (Figure 2.4b) of polytetrafluoroethylene, which was in contact with the cathodes **18**.

The diaphragm was fabricated by joining together two window frame sheets **1** (Figure 2.4a) by hot-pressing overlapping strips of a fluorinated ethylene-propylene copolymer fused at or near the edges of a starch-filled polytetrafluoroethylene sheet (2 mm thick). The diaphragm **6** was, in turn, attached to upper and lower supports (not shown) made of a fluorinated ethylene-propylene copolymer by hot-pressing strips of fluorinated ethylene-propylene (previously fused to the upper and lower edges of the diaphragm) to the supports. The anode-cathode gap was 13 mm. The starch was extracted from the diaphragm electrolytically in situ in the cell at a current density of 2 kA/m^2 anode surface.

The cell was fed with sodium chloride brine (300 g/ℓ NaCl) at a rate of 5 ℓ/hr, and the cell was at a current density of 2 kA/m^2. The cell operating voltage was 3.2 V. The chlorine produced contained 97% by weight of Cl_2 and 3% by weight of O_2. The sodium hydroxide produced contained 10% by weight of NaOH. The cell operated at a current efficiency of 96%.

Monopolar Filter Press Cell

T.W. Boulton and B.J. Darwent; U.S. Patent 4,204,939; May 27, 1980; assigned to Imperial Chemical Industries Limited, England, describe a monopolar filter press electrolytic cell for use in the electrolysis of alkali metal halide brine to produce cell liquor, halogen and hydrogen.

The cell comprises a plurality of anode plates and cathode plates and a hydraulically permeable diaphragm positioned between each adjacent anode plate and cathode plate, and at least one spacing plate of a nonconducting material positioned between each anode plate and adjacent diaphragm and between each cathode plate and adjacent diaphragm.

The anode plates, cathode plates and spacing plates are provided with at least two openings in the faces of the plates which, in combination, define a first compartment lengthwise of the cell separated from the second compartment, the spacing plates between the anode and adjacent diaphragm being provided with at least one passage which permits brine to pass between the first compartment and the anolyte compartments and which permits halogen to be released from the anolyte compartments to the first compartment.

The spacing plates between the cathodes and adjacent diaphragms are provided with at least one passage which permits cell liquor and hydrogen to pass from the catholyte compartments to the second compartment, the anode plates and cathode plates being made, in part, of a nonconducting material so that the first and second compartments are electrically insulated from one another.

Example: A diaphragm cell shown in Figure 2.5 was provided with four titanium louvered anode plates **1** (each 0.75 mm thick), coated with a mixture of ruthenium oxide and titanium dioxide, four mild steel louvered cathode plates **2**

(each 0.75 mm thick), and seven electrostatically-spun polytetrafluoroethylene sheet diaphragms (3 mm thick). The length of the louvers of the anode and cathode plates which follow the direction of current flow was 15 cm. The distance between diaphragm surfaces in the anolyte (or catholyte) compartments was 6 mm. The spacing plates **4** and frames **7, 8** were fabricated in polypropylene and the diaphragm plates **6** were fabricated in synthetic rubber.

Figure 2.5: Monopolar Filter Press Cell

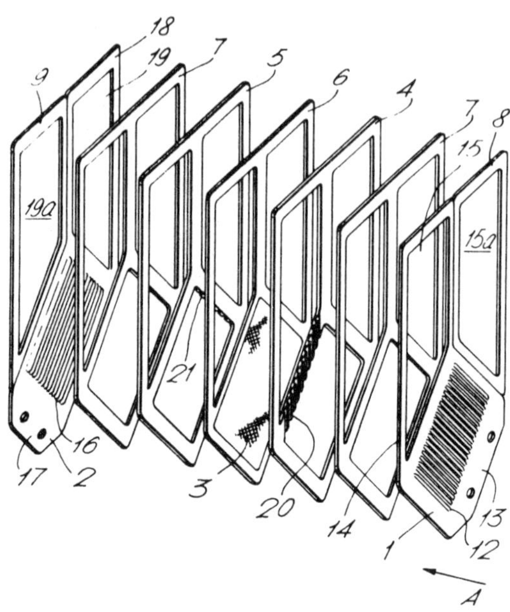

Source: U.S. Patent 4,204,939

The cell was fed with sodium chloride brine (300 g/ℓ NaCl) at a rate of 5 ℓ/hr, and a current of 480 A (corresponding to a current density of 3 kA/m^2) was passed through the cell. The cell operating voltage was 3.5 V. The chlorine produced contained 95% by weight of Cl_2 and 5% by weight of O_2. The sodium hydroxide produced contained 10% by weight of NaOH. The cell operated at a current efficiency of 86%.

Lightweight Filter Press Cell

K. Sato, Y. Sajima, T. Kuno and H. Ohbe; U.S. Patent 4,069,129; January 17, 1978; Asahi Glass Company, Ltd., Japan, provide a filter-press type electrolytic cell which is easily processed and prepared at low cost and low weight. The cell comprises a hollow member having therein a passage for liquid or gas. It comprises alternatively arranged frames and diaphragms fastened together to form alternating anolyte compartments and catholyte compartments. The frames comprise hollow members having an inlet or an outlet at the outer surface and holes at the inner surface. The appropriate electrolyte is passed into the anolyte and catholyte compartments respectively formed in the frame and the electrolyzed product is discharged from the anolyte or catholyte compartment.

The flow of the solution in the electrolytic cell of Figure 2.6b, using the frames of Figure 2.6a, will be illustrated with reference to Figure 2.6c, which shows the structure of the solution in the anolyte compartment.

Figure 2.6: Filter Press Cell

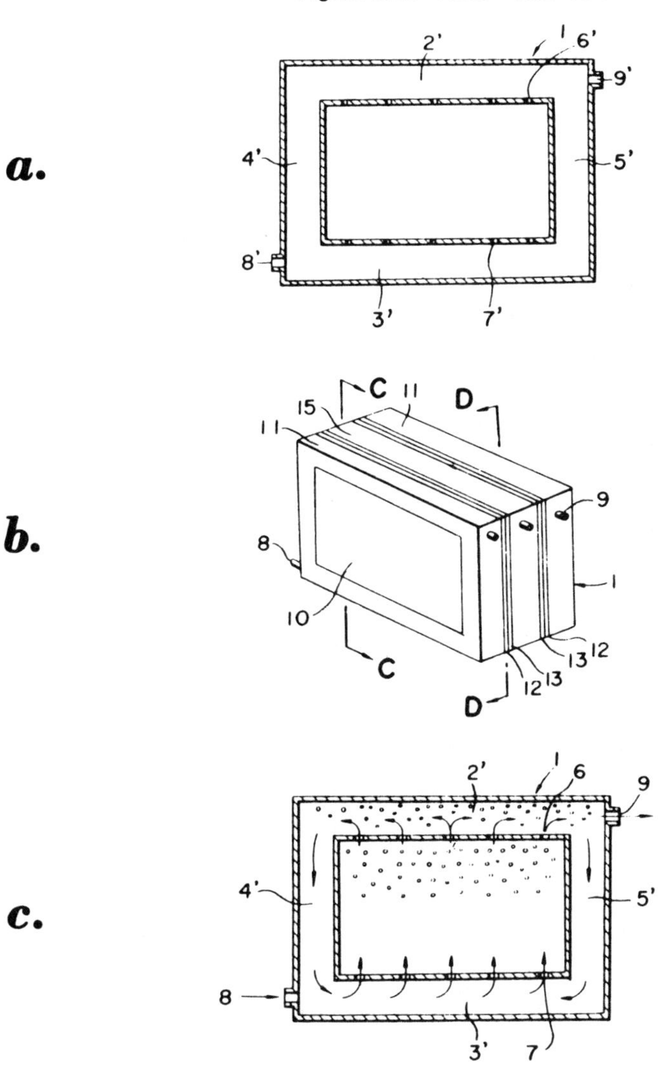

(a) Sectional view of the frame
(b) Schematic view of cell
(c) Sectional view along line **C–C** of (b)

Source: U.S. Patent 4,069,129

The catholyte compartment has the same structure except for the nature of the electrode. For example, an aqueous solution of sodium chloride is fed through the inlet **8** into the hollow zone **3'** corresponding to the lower part of the frame **15** for the anolyte compartment and is passed through the holes **7** into the anolyte compartment wherein electrolysis is conducted to generate Cl_2 gas.

The electrolyzed solution rises in the compartment under gas-lifting action, and is passed through the holes **6** into the hollow zone **2'** corresponding to the upper part of the frame **15** for the anolyte compartment. A part of the solution passes through the side hollow zones corresponding to the side parts **4'** and **5'** of the frame and is recycled into the anolyte compartment.

At the same time, water or a dilute aqueous solution of sodium hydroxide is fed from the inlet **8** into the hollow zone **3'** corresponding to the lower part of the frame **11** for the catholyte compartment and is passed through the holes **7** into the catholyte compartment wherein electrolysis is conducted to produce an aqueous solution of sodium hydroxide and to generate hydrogen gas.

The electrolyzed solution rises in the compartment under gas-lifting action and is passed through the holes **6** to the hollow zone **2'** corresponding to the upper part of the frame **11** for the catholyte compartment. A part of the solution passes through the side hollow zones **4'** and **5'** corresponding to the side parts of the frame and is recycled to the catholyte compartment.

The feed flow with the hollow member is remarkably slow and direct flow from the inlet **8'** to the outlet **9'** is usually prevented by appropriate selection of the size of the holes on the hollow members. Direct flow can be prevented by disposing appropriate members inside the hollow member. The recycle of the electrolyzed solution can also be effected by using an outer connecting pipe as well as the inner communicating hollow members.

Four hollow members made of titanium, having a cross section square of 70 x 70 mm and a 3 mm thickness, were assembled to form a rectangular frame (1 mm high and 2 m long) as shown in Figure 2.6a. An inlet and outlet for liquid and gas were formed in the frame and an anode was disposed in the frame to form a frame for the anolyte compartment.

Four hollow members made of stainless steel were assembled in the same structural form, and a cathode disposed in the frame to form a frame for a catholyte compartment. The inner surface of the upper hollow member had 17 holes (20 mm in diameter). The inner surface of the lower hollow member had 32 holes (9 mm in diameter). The frame for the anolyte compartment, a gasket made of natural rubber, a fluorine-type resin cation exchange membrane, and the frame for the catholyte compartment were serially arranged and fastened to form an electrolytic cell as shown in Figure 2.6b.

An aqueous solution of sodium chloride (315 g/ℓ) was fed at a flow rate of 0.1 m^3/hr to the anolyte compartment, wherein chlorine gas was generated at a rate of about 10 m^3/hr. The chlorine gas was discharged together with the diluted solution (electrolyzed–210 g/ℓ of NaCl aqueous solution) from the anolyte compartment.

The diluted solution was recycled through the vertical hollow members at a flow rate of about 3 m^3/hr. On the other hand, water was fed at a flow rate of 0.014 m^3/hr to the catholyte compartment, wherein hydrogen gas was generated at a rate of about 5.5 m^3/hr.

The hydrogen gas was discharged together with the resulting aqueous solution of sodium hydroxide (500 g/ℓ of NaOH aqueous solution) with a flow rate of 0.022 m^3/hr. The aqueous solution of sodium hydroxide was recycled through the vertical hollow members at a flow rate of about 2 m^3/hr.

The flows were effected by gas-lifting action. The electrolysis was continuously conducted for 1 month under a current density of 20 A/dm^2 and had a voltage of 4.0 V.

Automatic Response to Variation in Current Density or Feed Rate

L. Bourgeois; U.S. Patent 4,144,161; March 13, 1979; assigned to Solvay & Cie, Belgium has developed a diaphragm cell devised for reacting automatically and rapidly to a variation in the electrolyzing current density and/or in the electrolyte feed rate, so as to maintain the relationship between the rate of flow of the electrolyte through the diaphragm and the electrolyzing current density substantially constant at all times.

The cell comprises an anode compartment and a cathode compartment separated by a diaphragm, an inlet pipe and an outlet pipe for an electrolyte, means for controlling the feed of electrolyte, an outlet pipe for a gas produced in the anode compartment or in the cathode compartment, and a valve of variable aperture placed in the gas outlet pipe. Means for operating the valve comprises a float placed in a vessel that is in communication at its upper part with the gas outlet pipe, upstream of the valve, and at its lower part with the compartment that is connected to the gas outlet pipe.

The valve is designed so that in normal operation it occupies a position in which it partially constricts the gas outlet pipe, the cell being fed with electrolyte at a nominal rate. The valve is joined to the float by a connecting device designed for progressively opening or closing the valve as the float falls or rises in the vessel. Thus, if the current density falls, the gas pressure in the compartment where it is produced by electrolysis drops immediately and, as a consequence, the rate of flow of the electrolyte through the diaphragm changes immediately and proportionally.

While the rate of feed of electrolyte into the cell is being altered so that it matches the new value of the current density, the level of electrolyte in the compartment will fluctuate thus operating the valve so that this level will finally become stabilized at a position which will depend on the permeability of the diaphragm and the gas pressure in the compartment, upstream of the valve.

The cell as shown in Figure 2.7 comprises a foundation **1** forming the so-called base of the cell and supporting, at its periphery, a rectangular casing **2** made of steel, closed by a cover **3**. Within the casing **2**, cathodes **4** alternate with rows of substantially vertical and parallel anode plates **5** passing through the base of the cell and connected beneath the base to a current lead-in (not shown).

Figure 2.7: Diaphragm Cell Devised for Reacting Automatically to Current or Feed Rate Changes

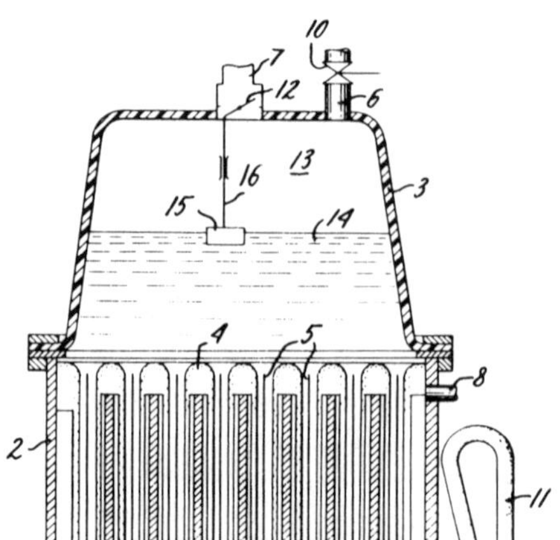

Source: U.S. Patent 4,144,161

The anodes **5** are, for example, constituted by plates of graphite or, preferably, plates of titanium carrying on their two surfaces a coating such as a metal of the platinum group.

The cathodes **4** are formed by a steel lattice, fixed to the walls of the casing **2** and shaped so as to form cathode pockets extending between the anodes **5**. The cathode lattice **4** is covered entirely on its surface facing towards the anodes by a diaphragm (not shown), which thus separates the cell into a cathode compartment and an anode compartment.

The anode compartment is in communication, by way of the cover **3**, with a pipe **6** for feeding in a sodium chloride brine to be electrolyzed and with a pipe **7** for removal of chlorine produced at the anodes **5** during electrolysis.

The cathode compartment is in communication, through the wall of the casing **2**, with a pipe **8** for removal of hydrogen produced at the cathodes **4** during electrolysis and with a pipe **9** for removal of a caustic liquor.

A valve **10**, placed in the brine feed pipe, allows the rate at which brine is fed into the cell to be regulated, while an inverted U-tube **11**, extending from the pipe **9**, allows the level of the catholyte in the cathode compartment to be adjusted by rotating the tube **11** around the axis of the pipe **9**.

A valve **12**, placed in the chlorine conduit **7**, allows adjustments of the pressure of chlorine within the vessel **13** beneath the cover **3**, above the level **14** of the brine in the anode compartment.

The valve **12** is connected to a float **15** by an actuating linkage **16**, so that opening or closing of the valve **12** is automatically brought about by falling or rising of the brine level **14**.

The position of the valve **12** is set so that it partially closes the pipe **7** during nominal operation of the cell, so that the vessel **13** is then maintained under a pressure of chlorine. This allows the operation of the cell to be easily adapted to a fall in the electrolyzing current density so as to maintain substantially constant the ratio between the rate of flow of brine through the diaphragm and the current density.

In order to accomplish this, the fall in current density causes an immediate reduction in the pressure of chlorine within the vessel **13** and, as a consequence, a fall in the rate of brine flow through the diaphragm.

Reduced Gap Between Electrodes

An improved electrolytic cell developed by *J.T. Rucker and D.H. Porter; U.S. Patent 4,174,259; November 13, 1979; assigned to Hooker Chemicals & Plastics Corp.* comprises the combination of a wide electrode, or more closely spaced electrodes, to achieve a narrow gap between electrodes, the electrodes being of conventional, simple, nonadjustable construction, guide spacers separating the anode elements and cathode elements of the electrolytic cell, and a hard, dimensionally-stable diaphragm.

Example 1: A 30,000 A chlor-alkali electrolytic test cell [HC-3B (Hooker Chemicals & Plastics)] was assembled with conventional dimensionally-stable anodes and wide-fingered perforated plate cathodes to reduce the average anode-to-cathode gap from the normal gap of 11.0 mm to a gap of 7.8 mm. Guide spacers of the hairpin-type design were constructed of rod stock [a chlorinated polyvinyl chloride (Trovidur)], ⅛" (3.175 mm) diameter, and were placed over the anodes.

The dimensionally-stable, asbestos-base, polymer-bonded diaphragm, containing 5% of a copolymer of ethylene and chlorotrifluoroethylene (Malar) was deposited upon the cathode and the diaphragm was processed in the conventional manner.

The test cell was assembled, operated for approximately 6 weeks, disassembled, rediaphragmed and reinstalled to test the ease of disassembly and reinstallation. No unusual difficulty in assembly, disassembly or reassembly was found. Voltage savings of 120 mV were observed in the test cell by comparison with the control cell described below in Example 2.

Example 2 (Comparative): A test cell (HC-3B) with a standard anode-to-cathode gap of 11.0 mm was constructed and operated under conditions identical to those of Example 1, and was used for the purpose of comparison.

The comparison of the cells shows a considerable savings in voltage for the narrowed gap cells, resulting in substantial savings in cost of power consumed for the electrolysis of brine in chlor-alkali cells. The data did not change significantly after cell disassembly, rediaphragming and reinstallation after approximately 6 weeks of operation.

Electrode Assembly with Flexible Gas Baffle Conductor

An electrode assembly is disclosed by *M.S. Kircher and M.F. Engler; U.S. Patent 4,101,410; July 18, 1978; assigned to Olin Corporation* which includes a common electrode riser and two opposed working faces. A movable, electrically conductive, downwardly opening gas baffle is provided between the faces to connect the faces to the riser and to bias the faces away from one another. The baffle also directs flow of electrically produced gases between the faces from an upward to a primarily sideways direction parallel to the faces. A diaphragm can surround the electrode assembly, if desired.

A method of assembling a diaphragm-type electrolytic cell is disclosed. The method includes placing a low friction protective sheet over an outer end of one of a plurality of electrode units and forcibly interleaving the protected electrode units with a plurality of other electrode units so as to contact the other electrode units by compressive forces applied to the other electrode units through the protective sheet and then removing the protective sheet so as to allow the other electrode units to expand against the formerly protected electrode unit.

Further disclosed is an expandable electrode assembly having at least two opposed working faces and a biasing device attached to each of the working faces for biasing the faces away from each other so as to expansively bias the assembly. An adjustable contraction device is attached to the bias device in order to contract the bias device and move the faces toward each other into a first position spaced a selected distance from one another. A releasable restraint device is also provided in order to hold the faces in the first position and selectively release the faces to move from the first position to a second position spaced a greater distance from one another than the selected distance. The restraint means can be a member which will corrode responsive to the presence of a corrosive medium thereabout.

Figure 2.8 is a side elevational view of an individual metal anode having vertically inclined conductor bar **52** and having spring conductor **42** vertically inclined in conformance with inclined conductor bar **52**. Conductor bar **52** passes outwardly from between the surfaces and can be provided with suitable threads **60** or other attachment means for connection to the anode support. Conductor bar **52** and spring conductor **42** can outwardly terminate at a point between the interior surfaces at a point spaced inwardly from outer end **51** of the metal anode so as to provide an outer passageway **61** beyond the outer end **62** of conductor bar **52** and spring conductor **42** associated therewith. This passageway **61** serves to allow upward liquid flow between the interior surfaces outwardly from ends **62** in order to avoid blockage or collection of gases.

Example: An anode of 24 ft^2 of planar electrode surface is constructed in the manner shown in Figure 2.8. The two 36" x 48" planar surfaces are fabricated from flattened expanded titanium mesh approximately 0.072" thick. The titanium clad copper rods or conductor bars are 1" diameter with 0.040" thick titanium-cladding. The conductor bars are horizontal at the attachment end, and slope downward at a 1 to 4 slope toward the opposite edge. The conductor bars terminate 4" from the opposite edge to provide a passageway for electrolyte rise upward between the titanium sheets. Attachment of the titanium sheets to the 1" diameter conductor rods is made with a 0.072" thick titanium strip 5" wide which is formed in the shape of an inverted U passing over the upper surface of the conductor bars. Connection between the strip or spring conductor and the interior surfaces of the expanded titanium mesh and between the spring conductor and the upper surface of the conductor bar is completed by welding.

Figure 2.8: Side View of Expandable Electrode

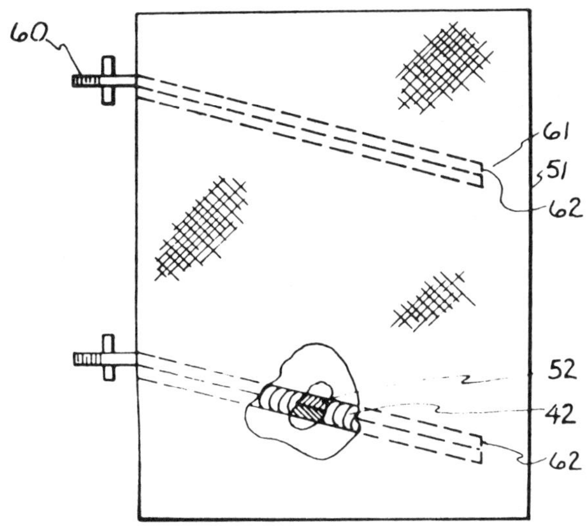

Source: U.S. Patent 4,101,410

In operation, a sodium chloride brine solution is fed into a cell containing the abovedescribed electrode surface to produce a caustic product, a chlorine gas product, a hydrogen gas by-product and a spent brine solution. Chlorine bubbles form on the outside surfaces of the titanium mesh but are forced through the openings in the mesh to a chamber between the titanium mesh sheets because the diaphragm and a surrounding spacer netting on the outside thereof obstruct upward passage of the gas.

On the inside of the titanium mesh sheets, the rising gas bubbles collect under the inverted U-shaped spring conductors and travel upward along the underside of the spring conductors to a channel provided at the connection end of the anodes. The rising gas tends to create a gas lift effect which induces electrolyte flow upward through the chamber. The spring conductors, in addition to collecting and transporting chlorine gas and conveying current to the anode surfaces, also maintain a pressure to hold the anode surfaces against the netting, diaphragm and cathode.

Recycling of Electrolyte

An alkali chlorine diaphragm cell such as is constructed in a conventional manner whereby evolved gaseous chlorine rises in the pool of alkali metal chloride solution is provided by *V. DeNora and O. de Nora; U.S. Patent 4,138,295; February 6, 1979; assigned to Diamond Shamrock Technologies SA, Switzerland* with recycling means to recycle the solution from an upper level of the solution to a predetermined lower level of the solution adjacent the cell bottom. A plurality of spaced conduits are preferably provided for this purpose.

The downward recirculation of the electrolyte through these conduits is induced by the upward movement of the electrolyte caused by the gaseous chlorine rising in the electrolyte outside the conduits. These conduits advantageously are located at or near anode surfaces and prevent or restrain lateral movement of recycled electrolyte until the recycled electrolyte reaches such lower level usually adjacent the bottom of the anolyte chamber. The circulation is preferably conducted as a plurality of spaced downward streams, each stream being between a pair of cathode elements and discharging into the space between an anode element and at least one of the cathode elements of the pair.

A cell provided with recycling means is shown in Figures 2.9a and 2.9b. In operation, the alkali metal chloride solution is fed into the top of the anolyte chamber by conventional feed means **23** and the electrolyte level in the chamber is maintained at a level above that of the cathode screens and the anodes, for example, as indicated at **24**.

The electrolyte level is held below the top of the cell cover **4** but above the electrodes so as to provide a gas space **26** above the electrolyte for collection of chlorine gas.

As the brine solution is fed into the anolyte chamber, a direct current electric potential of about 3 to 5 V is established between the anode base **1** and the cathode assembly **3**. The cell can **2** is provided with a copper grid bar **30** which is in contact with and extends around the cathode assembly and is connected to the negative pole of the dc source.

Because of the difference in hydrostatic pressure between the anolyte chamber and the hollow interior of the catholyte chamber **8** and the cathode fingers **12**, brine flows through the diaphragm and the cathode screen into the chamber **8**. Sodium hydroxide and hydrogen are formed at the cathode surfaces and carried into the interior of the fingers **12** and to chamber **8**. Hydrogen is withdrawn from the top of the chamber through a port or ports diagrammatically illustrated at **9**. Sodium hydroxide is withdrawn through conventional outlets **38** adapted to control liquid level within the catholyte chamber **8** and the cathode fingers at a convenient level.

Chlorine gas is evolved on the anode surfaces **16** in the form of bubbles which rise as a bubble stream extending from the bottom to the top of both the interior and exterior anode surfaces **16** as shown by arrows A in Figures 2.9a and 2.9b. These bubbles ultimately rise to the gas space **26** where chlorine gas is collected and withdrawn through the conventional outlets **28**.

When the anode is of screen or other perforated or reticulated structure, some of the evolved chlorine gas is diverted into the interior of the anodes **6**. By providing the anode with vanes, fins or other channeling means (not shown), a large part of such gas may be diverted to the interior of the anode **6**.

Figure 2.9: Cell with Recycling Means

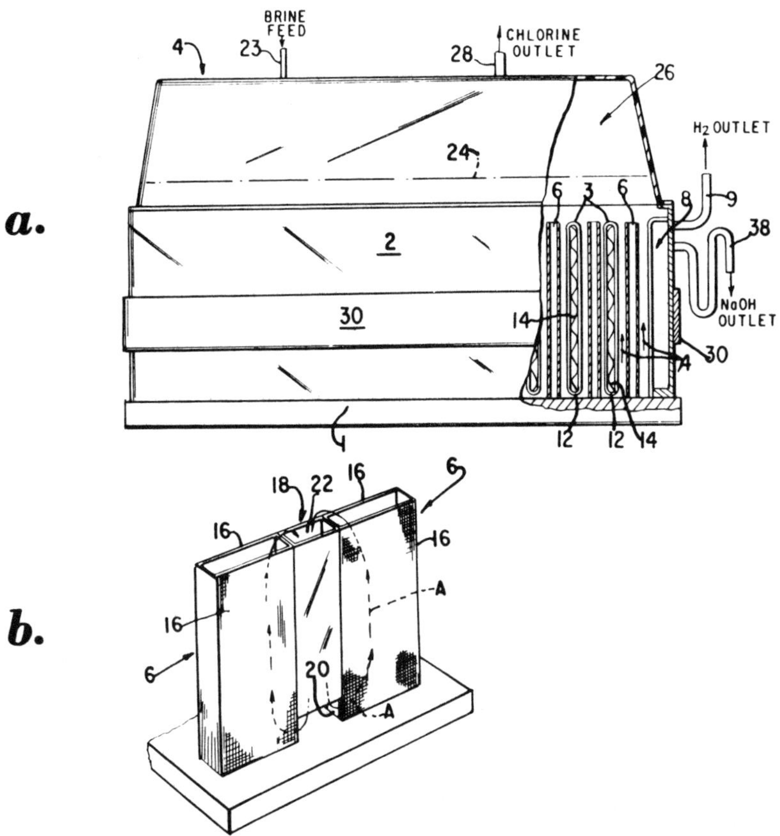

(a) Diagrammatic side elevational view, partly in section

(b) Diagrammatic fragmentary view of part of the cell base

Source: U.S. Patent 4,138,295

As the bubbles rise through the relatively narrow space between anode and cathode surfaces and within the anode interior, chloride solution (brine) is caused to move upwardly by gas lift. This brine solution ultimately rises to the top of anode **6** and the upward brine flow produces a downward brine flow through the conduits **18** ultimately delivering brine to the bottom of the conduits and out through opening **20** adjacent the bottom of the space between the anode and cathode as indicated by arrows **A**.

This circulation produces a positive transfer of electrolyte from the top to the bottom and bottom to the top of the electrolyte pool thus reducing the possibility of localized sodium chloride depletion in localized areas between electrodes.

Tandem Operation of Diaphragm and Membrane Cells

A method for economically producing a concentrated alkali metal hydroxide solution by electrolytic means is described by *J. Rutherford and C.F. Eggers; U.S. Patent 4,147,600; April 3, 1979; assigned to Hooker Chemicals & Plastics Corp.* The method involves the tandem operation of diaphragm and membrane electrolytic cells. More particularly, the cell liquor from the diaphragm cell or cells is utilized in place of all or part of the water usually utilized in the catholyte compartment of the membrane cell or cells.

In Figure 2.10, the points of addition and withdrawal of typical and preferred reactants and products are illustrated. Diaphragm electrolytic cell **11** includes outer wall **13**, anode **15**, cathode **17**, and conductive means **19** and **21** for connecting the anode and the cathode respectively to sources of positive and negative electricity. Inside the cell wall, permeable diaphragm **23** separates the cell into an anode or anolyte compartment **25** and a cathode or catholyte compartment **27**. An aqueous solution containing about 300 g/ℓ of sodium chloride, preferably acidic, is fed into the anolyte compartment **25** through line **29**.

Figure 2.10: Tandem Operation of Diaphragm and Membrane Cells

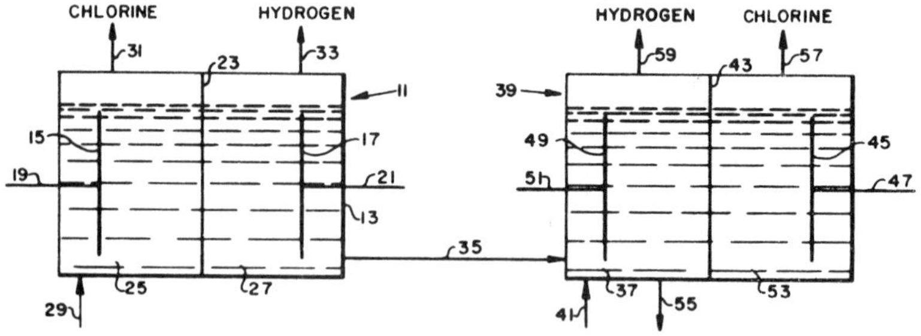

Source: U.S. Patent 4,147,600

During electrolysis, chlorine gas is removed from above the anolyte compartment through line **31**, and hydrogen gas is removed from above the catholyte compartment through line **33**. A solution containing about 140 g/ℓ sodium hydroxide and about 175 g/ℓ sodium chloride is withdrawn from the catholyte compartment through line **35** and fed into the catholyte compartment **37** of membrane cell **39**. Water flux may also be added to compartment **37** through line **41** to maintain the desired flow across the membrane and control the concentration of caustic in that compartment, thereby maintaining a high current efficiency by limiting the back migration of hydroxyl ions to the anolyte compartment through membrane **43**.

In membrane cell **39**, anode **45** is connected to a source of positive electric potential by conductor **47**, and cathode **49** is similarly connected by corresponding conductor **51**. A cation-active permselective membrane **43** separates catholyte compartment **37** and anolyte compartment **53**. A concentrated sodium hydroxide solution, 200 g/ℓ or more, is produced in catholyte compartment **37** and is withdrawn from the compartment through line **55**. Chlorine and hydrogen are withdrawn from the anolyte and catholyte compartments through lines **57** and **59**.

Example: An aqueous brine feed containing about 321 g/ℓ sodium chloride at about 60°C was fed into the anolyte compartment of a diaphragm electrolytic cell, designated as an H-4 cell. The cell was equipped with dimensionally stable anodes having a substrate of titanium with a coating of platinum group metals and platinum group metal oxides. The cell utilized a steel cathode and a deposited asbestos diaphragm. A current load of 33.6 kA was utilized to decompose the sodium chloride solution. A current efficiency of 90.5% was maintained. A cell liquor at about 90°C comprising 140 g/ℓ sodium hydroxide, 175 g/ℓ sodium chloride, and 913 g/ℓ water was removed from the catholyte compartment.

The diaphragm cell liquor was then filtered and fed at the rate of 0.15 gpm into the catholyte compartment of a membrane electrolytic cell. The membrane cell is the type designated as an MX cell. The cell was equipped with anodes and cathodes fabricated of similar materials as the corresponding components of the diaphragm cell discussed above. The cell was equipped with a permselective membrane of the PSEPVE-type. A water flux of about 1.6 gph was added to prevent salting out. The membrane cell was operated at a current density of 2 kA, about 1.16 A/in^2 of anode area. The anolyte temperature was 71°C.

The sodium hydroxide content of the cell liquor from the membrane cell varied over a range from about 200 g/ℓ, at start up, to about 340 g/ℓ, under stabilized operating conditions. The sodium chloride content similarly varied from about 55 to 140 g/ℓ. The tandem operation was carried out over a period of about 550 hours.

Vacuum-Assisted Assembly

Apparatus and method are disclosed by *S.J. Specht; U.S. Patent 4,078,987; March 14, 1978; assigned to Olin Corporation* for facilitating electrode installation in diaphragm-type electrolytic cells. The apparatus includes a vacuum generator for pulling the diaphragm against a first electrode during insertion between two other spaced electrodes.

The assembly is described with reference to Figures 2.11a through 2.11f. The cell **10** is basically assembled in three parts, first the anode assembly **20**; second the cathode assembly **22** together with catholyte outlet section **14**, caustic outlet **16** and diaphragm assembly **24**; and third cell body **18** together with chlorine outlet **12**, drain **15** and brine inlet **17**. These three parts are then assembled to form the electrolytic cell. A part of this assembly requires that the cathodes be inserted between the anodes. Alternatively, the anodes may be inserted between the cathodes. Such insertion involves the movement of at least one of the electrodes past the diaphragm surface, such as shown by arrows **80**, **82** and **84** of Figure 2.11e.

In order to minimize the damage resulting from the movement of the electrodes past the diaphragm, fluid is withdrawn by vacuum generator assembly **26** from the interior of cathode assembly **22** so as to produce a vacuum, or area of less than ambient pressure, within the cathodic side of cell **10** so as to force diaphragm assembly **24** tightly against the wire mesh surfaces **56** and **58** of cathode assembly **22**. A portion of the preferred expandable cathode assembly is shown in Figure 2.11e in the contracted position which results from the low pressure between mesh surfaces **56** and **58** relative to the ambient pressure existing between anodes in the anodic portion of the cell.

Figure 2.11: Vacuum-Assisted Assembly

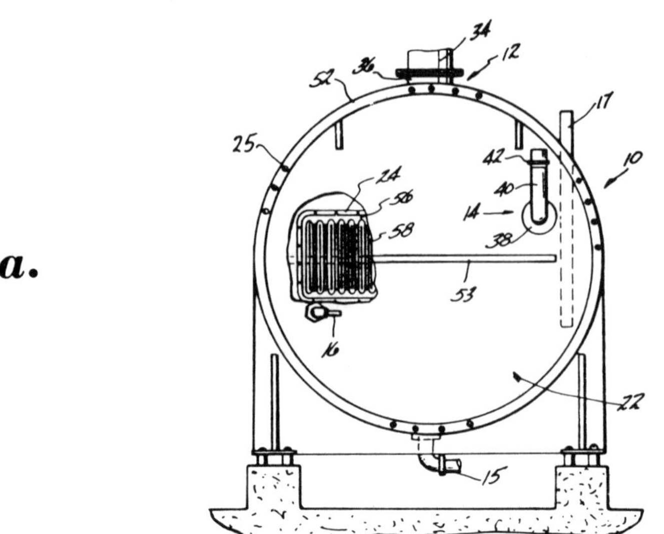

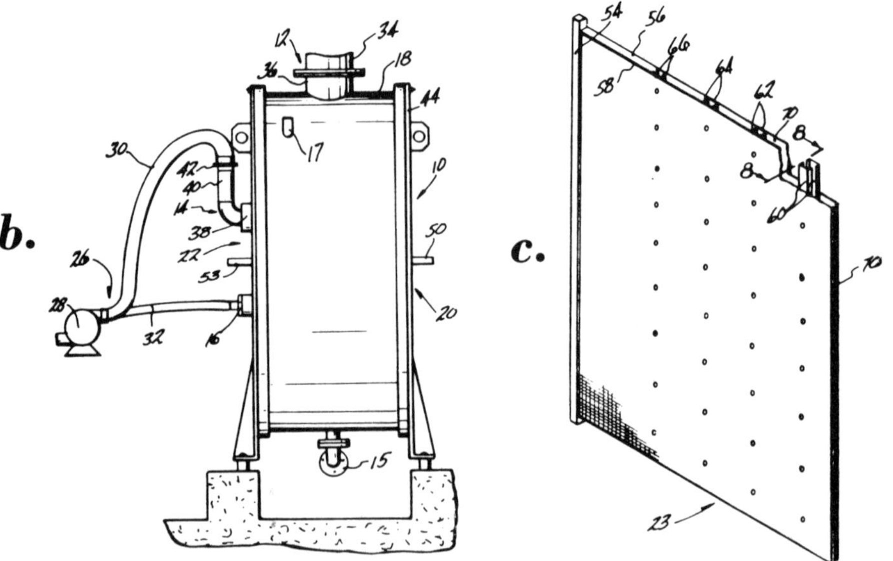

(continued)

Figure 2.11: (continued)

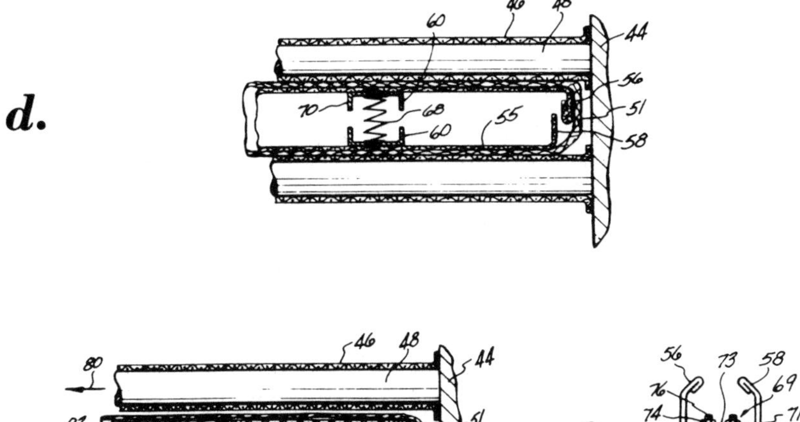

(a) Front view of cell assembly
(b) Right side elevational view of cell assembly, showing vacuum generator means
(c) Isometric view of expandable cathode
(d) Top partial sectional view taken horizontally through center of cell, showing cathode in expanded position
(e) Same as Figure 2.11d but cathode is in contracted position
(f) Side sectional view of a keeper assembly taken along line **8-8** of 2.11c

Source: U.S. Patent 4,078,987

This contraction of the wire mesh surfaces **56** and **58** toward each other is allowed by the support ribs **60, 62, 64** and **66** which are outwardly biased by spring **68** so that when the vacuum generator **26** is disconnected and the pressure on the cathodic side of the cell **10** is allowed to rise to ambient pressure, the wire mesh surfaces **56** and **58** are expanded by the force of springs **68** upon ribs **60, 62, 64** and **66**. This expansion continues until either keeper pins **76** reach the outer edges of opening **73** in the keeper plate or the wire mesh surfaces **56** and **58** are restrained by diaphragm assembly **24** coming into contact with an adjacent anode. If the latter occurs, the anode to cathode gap is set at the thickness of the walls **55** of the fingers **51** of diaphragm assembly **24,** and the springs **68** operate to maintain that gap until such time as keeper pin **76** contact the outer edges of openings **73** after which the width of opening **73** controls the anode to cathode gap.

Preferably, the keeper plate opening **73** will be sufficiently wide that the pins **76** will only contact the outer edges of opening **73** prior to assembly of cell **10,** so that at all times after reexpansion of surfaces **56** and **58** relative to one another, the anode to cathode gap is controlled by springs **68** and is thus held constant at a predetermined distance the thickness of wall **55**.

SUPPORTS AND SEALS

Supporting Diaphragms in Close Contact with Cathodes

In operating a diaphragm cell, there is a tendency for diaphragms to become detached from the cathode surface, for example, because of expansion of the sheet diaphragm during warming-up of the cell. Such a detachment of the diaphragm can result in a reduction in or partial blockage of the anolyte recirculation space between the diaphragm and the anode and/or trapping of gas generated at the cathode (e.g., hydrogen in a chlorine/caustic soda cell) between the diaphragm and the cathode.

A.M. Couper, P.J. Davies, and J.J.H. Krause; U.S. Patent 4,146,457; March 27, 1979; assigned to Imperial Chemical Industries Limited, England describe a device for supporting a sheet diaphragm on to a surface of a cathode. The device is made of a flexible insulating material and can be caused or allowed to expand into the space between the anode and the diaphragm so that the diaphragm is pressed into contact with the cathode.

Referring to Figures 2.12a and 2.12b, the diaphragm supporting device comprises a strip of flexible polymer **1**, for example, of polytetrafluoroethylene or polyvinylidene fluoride, suitably about 1 cm wide, which is provided with a plurality of holes **2** through which is threaded a titanium wire **3**.

Figure 2.12: Diagrammatic End Views of Diaphragm-Supporting Device

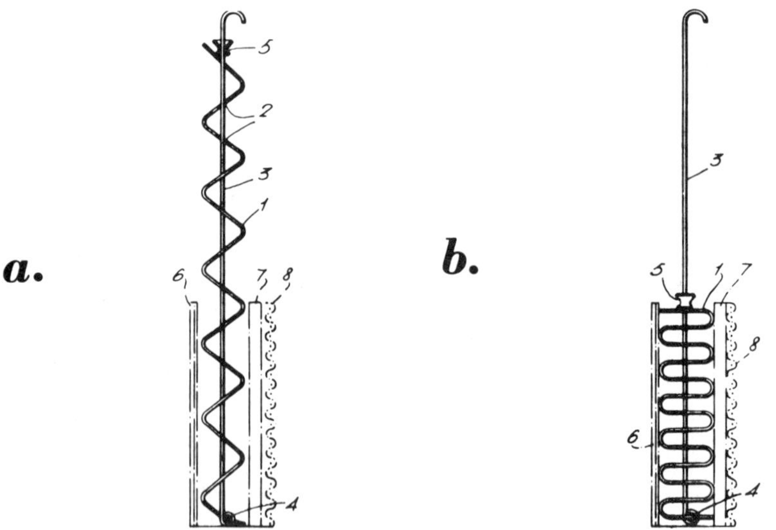

(a) Expanded (laterally compressed) position
(b) Compressed (laterally expanded) position

Source: U.S. Patent 4,146,457

Referring to the figures, the strip **1** is attached to the wire **3** at **4**. A clamping clip **5** is provided to limit a longitudinal movement of the strip **1** along the wire **3**.

Before assembling the supporting device into a diaphragm cell, the slip **5** is released and the strip is pulled along the wire into a longitudinally expanded position (as shown in Figure 2.12a). The supporting device is then inserted into a diaphragm cell (shown in Figures 2.12a and 2.12b) comprising an anode **6** (e.g., a titanium anode coated with an electrocatalytically active coating such as a mixture of ruthenium oxide and titanium dioxide), a diaphragm **7** (e.g., of asbestos, polytetrafluoroethylene or polyvinylidene fluoride) and a cathode **8** (e.g., of mild steel gauze). The clip **5** is then released and the strip is pushed down until it is laterally urged into engagement with the surfaces of the anode **6** and the diaphragm **7**, thereby pressing the diaphragm onto, and maintaining it in contact with, the cathode **8**.

Vented Diaphragm Support

J.A. Wood and S.J. Specht; U.S. Patent 4,196,071; April 1, 1980; assigned to Olin Corporation have developed a perforated support for helping prevent damage to a nonadherent membrane in a chlorine gas-producing cell configuration having downwardly open, gas-trapping pockets adjacent such a membrane in the upper end of a chlorine-producing anolyte chamber. The cathode chamber can be at a greater pressure than the anode chamber to force the diaphragm against the support.

Figure 2.13a is a front view of a pair of casings **10** enclosing a pair of electrodes on all but the rear side. The electrodes cannot be seen, since they are enclosed by casings **10** which have a closed front end **12**. An attachment assembly **27** connects casing **10** to a backplate **40** (see Figures 2.13b or 2.13c) which can be of any conventional design, such as a disc or rectangular plate or other shape.

Attachment assembly **27**, which is shown in greater detail in Figure 2.13b, comprises inner portion **34**, an inner gasket **42**, an outer gasket **44**, clamping flange **30**, outer portion **36** and a plurality of bolts **38**. Tabs **16** are twisted and are covered in part by a frame-like perimeter clamp **28** which forms a part of assembly **27**. Clamp **28**, as illustrated in Figure 2.13b, comprises inner portion **34**, outer portion **36** and bolts **38**. Portions **34** and **36** are held together by bolts **38**. Bolts **38** are tightened so that clamp **28** holds and seals tabs **16** against a ventilated clamping flange **30** (described below) projecting from backplate **40**. A cutaway view shows clamping flange **30** to be partially perforated for ventilation purposes described below.

Figure 2.13b is a vertical cross section through attachment assembly **27** of Figure 2.13a. Clamping assembly **27** comprises a clamping flange **30**, clamp **28**, gaskets **42** and **44** and support lip **37**. Clamping flange **30** is preferably attached to anode backplate **40** adjacent chlorine outlet **48**. Gas, liquid-gas foam or gas-containing liquid is collected in the space **56** between lip **37** and backplate **40** and flows to outlet **48** and is thus removed from the cell as a product of electrolysis. Clamping flange **30** is a T-shaped bar frame projecting inwardly from backplate **40**, with the base of the T attached to backplate **40** although other shapes could also be used. The perforated support lip **37** serves as a support member to underlie and support transition region **15** of casing **10** which extends downwardly from assembly **27** to edge **18**. Tab **16** is twisted about 90° so that its upper end is vertical and lies against the vertical upper end of transition region **15**. Tab **16** is held in this twisted position by gasket **42** which presses against tab **16** and gasket **44** which presses against transition region **15**. Gaskets **42** and **44** and the upper ends of tab **16** and transition region **15** are clamped between portion **34** and upper lip **31** of clamping flange **30** by bolt **38**. However, an unexpected number of pinholes were found to develop in region **15** until lip **37** was perforated.

In this position, region **15** is only loosely supported by lip **37** and there was unexpectedly determined to be a tendency for gas, e.g., chlorine, generated by electrolysis to accumulate in the space **54** between the inner surface **39** of lip **37** and the anodic side of region **15**. The deterioration of region **15** was determined to be caused by chemical attack and drying out of region **15** rather than physical stresses as would normally be thought in a clamping region. This accumulated gas under region **15** is unexpectedly found to dry and chemically attack region **15**. Therefore, a multiplicity of gas ventilation perforations **46** are made in lip **37** from space **54** to space **56** to ventilate and allow liquid-gas two phase flow through space **54** so as to keep region **15** wet.

Figure 2.13: Vented Diaphragm Support

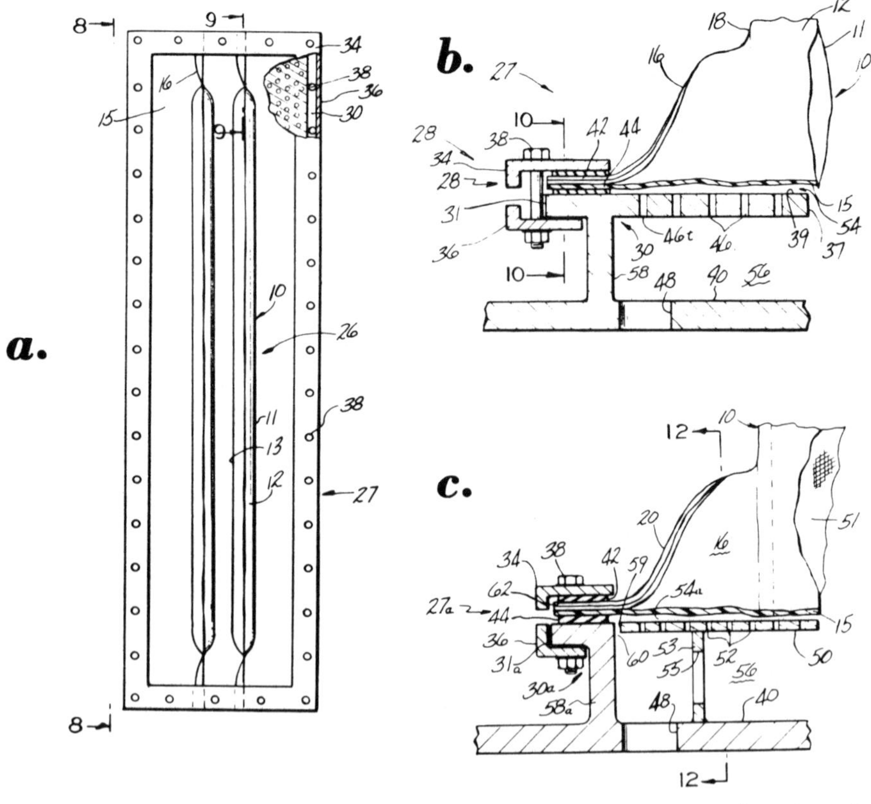

(continued)

Figure 2.13: (continued)

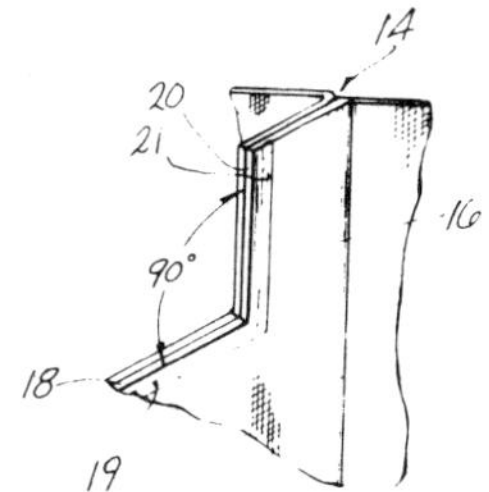

(a) Front exterior end view of electrodes covered by diaphragm casing and attached by disclosed clamping system
(b) Partial vertical side section of Figure 2.13a taken along line **9–9**, modified to show T-shaped clamping flange
(c) Partial vertical side section of Figure 2.13a taken along line **9–9**, modified to show preferred diaphragm casing attachment assembly
(d) Details of one tab on the diaphragm casing

Source: U.S. Patent 4,196,071

When chlorine outlet **48** is under a vacuum negative gauge pressure, as is the case in many conventional cells, perforations **46** help region **15** to be pushed by the higher pressure in the cathode chamber against surface **39** and thus give increased resistance to both gas accumulation in and gas or liquid flow through space **54** and reduce the volume of space **54**. Whether or not the chlorine outlet is at a negative gauge pressure, if the cathode chamber is filled to a higher level with catholyte than the anode chamber is with anolyte, a pressure differential or catholyte head will be created and will force region **15** against lip **37**.

If plate **37** is not perforated, this pressure may tend to trap gas under region **15** and prevent liquid from flowing into space **54** thus drying and chemically damaging region **15**. Also, the bottom edges of gaskets **42** and **44**, especially gasket **44**, are located approximately even in height with the upper edge of the uppermost perforation **46t** so that only a negligible gas pocket, if any, will remain above perforation **46t** and so that gas flow through perforation **46t** will keep any such negligible pocket well circulated.

Figure 2.13c is a modified form **27a** of the assembly **27** of Figure 2.13b. Support lip **37** is omitted and a perforated support plate **50** substituted therefor. Assembly **27a** comprises an L-shaped clamping flange **30a** with a vertical upper lip **31a** and a horizontal base **58a**. Base **58a** lies horizontal and connects lip **31a** to vertical backplate **40**. The upper end **59** of plate **50** is downwardly spaced from the horizontal base **58a** of flange **30a** so as to define a gas passageway **60** between base **58a** and end **59**. Plate **50** is provided with perforations **52** from space **54a** to space **56**. Perforations **52** correspond to perforations **46** of lip **37**.

Plate **50** is attached to backplate **40** by a perforated plate **53** having a vertical opening **55** therethrough so as to not restrict upward flow through space **56** to chlorine outlet **48**. Plate **53** is preferably perpendicular to plate **50** and backplate **40** and thus parallel with base **58a**. Plate **50** is preferably a titanium plate and can be in electrical contact with a current source so that plate **50** serves as a gas-evolving anode and thus makes effective use of the transition region **15** of the membrane.

In fact, it is most preferred to have support plate **50** be made of the same foraminous mesh material, such as, for example, TiO_2-RuO_2 mixed crystal coated (Beers coating) titanium mesh, as anode **51** or even be a part of anode **51**. In such a case, plate **50** would have literally thousands of perforations **52**. A corresponding cathode could be positioned on the opposite side of region **15** from plate **50** and in conforming structure to plate **50** and region **15** to assist in this regard.

Example 1: Two sheets **11** and **13** of a perfluorosulfonic acid membrane material (Nafion 391) are cut to provide a tab **16**. Nafion 391 is a homogeneous film 1.5 mils thick of 1,500 equivalent weight perfluorosulfonic acid resin and a homogeneous film 5 mils thick of 1,100 equivalent weight perfluorosulfonic acid resin laminated with a T-12 fabric of polytetrafluoroethylene. The outer edges of each of the two sheets are joined by heat sealing at a temperature of 230°C, a pressure of 3.0 kg/cm^2 and a dwell time of 4.5 seconds on a thermal impulse heat sealing machine to form a closed end **12**.

The sheets are then sealed by seal **19** linearly along each side to form closed edges **18** up to the edge of the tab using the same heat sealing conditions as above. Heat seal **21** is then applied to the tab portion at an angle of about 90° from the linear seal along each side to complete casing **10**. Seal **21** is applied so that it interconnects with seal **19** along the major portion of the side edge. The tab portion **16** is approximately 3" long. The casing is installed on an anode used in a cell for the electrolysis of sodium chloride in the production of chlorine and sodium hydroxide.

To provide a flat surface for sealing the casing along the top and bottom edges, the tabs are twisted and a clamp applied, as shown in Figure 2.13c. The clamp assembly includes a titanium mesh support plate having thousands of gas ventilation passageways therethrough. During electrolysis, the casing is found to be leak-proof during tests in excess of 16 weeks and expanded or contracted with changes in cell operating conditions without placing a detrimental mechanical stress on the separator material.

Example 2 (Comparison): Another casing **10** identical to the above Nafion 391 casing was fabricated by the same procedure but was instead attached to a similar anode backplate by the attachment assembly **27** of Figure 2.13b, except that perforations **46** were deleted so that a restricted flow gas pocket was created between flange **30** and transition **15**.

The cell was operated for 16 weeks to produce chlorine gas within casing **10**. At the end of the 16-week period, there was found to be considerable leakage between anode and cathode and upon disassembly of the cell, numerous holes were found in region **15** adjacent flange **30**. From the size and shape of the holes which were then analyzed, it was determined that chemical attack had caused the holes.

Cell Liner and Seal Device

A cell liner for an electrolytic cell which covers the base plate of the cell is adapted by *R.F. Anderson; U.S. Patent 4,081,348; March 28, 1978; assigned to The B.F. Goodrich Company* to provide an effective seal between the base plate and the cover of the cell. Around the perimeter of the liner extends a sealing device having two parallel ribs defining a channel for receiving an elongated elastomeric extrusion of circular cross section. The extrusion is adapted to sealably contact the sealing rim of the top cover.

Figure 2.14a represents an electrolytic cell **10** of the diaphragm-type used in the chlor-alkali industry for the production of chlorine. The enclosure or cover member **26** has a lower edge or rim **25** which seals the cell along the outer margins of base plate **20**.

Figure 2.14: Cell Liner and Seal Device

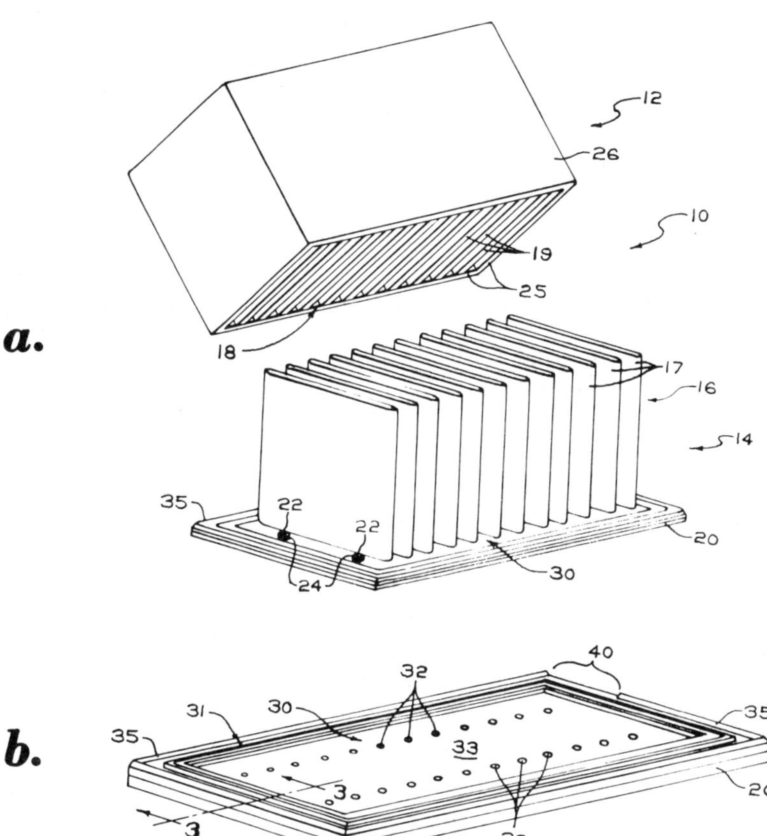

(continued)

Figure 2.14: (continued)

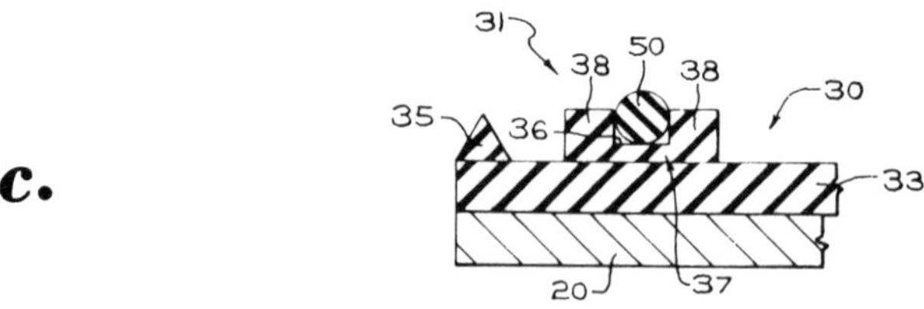

(a) Isometric view of electrolytic cell
(b) Isometric view of cell liner used in Figure 2.14a
(c) Cross-sectional view of a portion of the peripheral margin of the cell liner of Figure 2.14b

Source: U.S. Patent 4,081,348

The cell liner **30**, which covers base plate **20**, is more clearly shown in Figure 2.14b. The liner **30** is basically a vulcanized rubber sheet **33** sized to completely cover the upper surface of base plate **20**. The central zone of the sheet **30** includes suitably arranged openings **32** for passage therethrough of anode stems **24** which secure and electrically connect the anode assembly **16** to base plate **20**. The stem collars **24** hold the edges of opening **32** against base plate **20** in a sealed relationship.

Adjacent the outer margins of sheet **33**, the cell liner includes an outer upwardly projecting elastomeric rib **35** extending almost entirely around the sheet **33**. This rib, sometimes known as a drip strip, is adhered in suitable fashion to the surface of sheet **33** and functions to direct any spillage, leakage or condensate to an open area **40** for collection by suitable means (not shown). The drip strip **35**, as shown in Figure 2.14c, can be triangular in cross section, or may be some other desired contour such as semicircular, or rectangular.

The cell liner **30**, as seen in Figure 2.14c, further includes a sealing device **31** adhered to the surface of sheet **33** disposed in spaced relation to and inwardly of drip strip **35** and extending substantially parallel thereto. The device **31** includes a sealing channel **37**, generally of U-shaped cross section having spaced upwardly projecting legs **38** and an intermediate seat area **36**. The channel **37** could be a one-piece member as shown or could be sections of different thickness rubber mutually adhered to achieve the channel contour. The spacing between legs **38** of channel **37** should be less than the width of sealing rim **25** of cover **26** and preferably is such to snugly receive an elastomeric sealing strip or extrusion **50** of such cross section that it will project slightly above the upper surfaces of legs **38** of channel **37** when placed in channel member **37**.

The extrusion **50** is preferably of circular cross section as shown in Figure 2.14c, but can, if desired, be an alternative contour such as semicircular. The extrusion **50** is of sufficient length to extend within and throughout the entire length of channel member **37** which is shown endless in Figure 2.14b. While the channel member **37** is adhered to the cell liner, the extrusion is not adhered to the cell liner either directly or indirectly as by adherence to channel seat **36**. Thus, when need to change or replace a seal arises, it is needed only to lift the extrusion **50** from channel **37** and easily replace it with a fresh one.

Sealing Member Resistant to Heat and Chemicals

H.S. Custer and B.H. Oliver; U.S. Patent 4,098,670; July 4, 1978; assigned to The Goodyear Tire & Rubber Company provide a heat resistant and chemically resistant polymeric sheet which has relatively low compression set properties and which will retain a relatively high resistance to compression set after long periods of use as a seal and cover for the cell base of an electrolytic cell of the diaphragm-type.

Example: Two flexible vulcanized elastomeric sheets **1** of the type shown in Figure 2.15 were manufactured having the following Compositions A and B:

Components	Parts by Weight A	Parts by Weight B
Ethylene propylene terpolymer (Nordel 1470)	100.00	-
Ethylene propylene terpolymer (Nordel 1070)	-	100.00
Carbon black	90.00	90.00
Plasticizer	50.00	50.00
Zinc oxide	5.00	5.00
Stearic acid	1.00	-
Zinc stearate	-	1.00
Antioxidant	2.00	2.00
Sulfur	0.10	-
4,4'-dithiodimorpholine (Sulfasan R)	1.50	-
Zinc dibutyldithiocarbamate (Butyl Zimate)	4.00	-
Mixture of tetramethyl and tetraethyl thiuram disulfide (Methyl, Ethyl Tuads)	2.50	-
Bis(tert-butylperoxyisopropyl)benzene (Vulcup R)	-	6.00
Total	256.10	254.00

In the above compositions, the plasticizer used was paraffinic oil and the carbon black used was the fast extruding furnace type. The vulcanizing agents used in Composition A were of the sulfur donor type and in Composition B were of the peroxide type.

The above formulas are expressed in proportions on the basis of parts by weight based on the weight of the ethylene propylene terpolymer rubber.

The above compositions were prepared in the following manner. All the compounding ingredients, except the vulcanizing agents, were added to a Banbury mixer and mixed to produce a nonproductive stock. The vulcanizing agents were then added to the nonproductive stock in the Banbury during a second pass mixing procedure.

The above compositions were then processed on a roller die extruder and each was formed into a rectangular sheet or web of material having a gauge or thickness of ¼" (6.35 mm). Each sheet was next rolled into a polyethylene liner. Each roll thus formed was cured in a rotocure at a temperature of 360°F for 18 minutes. After vulcanization, each sheet **1** is prepared for installation in an electrolytic cell **2** of the type shown in Figure 2.15 by being cut and trimmed to size. Each sheet has, for example, a width of 45" (1,143 mm), a length of 69" (1,771.7 mm) and a gauge of ¼" (6.35 mm). Holes **3** for receiving the anode risers **16** are punched in this sheet with the holes **3** being disposed, for example, in equally spaced rows with an equal number of holes in each row. The holes **3**, for example, may have a diameter of about 1.25" (31.75 mm).

Each sheet is then provided with extruded strips **5**, **5'** and **6** of neoprene rubber adhered to the sheet in the locations shown in Figure 2.15a by means of a commercially available adhesive. In this regard, however, it should be appreciated that the strip may also be of the same composition as the sheet **1** indicated above with other suitable adhesives being used.

Figure 2.15: Use of Sealing Member

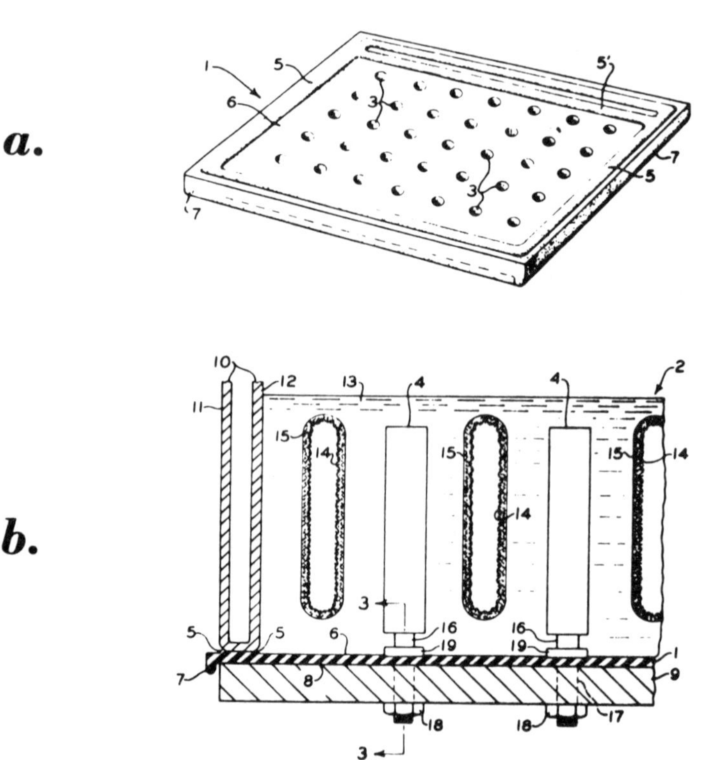

(a) Perspective view of sealing member
(b) Partial side elevational view of cell showing sealing member in installed position

Source: U.S. Patent 4,098,670

Each sheet is then installed in an electrolytic cell **2** as shown in Figure 2.15b of the type described in U.S. Patent 2,987,463 (Hooker type S-3) except having a cell base construction as described in U.S. Patent 3,591,483. The installation includes placing the sheet **1** on a copper base plate **9** so that the holes **3** in the sheet **1** are aligned with an equal number of holes **17** in the base **9**. Then, dimensionally stable anodes **4** are set in place with anode risers **16** having a diameter of about 1.25 of an inch (31.75 mm) being inserted through the anode-receiving holes **3** of the sheet **1**. The anode risers **16** are bolted to the bottom side of the copper base plate **9** and the collar or flange **19** of the riser **16** is seated against the upwardly disposed surface **6** of the sheet **1** with the combined diameter of the flange **19** and riser **16** being about 2" (50.80 mm).

At the bottom of the cell base **9**, the nut **18** is tightened so that the sheet **1** is compressed from a diameter of about ¼ of an inch (6.35 mm) to a diameter of about 3/16 of an inch (4.76 mm). After the installation of the cell base **9**, the cell can **10**, cathodes **14** and diaphragms **15** are fitted in place to complete the assembly of the cell **2**.

The cell **2** is then fed with a brine solution **13** and operated under general conditions of the type described in the example provided in U.S. Patent 3,591,483, Column 8. After several months of service, the sheets have shown no visible signs of heat cracking and the compression seal between the flange **19** and the surface **6** of each sheet **1** has been adequately maintained.

TESTING METHODS

In Situ Reference Electrode

A method and apparatus for measuring the overvoltage of a gas producing electrode of a chlor-alkali diaphragm cell is disclosed by *H.C. Kuo, G.W. Geren, T.E. Corvin and B.K. Ahn; U.S. Patent 4,163,698; August 7, 1979; assigned to Olin Corporation*. The method includes positioning an exposed metal tip of a reference electrode from about 0.2 to 1.0 mm away from the gas producing electrode within a stream of gas produced by the gas producing electrode. The apparatus includes the reference electrode so positioned. Platinized platinum wire and a RuO_2-TiO_2 coated Ti wire are preferred as the reference electrodes for cathode and anode, respectively.

As shown in Figures 2.16a and 2.16b, reference electrode assembly **26** comprises insulating tube **20**, reference wire **22**, stem wire **24**, fasteners **28**, tip seal **32** and a voltmeter **25**. Insulating tubing **20** is a chemically-resistant, nonconductive tubing such as polyethylene, polypropylene or tetrafluoroethylene which serves to insulate reference wire **22** from electrode **16**. Reference wire **22** is a small gauge, preferably 20 to 22 gauge, wire of a material suitable for use as a reference electrode.

The in situ reference electrode **26** can be assembled by placing a small gauge, such as for example a 20 to 24 gauge, wire of material suitable as a reference electrode in a chemically resistant tubing, leaving tip **30**, approximately 0.3 to 0.7 cm of the wire, exposed. A platinized platinum wire has been found to be particularly suitable as the small gauge wire in an in situ reference electrode for measuring hydrogen overvoltage of a hydrogen producing cathode of an electrochemical cell. A small gauge (e.g., 20 to 24) conductive wire comprised of a valve metal selected from the group consisting of tantalum, titanium, zirconium, bismuth, tungsten, niobium and alloys thereof, such as for example a titanium wire, coated with a mixed crystal material consisting essentially of at least one oxide of a film-forming metal and at least one oxide of a platinum group metal, such as for example a titanium dioxide (TiO_2) and ruthenium dioxide (RuO_2) coating, according to U.S. Patent 3,632,498, is suitable as the small gauge wire in an in situ reference electrode for measuring chlorine overvoltage of an anode in a chlorine gas producing electrolytic cell.

Figure 2.16: In Situ Reference Electrode

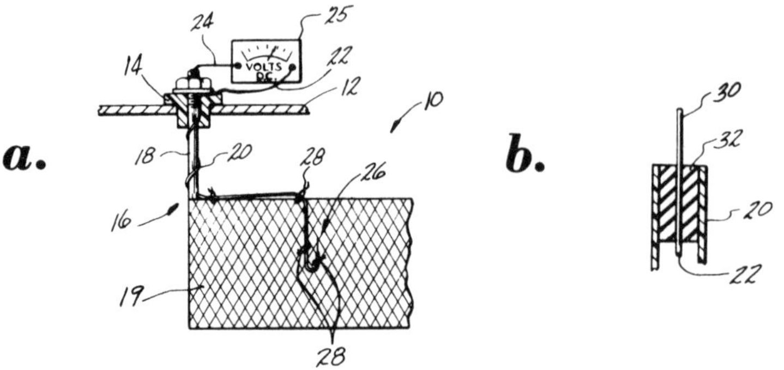

(a) Schematic, vertical, cross-sectional view through a diaphragm cell
(b) Cross section through tip of reference electrode

Source: U.S Patent 4,163,698

Wire **22**, whether for use as a cathode or anode reference electrode, is placed in the chemically resistant insulating tubing **20** with about a 0.3 to 1.0 cm tip **30** of wire **22** exposed. The tip **30** is then sealed with a chemically resistant material, such as a silicon rubber or epoxy. This insulated wire is then attached to face **19** of electrode **16** adjacent the center of face **19** or elsewhere.

It is preferred to place the exposed tip **30** parallel to and from a lower limit of about 0.2 to an upper limit of about 1.0 mm away from the surface of electrode **19**. The lower limit is set so as to avoid shorting between electrode **19** and tip **30** and to prevent bubbles being trapped between electrode **19** and tip **30**. The upper limit prevents tip **30** from being outside the gas stream generated by electrode **19**. It is preferred that tip **30** be pointed upwardly so that gas bubbles have less tendency to collect on tip **30** than would be the case if the tip were pointed down, since pointing tip **30** down would leave the seal **32** and end of tube **20** as downward facing ledges to hold or slow gas bubble release.

Tube **20** can be attached to the cathode by any corrosion resistant fastening means, such as for example a tetrafluoroethylene tie. Tube **20** can be either spaced from or in contact with the surface of electrode **16** since tube **20** is of insulating material. The end of wire **22** opposite tip **30** is also exposed so that a connection can be made to a first input terminal of a dc voltmeter **25** to measure the difference between the potential of wire **22** and electrode **16**. Stem wire

24 serves to connect a second input terminal of dc voltmeter **25** to electrode **16** for purposes of making such a comparison. Voltmeter **25** can be any conventional readily available dc millivoltmeter and need not be an elaborate or sophisticated device as is conventionally needed in order to get a reliable overvoltage reading with a Lugin type measurement method. The overvoltage is read directly from a relatively inexpensive dc voltmeter **25** under normal operating conditions and is thus particularly suited for incorporation into production type diaphragm or membrane cells for which overvoltage has not previously been monitored on a regular basis, even though such information is quite valuable as an indication of the performance of the catalytic metal coatings normally used on electrodes of such cells.

Example: A 20-gauge titanium wire is coated with a titanium oxide (TiO_2)-ruthenium oxide (RuO_2) coating by using the following procedure:

(1) The titanium wire was etched in a solution of 99 parts by volume concentrated (36%) HCl and 1 part by volume concentrated (50%) HF for 1 minute;

(2) The wire was removed from the HCl-HF solution, etched in HCl (36%) for 2½ hours at room temperature and then removed;

(3) The wire was soaked in deionized water for 16 hours (i.e., overnight) until just before coating;

(4) The wire was removed from the deionized water and etched for 8 minutes in concentrated (36%) HCl just before coating and then rinsed with deionized water;

(5) A ruthenium chloride ($RuCl_3$) plating solution having, per gram of $RuCl_3$, 6.2 ml butanol, 0.4 ml concentrate (36%) HCl and 6 ml (5.46 grams) tetra-n-butyl-orthotitanate was brushed on the wire;

(6) The solution covered wire was then heated in air (fired) at 400°C for 5 minutes;

(7) Steps 5 and 6 above were repeated four times; and

(8) The coated wire was then heated in air at 400°C for 6 hours and allowed to air cool to room temperature to give the coating greater stability.

The coated wire was then placed in a polyfluorinated hydrocarbon insulating tube leaving a tip of about 1 cm of coated wire exposed. The tube was sealed at this tip with a silicon rubber glue to prevent leakage.

This insulated coated wire was then attached as an in situ reference electrode to an anode of a chlor-alkali electrolytic cell (see Figure 2.16a) in a position parallel to the surface of the anode and with a fixed gap of 0.5 mm between the tip and the anode surface.

To monitor anode overvoltage, a dc voltmeter was attached across the coated titanium wire and the anode and the chlorine overvoltage of the anode was read directly from the voltmeter under normal operating conditions.

The readings were found to be in conformity with readings taken from a conventional Luggin capillary probe.

Method to Determine Suitability of a Diaphragm

C.K. Bon; U.S. Patent 4,090,924; May 23, 1978; assigned to The Dow Chemical Company has developed an electrical method to determine the suitability of a diaphragm for use in an electrolytic cell. The method comprises inserting the diaphragm between a primary anode and a primary cathode immersed in an electrolyte and then impressing a known direct current electromotive force between the electrodes. The change in electrical properties across the electrolyte resulting from insertion of the diaphragm is determined. Such change is indicative of the suitability of the diaphragm for use in an electrolytic cell and can be a measure of diaphragm uniformity.

Referring to Figure 2.17, primary electrodes, such as a primary anode **60** and a primary cathode **61** are immersed in an electrolyte **62** and connected to a power source **64**.

Figure 2.17: Apparatus for Determining Suitability of a Diaphragm

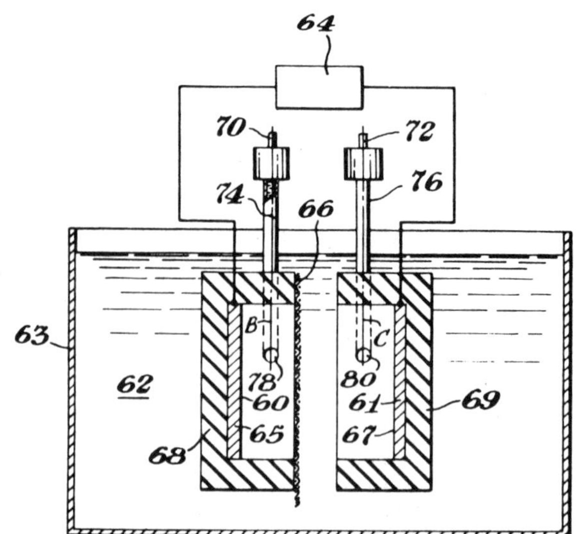

Source: U.S. Patent 4,090,924

Two auxiliary calomel measuring electrodes **70** and **72** are connected to the electrolyte **62** by a first salt bridge **74** and a second salt bridge **76**. Orifices **78** and **80** and salt bridges **74** and **76**, respectively, pass through the retaining members **68** and **69** at a predetermined position between the primary electrodes **60** and **61** and communicate with the electrolyte **62**. The orifices **78** and **80** are positioned apart to define a predetermined distance, for example, ¾", of electrolyte **62** between the center of such orifices as represented by center lines **B** and **C**.

A direct current electromotive force is impressed between the primary anode **60** and the primary cathode **61** to produce a constant current flow between such primary electrodes.

The diaphragm **66** is positioned in electrolyte **62** between the primary electrodes **60** and **61** and the salt bridge orifices **78** and **80** to the measuring electrodes **70** and **72** to thereby alter the electrical resistance between the measuring electrodes. At a constant known current, the change in voltage across the predetermined portion of electrolyte **62**, as measured by the measuring electrodes **70** and **72**, is an amount characteristic of the porosity and surface characteristics or effectiveness of the diaphragm in an electrolytic diaphragm cell.

BRINE FEED, ADDITIVES, AND CORROSION PREVENTION

Brine-Introduction Tube Mounted Within Larger-Diameter Tube

A.J. Schweickart; U.S. Patent 4,149,955; April 17, 1979; assigned to BASF Wyandotte Corporation has modified a brine-feeding structure so that the brine-introduction tube is mounted within a larger-diameter tube sealingly mounted in the brine-introduction opening in the side of the cell top, with an orifice member preferably being mounted within an end of the brine-introduction tube nearer to the cell. This makes it possible to avoid unwanted variations in the brine flow rate and to see instantly whether an orifice member is clogged.

In Figure 2.18, there is shown a cell top **2** having a vertically extending side portion **4**, which has therein an opening **6**. There is provided a first tube **8**, preferably of glass, and having an outside diameter of over 25 mm. Surrounding the tube **8** is a rubber stopper **10** which serves as a means for sealingly mounting the tube **8** within the opening **6**.

Also shown is a second tube **12**, also preferably of glass, although either of the tubes **8** and **12** may be made of any transparent material which is adequately strong and resistant to the action of brine. The tube **12** is of 10 to 26 mm in outside diameter and is surrounded by a rubber stopper **14** or other suitable means for permitting it to be mounted in the end **16** of the tube **6** which is located exteriorly of the cell **2**.

There is also provided an orifice member **18**, having around it a rubber packing **20**, or other means whereby it is sealingly mounted within the tube **12**, and within the end **22** of the tube **12** which is proximal to the cell **2**. The orifice member **18** may be made of any material satisfactorily resistant to the action of brine; satisfactory results have been obtained with the use of an orifice member made of tantalum metal.

In a typical case, the orifice member has an outside diameter of 0.27" to 0.44" (6.9 to 11.2 mm) and an orifice opening 0.15" to 0.40" (3.8 to 10.1 mm) in diameter. With any usual head, such as 0.8 to 2.2 m, an orifice opening as mentioned above will usually provide a flow rate on the order of 2 to 15 ℓ/min. The desirable flow rate, within the foregoing range, depends upon the capacity of the individual cell involved; a flow rate of about 6 ℓ/min would not be unusual.

A flexible brine-supply line **24** is snugly fitted over the end **26** of tube **12** which is distal from the cell **2**. The cell top **2** fits upon the cell bottom **28**. The level of liquid **30** within the cell is also indicated.

The mode of operation of the apparatus disclosed above is self-evident. Brine is supplied as indicated at **32**, passing as a narrow stream **34** from the orifice member **18**. Inasmuch as the brine-supply line **24** does not cover the vicinity of the orifice member **18**, the operator can see instantly whether a stream **34** of flow is being produced, and if it is not, whether liquid is being supplied to the orifice member **18** or not.

The structure disclosed makes it possible to restore a cell to proper operation quickly and conveniently in the event that the orifice member **18** becomes plugged, or an adjustment in the rate of feeding brine to the cell is required. The stopper **14** is removed from the end **16** of the tube **8**, and if desired, the entire subassembly comprising the tube **12**, stopper **14**, orifice member **18**, and packing **20** may be detached from the brine-supply line **24** and replaced with a similar assembly containing an unclogged orifice of appropriate size. Alternatively, the orifice member **18** may be removed and replaced or be returned after cleaning.

Figure 2.18: Brine-Introduction Tube Mounted Within Larger-Diameter Tube

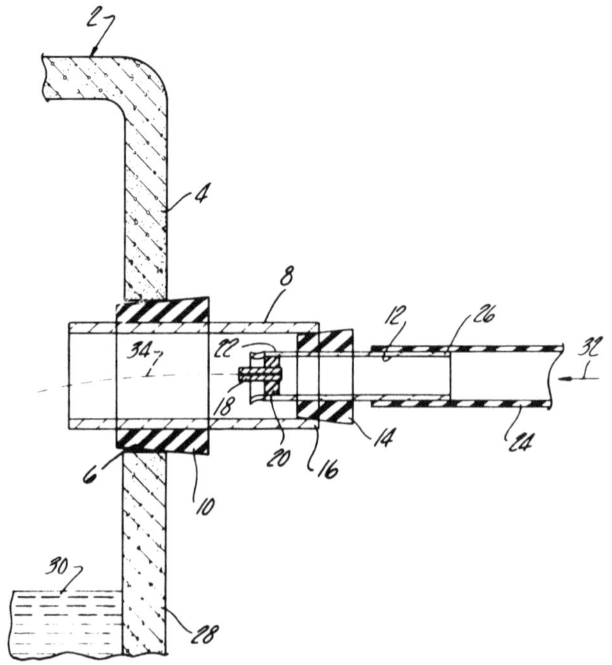

Source: U.S. Patent 4,149,955

Feed System for Slurries of Alkali Metal Chlorides

The apparatus and method developed by *A.C. Wiley, Jr. and O.C. Taylor; U.S. Patent 4,082,630; April 4, 1978; assigned to The Dow Chemical Company* provide a satisfactory means of supplying or feeding a finely divided solid, aqueous alkali metal chloride slurry to an anode compartment of an electrolytic cell without either causing deposition of the solid salt within conduits leading to the anode compartment or adversely affecting the cell performance by substantial over-saturation of the anolyte with the salt. Such over-saturation generally results in plugging of a diaphragm separating the anode and cathode compartments with a solid salt.

The improvement of this disclosure includes a means adapted to feed an aqueous slurry of the alkali metal chloride particulate to the anode compartment through at least one primary conduit and then through a secondary conduit at a linear velocity of at least about 6.5 fps. The secondary conduit has an interior cross-sectional area of up to about 0.2 in^2 and less than that of the primary conduit.

In operation of the aqueous slurry feed means of Figure 2.19, finely divided sodium chloride particles are fed into a mixing container **30** together with water, and/or an alkali metal salt containing brine, and, optionally, aqueous salt slurry which was not passed into the anode compartments to be electrolyzed.

The water-salt feed is suitably mixed to form a generally uniform slurry composition by means of, for example, an impeller **32** in the container **30**. The slurry flows from the container **30** into first and second primary conduits or pipes **34** and then into secondary conduits **36**, which are physically connected to the anode compartments (not shown) within the cell series **37**. The respective lengths of the secondary conduits **36** between the primary conduits **34** and the anode compartments can be suitably adjusted to provide a substantially uniform volume of slurry through the secondary conduits in a unit time period. A shorter secondary conduit length results in a greater slurry flow than does a longer secondary conduit, when the pressure within the primary conduit **34** is maintained constant.

It has been found that the slurry can be passed into the anode compartments on a substantially continuous basis without plugging the diaphragm when the secondary conduit **36** has a cross-sectional area of up to about 0.2 in^2. Preferably, the secondary conduit **36** is a tube with an inside diameter of from about $^3/_{16}$" to $^5/_{16}$". It is essential that the primary conduits **34** have a cross-sectional area greater than that of the secondary conduits **36**. Preferably, each position of the primary conduit **34** has an interior cross-sectional area at least equal to the total interior cross-sectional areas of the secondary conduits **36** fed from that portion of the primary conduit. Generally, the primary conduit **34** will be a pipe with an inside diameter of from about 1½" to 3".

That portion of the aqueous slurry which is not passed into the electrolytic cells **10** through the secondary conduits **36** from the primary conduits **34** can be returned to the mixing container **30** through a return conduit or pipe **38**.

Example: A plurality of electrolytic cells, electrically connected in series, for producing gaseous chlorine and sodium hydroxide from an aqueous sodium chloride solution was operated by feeding a sodium chloride brine containing finely divided solid sodium chloride particles to anode compartments within the series. The brine was mixed with sodium chloride particles with a size of about 100 mesh by means of an impeller in a mixing container spaced apart from the electrolytic cells. The mixture or slurry contained about 5 wt % solid sodium chloride particles.

Figure 2.19: Schematic Illustration of Electrolytic Cell System

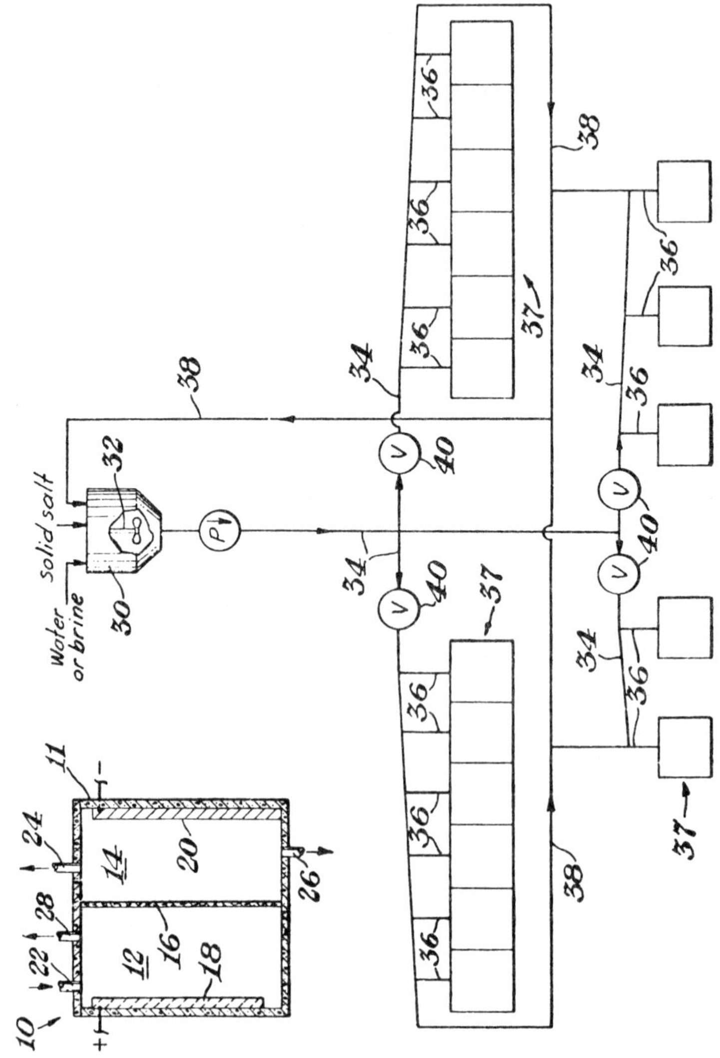

Source: U.S. Patent 4,082,630

The slurry was pumped at a pressure of 40 lb/in^2 (gauge) to anode compartments within the series through corrosion-resistant feed pipes with an inside diameter of about 1½" to 2" and polytetrafluoroethylene tubes with an inside diameter of 3/16". The linear velocity of the slurry through the primary conduit was about 6.7 fps. Substantially the same volume of slurry was fed to each of the anode compartments by suitably adjusting the length of the tubes extending to the anode compartments from the feed pipes.

The electrolytic cell was operated by established procedures to produce chlorine and sodium hydroxide. With the slurry feed rate to the anode compartments adjusted to provide an amount of sodium chloride feed substantially equal to that being electrolyzed, there was no evidence of detrimental deposition of solid salt within the slurry feed conduit system or plugging of the diaphragms. Excess slurry in the feed pipes was returned to the mixing container and recirculated through the system.

Transition Metal Compounds Added to Catholyte Liquor

A. Martinsons and H.B. Johnson; U.S. Patent 4,105,516; August 8, 1978; assigned to PPG Industries, Inc. have found that, when the compound of a transition metal is added to the catholyte liquor while an electrical current is caused to pass from the anode of the electrolytic cell to the cathode of the electrolytic cell, the cathode component of the cell voltage is found to be reduced, for example, from about 1.45 volt SCE before addition to about 1.25 volt SCE after addition. The exact mechanism for attaining this cathode voltage reduction is not clearly understood but it is believed that the transition metal deposits on the cathode while chlorine is being evolved at the anode, thereby maintaining a clean transition metal surface of high surface area on the cathode during electrolysis.

Example: A test was conducted to determine the effect of Fe^{++} addition on the cathode hydrogen evolution potential of a laboratory chlor-alkali diaphragm cell. The cell had a 5" x 7" (12.7 x 17.8 cm) ruthenium dioxide-titanium dioxide coated titanium mesh anode spaced from a 5" x 7" (12.7 x 17.8 cm) etched, expanded iron mesh cathode. An asbestos paper diaphragm was interposed between the anode and the cathode.

An aqueous solution of $FeCl_2 \cdot 4H_2O$ was added directly to the catholyte compartment of the cell. Electrolysis was carried out at a current density of 190 A/ft^2 (0.20 A/cm^2). The brine feed contained 315 g/ℓ of sodium chloride. The catholyte liquor contained 160 g/ℓ of sodium chloride and 120 g/ℓ of sodium hydroxide. The iron chloride feed to the cell was through a feed line directly to the catholyte compartment. The results are shown below:

Days of Operation	. . Amount of Iron Added . . . (g/ft^2)	(meq/cm^2 day since last addition)	. . . .Cathode Voltage (volts) . . . . Before Addition	After Addition
1	4	1.54×10^{-1}	1.390	1.298
8	2	9.6×10^{-3}	1.341	1.295
11	1	1.29×10^{-2}	1.310	1.287
16	0.5	3.86×10^{-3}	1.316	1.304
21	0.5	3.86×10^{-3}	1.335	1.295
24	1	1.29×10^{-2}	1.332	1.290

Prevention of Equipment Corrosion

The alkaline liquors produced by the electrolysis of concentrated sodium chloride brines in diaphragm cells generally contain between 9 and 13% by weight of sodium hydroxide and between 14 and 17% by weight of sodium chloride. As they leave the electrolytic cells, these hot alkaline liquors are usually transferred by way of metallic conduits to storage vessels and/or evaporators where they are concentrated in respect to sodium hydroxide while precipitating sodium chloride.

In view of the highly corrosive character of hot alkaline liquors of high sodium chloride content, and particularly of liquors saturated or supersaturated with sodium chloride, that are treated in the evaporators, these evaporators are generally made of corrosion-resistant metals or alloys, for example, of nickel or stainless steel.

The use of corrosion-resistant metals or alloys severely burdens the cost of equipment handling alkaline liquors.

A method is disclosed by *G. Sluse and F. Dujardin; U.S. Patent 4,110,181; August 29, 1978; assigned to Solvay & Cie, Belgium* for handling corrosive concentrated aqueous solutions of alkali metal hydroxides containing major amounts of alkali metal halides in equipment made of ordinary or mild steel. Corrosion is prevented by a combination of controlling the amount of alkali metal hydroxide to less than 500 g/kg of solution; maintaining the temperature of the solution below 130°C; and providing an anodic potential of between -1,000 and -200 mV with respect to the potential of the saturated calomel electrode.

Example 1 (Comparative): This test was carried out on bars of extra mild steel immersed in an alkaline liquor saturated with sodium chloride, with no protection of the bars whatever against corrosion.

A bar of extra mild steel was immersed in an alkaline liquor containing 400 g of sodium hydroxide per kg and heated to 60°C. At the end of the test there was recorded for the bar, a loss in weight corresponding to 1.5 $g/m^2/day$. After pickling, the loss in weight had increased to 1.6 $g/m^2/day$.

Example 2: For this test, there was employed an electrochemical cell comprising a container made of polytetrafluoroethylene for the liquor, a cathode in the form of a platinum wire and an anode consisting of the bar under test. By means of a potentiostat, a fixed potential with respect to a calomel/saturated KCl reference electrode communicating with the solution by way of a bridge was impressed on the anode bar. The same conditions of temperature and concentration of the liquors as in Example 1 were used.

Example 3: The conditions of Example 1 were repeated, applying additionally a potential of -800 mV to the bar. At the end of the test, the loss in weight suffered by the bar was imperceptible before pickling; it had increased to 0.3 $g/m^2/day$ after pickling.

MERCURY CATHODE CELLS

ADJUSTMENT OF ANODE-CATHODE SPACING

Monitoring a Plurality of Control Parameters

Although it is known to be desirable to reduce the anode-cathode spacing to reduce power consumption, such a reduction can increase the hazards of cell shorts. These are undesirable in that they often require stopping and restarting of the electrolysis reaction, they cause damage to the anodes, and they are accompanied by undesirable alternative and dangerous hydrogen generation which reduces current efficiency and may result in explosions and cell damage.

E.L. Hixson and C.M. Talkington; U.S. Patent 4,069,118; January 17, 1978; assigned to Stauffer Chemical Company have designed an apparatus for controlling and monitoring an electrolysis reaction so as to reduce or avoid anode-cathode shorts, while minimizing power consumption through minimizing anode-cathode spacing.

A plurality of control parameters are utilized in the controlling and monitoring functions. One control parameter is current, which is analyzed to detect current oscillations, which are precursors of shorts. When current oscillations are detected, the anode-cathode spacing is increased until stable current is regained. Other parameters are excursions of current and voltage above or below predetermined upper and lower limits.

In Figure 3.1, an array of 52 electrolysis cells is depicted. Each cell includes 16 anode units. For example, the cell **1** includes the units **A1,1** through **A1,16**; the cell **2** includes the units **A2,1** through **A2,16**, etc. Each cell receives the 16 anode buses **32-1** through **32-16**. The cell units which connect to anode bus **32-1** are the units **A1,1** through **A52,1**.

Each cell and each unit in a cell are connected to a cell instrumentation package (CIP). The 52 cells (cells **1** through **52**) are connected to the cell instrumentation packages **CIP-1** through **CIP-52**, respectively. The anode buses **32** and the cathode buses **33** connect to the power supply **7**.

Figure 3.1: Digital Control System

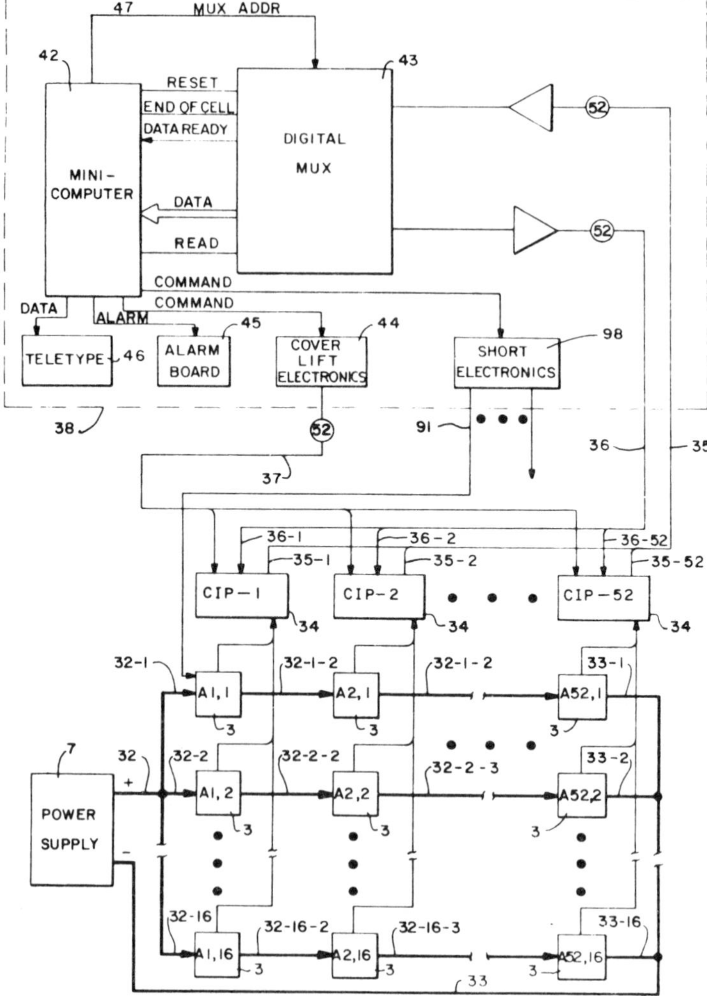

Source: U.S. Patent 4,069,118

Each cell instrumentation package **34** has an output data line (typically two wires per line) which forms 1 of the 52 lines in the data bus **35** to the data processor **38**. Each CIP **34** receives an address line (typically two wires per line) which is 1 of the 52 lines in the address bus **36** from data processor **38**. Each of the CIPs also receives a cover lift control line from the 52-line cover lift bus **37**.

The data bus **35** and the address bus **36** are controlled by the data processing circuitry which includes a programmable computer **42** and a digital multiplexer **43**.

In one preferred embodiment, the computer **42** is a Nova 1200 computer and the digital multiplexer **43** is a Model 4100, both manufactured by Data General Corporation.

The digital multiplexer **43** is addressed by the 12-bit MUX ADDR bus **47** from the computer **42**. The bus **47** is divided into two 6-bit fields. The high order 6 bits address the 52 cell instrumentation packages. The lower order 6 bits of the MUX ADDR define the individual measurements for each of the cells. In each cell, a total of 33 measurements are made, specifically 16 voltage measurements, 16 current measurements and 1 temperature measurement. Since there are 33 measurements per cell and since there are a total of 52 cells, a total of 1,716 measurements are made for the 52 electrolysis cells.

The multiplex address on bus **47** sequentially steps through each of the 1,716 addresses. The multiplexer **43** responsively addresses the cell instrumentation packages and the 33 measurements associated with each package. The computer **42** and multiplexer **43** complete the addressing in a fixed cycle period. The cycle period is selected, in one embodiment, to be approximately 0.1 sec and, therefore, a sampling rate of 10 samples per second is established. With a sampling rate of 10 times per second, the 1,716 measurements are completed 10 times per second.

The sampling rate determines the highest frequency of oscillation which can be detected by the apparatus. A sampling rate of 10 samples per second will readily allow detection of oscillation frequencies up to 5 cycles per second. If higher frequencies are to be detected, then the sampling rate must be increased. For readily detecting frequencies of less than approximately 1 cycle per second, a sampling rate of 2 samples per second may be used.

Comparison with Predetermined Standards

An improved method and apparatus for adjusting the space between an adjustable anode and a cathode in an electrolytic cell is described by *R.W. Ralston, Jr.; U.S. Patents 4,098,666; July 4, 1978 and 4,155,829; May 22, 1979; both assigned to Olin Corporation* wherein current measurements and voltage measurements are obtained for conductors to the anode sets and compared with predetermined standards for the same conductors and anode sets. Measurement of deviation from the predetermined standards are used to determine the direction of anode adjustment. A digital computer operably connected to motor drive means adapted to raise or lower anode sets upon appropriate electric signals from the computer is a preferred embodiment.

Example: A horizontal mercury cathode cell for electrolyzing aqueous sodium chloride to produce chlorine containing 12 anode sets of 8 graphite anodes per set was equipped with the anode control system of Figure 3.2.

Current and voltage signals for all 12 anode sets were transmitted simultaneously to automatic control unit **6**, a digital computer, for about 5 seconds until about 180 readings of current and of voltage were received for each anode set. The average voltage, current and the difference between each current reading and the previous current reading was determined by the digital computer for the series of readings. The voltage coefficient was calculated for each anode set according to the formula: $Vc = V -3.1/kA/m^2$.

Figure 3.2: Apparatus for Regulating Anode-Cathode Spacing

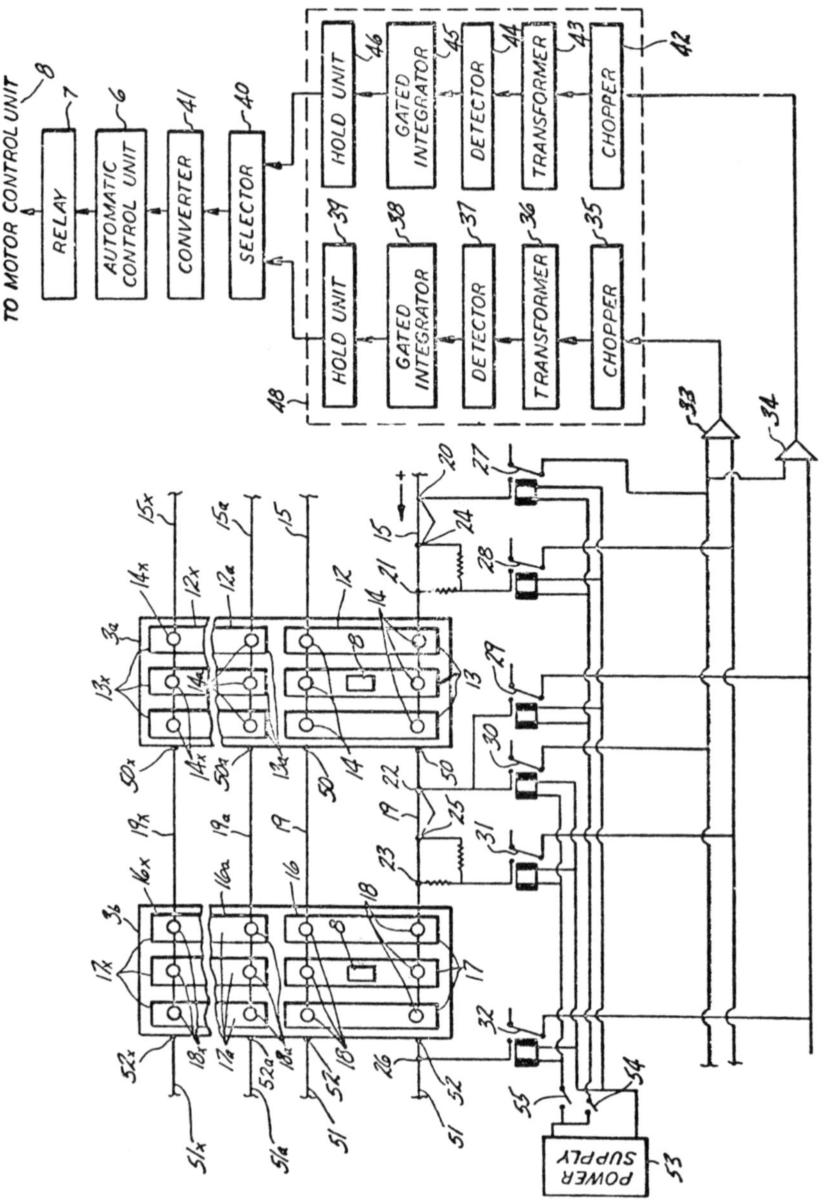

Source: U.S. Patent 4,098,666

Anode set **12**, with a cathode surface area of 2.4 m^2, was found to have a Vc of 0.128, based on an average voltage of 4.38 and an average current reading of 12.0 kA. When Vc was compared with its standard coefficient S, of 0.115, it was found to have a value above the deviation range k, where k was ±0.006. When the coefficient comparison determined that the value of Vc was above S by a value greater than k, a signal from the computer activated a relay which energized a hydraulic motor to lower anode set **12** to decrease the anode-cathode spacing by 0.3 mm. Following the decrease in anode-cathode spacing, the following sequence of operations were performed.

(1) A second set of about 15 measurements of current was taken for each conductor **15** to anode set **12** only and the difference between each measurement in each set was determined.

(2) The first analysis compared the initial increase in current after decreasing the anode-cathode spacing with the maximum increase prior to the adjustment and was found to be within the predetermined limits.

(3) A second set of about 15 current readings was taken and the second analysis for current fluctuation determined using the formula $\Sigma\Delta^2/N$. The fluctuation was found to fall within the predetermined limit of 0.5%.

(4) A third analysis showed that the time since lowering the anode had not exceeded a fixed limit.

(5) A fourth analysis revealed that the total increase in current did not exceed a predetermined limit of 7%.

(6) The last reading was found to be larger than the previous reading and steps (3) through (5) were repeated, with the same result. The latest reading was then found to be smaller than the previous reading indicating that the current to the anode set has stopped increasing.

Readings were then taken for all anode sets on the cell and the Vc calculated for each was found to have a value within 5% of the stored value, S. No further adjustments were made and the next cell to be adjusted was selected.

Method for Detecting Incipient Short Circuits

R.W. Ralston, Jr.; U.S. Patent 4,174,267; November 13, 1979; assigned to Olin Corporation discloses a preferred embodiment of the apparatus described in the preceding two patents in which potential incipient short circuits are detected and avoided by the improvement which comprises:

(a) Obtaining the difference between two successive current signals for a selected conductor and doubling the difference to obtain a total current difference, Δ_t;

(b) Obtaining the difference between two successive current signals for the conductor adjacent to the selected conductor and adding the difference for each adjacent conductor to obtain an adjacent total current difference Δ_a;

(c) Subtracting Δ_a from Δ_t to obtain a remainder, R, for the selected conductor; and

(d) Raising the selected anode when R exceeds Δ_t by more than about 0.5%.

Example: A group of horizontal mercury cathode cells for the electrolysis of sodium chloride were used in this example; each cell contains ten anode sets and each anode set contains five anodes. The anodes were constructed of titanium metal and partially coated with a noble metal compound. Each anode set was supplied with current by two conductors. The anode adjustment system of Figure 3.2 was installed on the cells. Upon selection of one cell for possible adjustment of the anode-cathode spacing, a series of 180 readings were taken simultaneously for all anode sets in the cell over a period of about five seconds.

The current measurement was obtained by measuring the voltage drop between two terminals spaced 30" apart on each conductor and the voltage measurement was obtained between two corresponding terminals on each conductor supplying current to the corresponding anode set for the next adjacent cell. Thus, a group of 180 current measurements and 180 voltage measurements were obtained for each of the two conductors supplying an anode set and for all ten sets in the cell.

Each group of measurements were signal conditioned and converted from analog to digital form and supplied to automatic control unit **6** (a digital computer) where the average total current and voltage measurements were calculated.

The voltage coefficient was calculated from the average total current and voltage readings obtained and then compared with a predetermined standard individually selected for each of the anode sets. Measurements of current and voltage taken for each set of anodes along with the calculated Vc (as defined in the preceding patents) and the predetermined standard Vc are given in Table 1.

Table 1

Anode	. Current in Kiloamperes .		. . . . . . .Voltage		Calculated	Standard
Set No.	Conductor A	Conductor B	Conductor A	Conductor B	Vc	S
1	6.86	6.38	4.44	4.47	0.154	0.150
2	7.15	7.93	4.41	4.55	0.137	0.130
3	7.71	7.92	4.44	4.48	0.131	0.130
4	7.40	7.74	4.46	4.48	0.136	0.130
5	7.51	7.44	4.46	4.48	0.138	0.130
6	7.88	7.31	4.46	4.51	0.137	0.130
7	7.47	7.47	4.48	4.46	0.137	0.130
8	7.25	7.75	4.48	4.47	0.137	0.130
9	7.57	7.38	4.41	4.48	0.135	0.130
10	6.96	6.16	4.41	4.40	0.149	0.140

Note: Average anode set current, 14.72 kA; average cell voltage, 4.46; average k = ±0.010.

The incipient short circuit values or remainder (R) were determined later by:

(A) Obtaining the difference between two successive current signals for a selected conductor and doubling the difference to obtain a total current difference, Δ_t;

(B) Obtaining the difference between two successive current signals adjacent to the selected conductor and adding the difference for each adjacent conductor to obtain an adjacent total current difference, Δ_a;

(C) Subtracting Δ_a from Δ_t to obtain a remainder (R) for the selected conductor which is added to the summation table;

(D) Repeating steps (A) through (C) 180 times during a 5 second period;

(E) Dividing the total R in the summation table by 180 to obtain an average R; and

(F) Raising the selected anode when the average R exceeds the average current in the above table for the selected conductor by more than about 0.5%.

Each end conductor (conductor A of anode set 1 and conductor B of anode set 10) was compared with double the adjacent conductor (conductor B of anode set 1 and conductor A of anode set 10, respectively). The values for R are presented below in Table 2.

Table 2

Anode Set No.	Incipient Short Value, R, in kiloamperes Conductor A	Conductor B
1	0.002	0.002
2	0.002	0.003
3	0.003	0.002
4	0.002	0.003
5	0.03	0.04
6	0.003	0.003
7	0.002	0.003
8	0.002	0.002
9	0.003	0.002
10	0.003	0.003

From the results of Table 1, it can be seen that none of the anode sets fell outside of the limits of k and, therefore, no adjustment of the anode-cathode spacing was required. However, Table 2 shows a later need to adjust anode set 5.

These measurements showed that there was an incipient short circuit about to occur at anode set 5, and anode set 5 was raised a distance of 0.3 mm. All values, R, were in the range of 0.002–0.003 following the raising of the anode set and normal operation of the cell was observed.

ELECTRODES AND OTHER CELL COMPONENTS

Vertical Liquid Electrode

D.J. Cotton; U.S. Patent 4,091,829; May 30, 1978 describes a method of creating, maintaining and controlling a vertical component of liquid surface by the use of capillary troughs.

By way of example, the method is demonstrated through its use to create a liquid electrode with an active vertical surface suitable for use in electrolytic cells, particularly in cells of the liquid metal electrode type and, more particularly in cells that utilize a mercury cathode in the electrolysis of molten metal halides, or their solutions, to produce halogen gas and a metal, its amalgam, its hydroxide, or its oxide.

Figure 3.3 illustrates a typical cross section normal to the longitudinal axis of a cathode-anode pair in such a cell. **1** is a capillary trough which is nonadhesive to mercury and which, at its top, has an opening that approximately equals twice the capillary constant of mercury (typically, 5.0±0.5 ml) in the cell environment. **3** is mercury, which, because of the size of the capillary trough opening, forms a meniscus that has an approximately circular cross section which projects above the trough edge.

Figure 3.3: Vertical Liquid Electrode

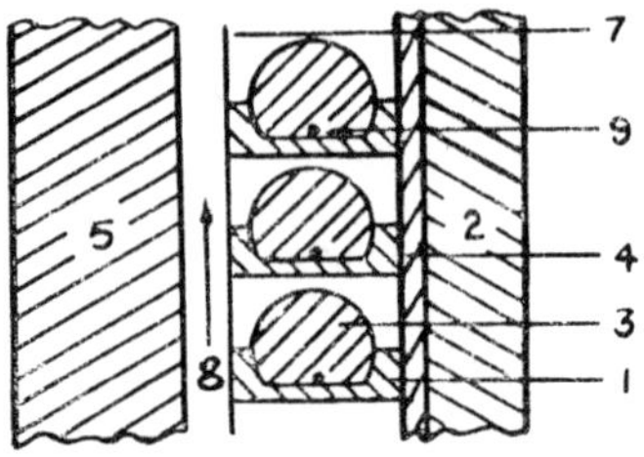

Source: U.S. Patent 4,091,829

The vertical component of the projecting surface area of the meniscus acts as a cathode surface in parallel to the surface of **5**, the anode. The capillary trough **1** is attached to a metallic support plate **2** covered with a layer of nonconductive material **4**. The trough has a flat exterior edge to which an optional diaphragm **7** is attached. The diaphragm **7** effectively divides the space between the cathode and the anode into two separate chambers so as to segregate the cathode- and anode-evolved gases. The electrolyte, aqueous alkali chloride, is forced to flow in the space between the cathode and the anode in the direction indicated by arrow **8**. A wire metal conductor **9** insures electrical continuity through the mercury.

Flexible Cover

R.F. Anderson; U.S. Patent 4,100,053; July 11, 1978; assigned to The B.F. Goodrich Company has developed an improved flexible sheet elastomer cover for sealing the electrolyte tank of a De Nora type chlorine generating cell.

Elastomer materials of different compositions are used for the opposite surfaces of the cover with the inside surface composed of a chlorinated polyethylene compound and the outside surface composed of an ethylene-propylene terpolymer compound.

Figure 3.4 illustrates an electrolytic cell (De Nora type) as including a tank **10** having a flat horizontal rim flange **10a** around the periphery of the open top. The cell is shown as partially filled with brine electrolyte **11** overlying the cathodic electrode **12** which consists of a thin layer of mercury at the bottom of the tank. The several spaced anodic electrodes **13** are made of carbon blocks or metal adjustably suspended from an overhead rigid supporting structure **14**, shown as a plate reinforced with I-beams, which is independently supported by vertical pillars **15** spaced uniformly around the tank.

The carbon or metal anodes **13** are secured vertically in an adjustable manner by terminal members including fastening means, preferably threaded shafts and locknuts, to facilitate maintaining a constant gap between the anodes and the mercury cathode.

Figure 3.4: De Nora Cell

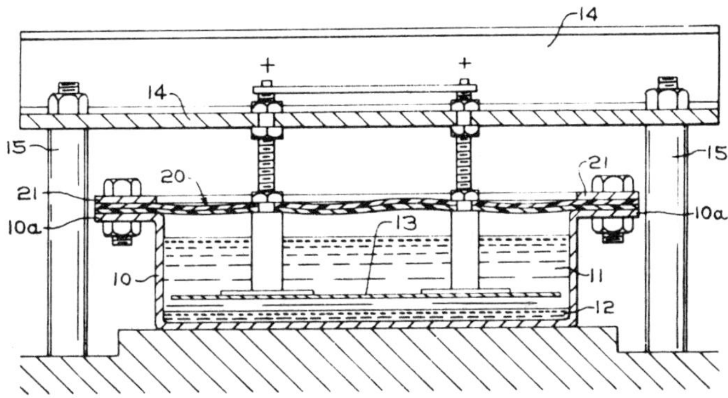

Source: U.S. Patent 4,100,053

The cell **10** has a flexible cover **20** which is sealed around the flange **10a** of the tank by being clamped intermediate a frame **21** and the flange **10a**. External securing means, which may be C-clamps but are shown here as machine bolts and nuts, with the bolts passing through the matching holes in the frame, cover and flange, are used to sealingly clamp the cover between the frame and the flange. The cover **20** is also provided with holes through which threaded portions of the terminals for anodes **13** extend with sealing between each terminal and the flexible cover being effected by the locknuts on the threaded portions, which are tightened sufficiently to clamp the cover between the nuts and integral shoulders on the terminals.

The flexible cover **20**, in its preferred embodiment, is a unitary sheet of two layers of different elastomeric materials integrally united by being vulcanized together. The upward or outside layer is of uniform thickness and is made of sulfur-curable terpolymer of ethylene, propylene and a nonconjugated diene (EPDM). The downward or inside layer is also of uniform thickness and is made of a chlorinated polyethylene composition.

The improved cell cover has exhibited a much longer, trouble-free life with fewer difficulties in cell operation, such as contamination of the electrolyte by deposits on the cover falling therefrom, than has heretofore been possible with prior cell covers. The longer life is due to the fact that the chlorinated polyethylene on the underside of the cover has high resistance to erosion and other deterioration by the hot chlorine gas, in contact with this surface of the cover, while the ethylene-propylene terpolymer resists deterioration by the high ozone content of the atmosphere in contact with the outer surface of the cover.

The following is an example of a suitable composition for the outside layer:

Ingredients	Parts by Weight
EPDM	180.0
Chlorosulfonated polyethylene	5.0
Zinc oxide	5.0
Methylated paraffinic resin	5.0
Stearic acid	1.0
Carbon black, FEF	160.0
Paraffin oil	45.0
Mercaptobenzothiazole	0.65
Tetramethylthiuram disulfide	0.90
Sulfur master batch, 80% sulfur	2.50

In the above composition, the EPDM was in the form of a blend of 160 pbw Epcar 5465 and 20 pbw Epcar 346 (both from B.F. Goodrich Company); the 5465 terpolymer has been identified as having a Mooney of 53 at 257°F (125°C) and a specific gravity of 0.90, while the 346 terpolymer has a Mooney of 30 at 212°F (100°C) and a specific gravity of 0.86.

The following is an example of a suitable composition for the inner layer:

Ingredients	Parts by Weight
Chlorinated polyethylene	80.0
Natural rubber	20.0
Magnesium oxide	10.0
Peroxide	8.0
Carbon black	75.0
Paraffin oil	10.0
Dioctyl phthalate	10.0
Epoxy polymeric plasticizer	10.0
Triallyl isocyanurate	1.0
Desiccant	12.0

The chlorinated polyethylene in the above compound [CM0136 (Dow Chemical Company)] is reported to have a specific gravity of 1.16, a chlorine content of 36%, and an average Mooney of 80 at 250°F (121.11°C).

DECOMPOSITION OF ALKALI METAL AMALGAMS

Electrolysis of aqueous sodium chloride solutions to produce chlorine and sodium hydroxide by the so-called mercury amalgam process is still widely used industrially as it presents several advantages over other existing processes, for example, those utilizing diaphragm or membrane cells.

In all the commercially known plants, the amalgam leaving the electrolysis cell is decomposed in a reactor provided with a catalytic filling; water and hydrogen and caustic soda produced by the decomposition process are recovered and mercury is recycled to the cell.

Unpurified Salt as Raw Material

One of the main factors affecting reliable operation and safety of the mercury amalgam process is the purity of the brine introduced into the cell as the level of impurities that can be tolerated in the process is very low. Quantities varying from 0.3 to 0.01% of impurities such as calcium, magnesium and iron are usually present in salt, while other heavy metals like Cr, V, Mo, Mn are often present in a concentration of about 0.01 ppm.

These impurities must be carefully removed from the brine since quantities higher than 0.01 ppm in the brine can cause hydrogen to evolve at the mercury cathode after an extended period of time and the Cl_2-H_2 mixture formed can explode with disastrous effects.

G. Bianchi, O. De Nora and P.M. Spaziante; U.S. Patent 4,166,780; September 4, 1979; assigned to Oronzio De Nora Impianti Elettrochimici SpA, Italy, provide a process wherein unpurified salt is utilized as raw material, and wherein expensive dechlorination and brine purification plants are no longer necessary.

The process comprises subjecting the amalgam leaving the electrolysis to decomposition to form mercury and an alkali metal hydroxide solution and subjecting the mercury to anodic polarization in an electrolyte with a counterelectrode maintained at a sufficiently negative potential to remove from the mercury at least a portion of metal impurities contained therein and recycling the purified mercury to the electrolysis cell. The metal impurities in the mercury are preferentially anodically dissolved in the electrolyte so that the level of impurities in the mercury will be held below the levels which would adversely affect the electrolytic reaction taking place in the electrolysis cell.

Figure 3.5a illustrates the mercury circuit in a chlorine plant wherein brine is electrolyzed in mercury electrolysis cell **1**. The amalgam leaving the cell **1** is introduced at the upper portion of denuder **2** which is filled with a static porous bed of catalytic material such as graphite granules.

Water is introduced by line **11** into the lower portion of denuder **2** and flows countercurrent to the amalgam during which sodium is stripped from the amalgam to form sodium hydroxide and hydrogen is evolved. The hydrogen is removed through outlet **13** and the sodium hydroxide solution is removed through outlet **12**.

The mercury from the bottom of denuder **2** is conducted by pump **3** to the electrolytic purification cell **4** and then back to electrolysis cell **1** which is provided also with brine inlet **16**, brine discharge **17** and chlorine outlet **18**. Electrolyte is added to purification cell **4** by line **14** and is discharged through outlet **15**.

In the preferred embodiment of the process, an electrolytic decomposition stage **5** is provided at the bottom of denuder **2** to eliminate any residual sodium in the mercury before the electrolytic purification step of cell **4**.

Figure 3.5: Electrolytic Decomposition of Amalgam

a.

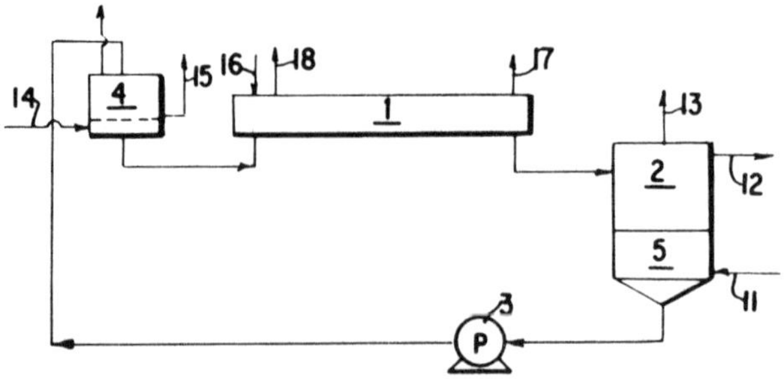

b.

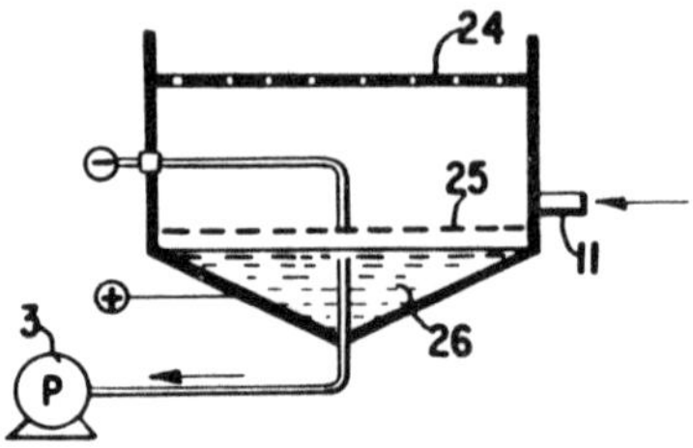

(a) Mercury circuit
(b) Schematic partial cross-sectional view of denuder

Source: U.S. Patent 4,166,780

In Figure 3.5b, the lower portion of denuder **5** is provided with an electrolytic decomposition zone below divider plate **24** in which a pool **26** of mercury collects in the denuder bottom. A counterelectrode **25** made of graphite, steel, nickel or other suitable, electrically conductive material is placed a certain distance from mercury pool **26** and the electrode **25** is cathodically polarized with respect to pool **26** by a floating direct electric current supply (not shown) whose positive pole is directly connected to pool **26**. The electrolyte for the decomposition stage is the water introduced by line **11** to form sodium hydroxide solution during its passage through the denuder.

Polyphase Packing

A polyphase amalgam denuder packing material disclosed by *O. De Nora, A. Nidola and P.M. Spaziante; U.S. Patent 4,161,433; July 17, 1979; assigned to Oronzio De Nora Impianti Elettrochimici SpA, Italy,* is comprised of a sintered mixture of at least one powdered valve metal boride and at least one valve metal carbide. The material remains active and stable under all operating decomposition conditions for a long duration. Examples of suitable valve metals are tantalum, titanium, zirconium, hafnium, niobium, tungsten and silicon and mixtures thereof.

Example: A cylindrical vessel with an inner diameter of 90 mm and a height of 100 mm was immersed in a thermostatic water bath maintained at 80°C and was provided with saddles of about 2" (5.08 cm) of the sintered material as specified below, and 100 ml of aqueous 50% sodium hydroxide. 30 ml of sodium mercury amalgam containing 4% sodium were then added at 60°C and the vessel was closed. Stirring was effected with a magnetic stirrer floating on the amalgam.

The hydrogen evolution rate reached a volume of 500 ml and the slope of the straight line obtained by plotting the amount of evolved hydrogen against time (square centimeter of mercury per minute) gave the decomposition rate which is related to the surface area of the sintered material. The rate is expressed as milliliters of mercury (at 25°C) per minute times square centimeters. The catalytic materials were aged in a laboratory denuder for 100 hours and the activity test was repeated. The results obtained with the disclosed material as well as with graphite, activated graphite, molybdenum and on 30% iron/70% vanadium alloy are reported below.

Sample Composition	Decomposition Rate ..... ml Hg at 25°C/min x cm^2....	
	Initial	After 100 Hours
ZrB_3 (60%) SiC (40%)	65	63
ZrB_2 (50%) ZrC (50%)	60	60
ZrB_3 (40%) ZrC (30%) SiC (30%)	63	61
ZrB_2 (40%) WC (10%) SiC (50%)	60	56
TiB_2 (50%) SiC (30%) C (20%)	68	67
ZrB_2 (50%) SiC (20%) C (20%) B_4C (10%)	68	66
ZrB_3 (40%) SiC (30%) C (25%) Ni (5%)	79	73
ZrB_2 (40%) SiC (30%) C (25%) Fe (5%)	75	68
Graphite	3	3
Graphite activated with $2NiO \cdot MoO_3$	10	10
Molybdenum 99.99%	60	40
Fe (30%)-V (70%) alloy	75	50

The results show that the decomposition rate for the disclosed materials is much greater than graphite and activated graphite and the stability of the material is much greater than molybdenum and the iron/vanadium alloys. This means that the process denuders may be much smaller than conventional denuders for the same capacity which permits a substantial reduction in the amount of mercury required for a mercury chlor-alkali plant with a considerable holdup savings.

Surface-Active Agent Embedded in Polymer Matrix

A method and composition are described by *H.M. Patel; U.S. Patent 4,105,441; August 8, 1978; assigned to Olin Corporation* for the decomposition of sodium-mercury amalgam in a reactor containing solid packing particles to form hydrogen, sodium hydroxide and denuded mercury.

The packing particles are comprised of a heterogeneous solid mixture of a matrix of a thermally stable polymer having embedded therein discrete particles of a surface-active composition. Suitable thermally stable polymers include polyphenylene sulfide and suitable surface-active compositions include carbon, iron, nickel, cobalt, vanadium and molybdenum. The proportion of the polymer in the packing generally ranges from about 20 to 80% by volume. Preferably, the matrix is prepared in porous form, which may also be coated with a surface-active metal.

Examples 1 Through 3: A decomposer packaging material was prepared in accordance with the following procedure. Polyphenylene sulfide resin [Ryton (Phillips Petroleum Company)], having an average diameter of ~150 μ and a melting point of ~288°C (550°F) was admixed with particles of iron (~150 μ) and particles of sodium carbonate (~150 μ) in the proportions set forth below in Example 1. This mixture was stirred to form a substantially uniform dispersion of each component within the mixture.

In a similar manner, two other batches of the same resin with other surface-active materials were prepared in proportions set forth below in Examples 2 and 3, respectively. Mixtures of ~500 cc were prepared in each instance.

In each example, the mixture was placed in a die having a length of ~3", a width of ~4.5" and a depth of ~3⁄16". The filled die was placed in a vice and compressed to a pressure of 1,500 psig for a few seconds to effect compression of the particles in the mixture. The compressed die was then placed in an oven and heated to a temperature of ~400°C for about ½ hour to effect curing of the resin component of the mixture. The die was removed from the oven, allowed to cool to room temperature, and the resulting product comprised of a polymer matrix having the surface-active ingredients uniformly dispersed throughout was removed from the die.

In each example, the cooled product removed from the die was cut into small pieces of ~1" x 2". These particles were immersed in hot water and agitated for a period of about 2 hours to effect leaching of the water-soluble salt component therefrom.

The particles were washed several times to remove the residue of the water-soluble salt and then dried to form a packing material comprised of a porous polymer matrix having particles of the surface-active component uniformly dispersed throughout.

These particles of packing composition were each tested to determine its effectiveness as a decomposition packing for sodium-mercury amalgam. A heated reactor provided with means to suspend packing particles above the bottom of the vessel, means to add sodium-mercury amalgam and means to remove hydrogen gas and denuded mercury and sodium hydroxide was used in this test.

A measured amount of sodium-mercury amalgam, having the sodium concentration set forth below, was preheated to ~80°C and fed to the reactor, followed by deionized water in the proportions and at the temperature set forth below in the respective examples.

About 10 pieces of the packing material being tested were placed on the suspending means and lowered into the reactor. After approximately 15 minutes of reaction, the denuded amalgam was removed from the reaction vessel and analyzed to determine the sodium concentration remaining. These analyses are set forth below, along with a calculation of the sodium stripping rate for each example.

For purposes of comparison, conventional decomposition packing comprised of small graphite pieces having an average diameter of ~6.5 mm and the same surface area as the process packing was treated in the same reactor and the results are shown below in Comparative Test A.

Compositions and Effectiveness of Compositions of Examples 1 Through 3 When Used as Decomposer Packing

	. . . . Example	. . . .		
Composition	**1**	**2**	**3**	**Test A***
Small graphite pieces conventional packing	0	0	0	100
Polyphenylene sulfide resin, %	40	40	40	0
Sodium carbonate, %	30	0	30	0
Iron, %	20	0	20	0
Cobalt, %	10	0	10	0
Graphite, %	0	30	0	0
Sodium chloride, %	0	30	0	0
Mercury amalgam, ml	330	350	380	400
Deionized water, ml	50	100	50	100
Reactor temperature, °C	80-83	86-88	85	80-82
Sodium weight % in amalgam				
at start	0.2050	0.2375	0.185	0.2300
at end	0.1025	0.1950	0.0925	0.1925
Difference in sodium concentration, %	0.1025	0.0425	0.0925	0.0375
Sodium stripping rate, ΔNa/min	0.00683	0.00283	0.00615	0.0025

*Comparative.

An analysis of results shows that the composition of Example 1 had a stripping rate of 2.73 times the rate of conventional graphite. In addition, the stripping rate of Example 2 was 1.13, and the stripping rate of Example 3 was 2.46 times the stripping rate of conventional graphite packing.

BIPOLAR CELLS AND ELECTRODES

CELL DESIGN AND PROCESS CONTROL

Electrolytic cells may be connected in series in a common housing with the anodes of one cell being electrically in series with the cathodes of the prior cell and mounted on the opposite sides of a common structural member. In this way the cathodes of one cell are in series with the anodes of the next adjacent cell in the electrolyzer and mounted on a common structural member, and the anodes of the cell are in series with the cathodes of the prior cell in the electrolyzer. Such a configuration is called a bipolar configuration.

An electrolyzer is an assembly of electrolytic cells in bipolar configuration. The common structural member is called a bipolar unit or bipolar electrode. The common structural member includes the backplate, the anodes of one cell in the electrolyzer and the cathodes of the next adjacent cell in the electrolyzer connected thereto. The electrolytic cell provided by the anodes of one bipolar electrode facing the cathodes of the adjacent bipolar electrode and facing each other so that electrolysis of electrolyte may be carried out therebetween is called a bipolar cell.

Bipolar electrolyzers provide economy of materials of construction and plant space. However, in order to take advantage of the apparent economies of bipolar electrolyzers, electrolysis should be conducted at high current densities, for example, above about 120 A/ft^2 or even above about 190 A/ft^2. When electrolysis is carried out at such current densities it is important that the electrical current flow through the electrolyzer with minimum electrical resistance between adjacent cells in the electrolyzer. It is also important that the seepage of electrolyte into the backplates be completely prevented.

Reduction of Titanium Hydride Formation by Changing Direction of Current

In early bipolar electrolyzers, the flow of electricity through the backplate was enhanced by providing metal to metal contact between the titanium of the anolyte surface of the backplate and the steel of the catholyte resistant surface of the backplate, for example, as in explosion bonded backplates.

In other bipolar electrolyzer designs, electrically conductive structures in the backplate carried the current from the cathodes through the backplate to the anodes connected thereto. One way this was accomplished was by the use of copper studs which extended through the backplate.

However, it was soon found that in bipolar electrolyzers having steel-titanium laminate backplates the atomic hydrogen generated on the steel cathodic surface of the backplate migrated through the steel toward the titanium member of the backplate. This resulted in the formation of titanium hydride at the interface between the steel and the titanium.

H. Cunningham; U.S. Patents 4,152,239; May 1, 1979; and 4,093,525; June 6, 1978; both assigned to PPG Industries, Inc. has found that if the flow of electrical current through the backplate can be caused to be lateral, i.e., perpendicular, to the overall flow of electrical current from the first anodic half cell of the electrolyzer, the hydrogen diffusion toward and into the titanium may be substantially reduced. The formation of titanium hydride is further diminished if the anodic member of the backplate is spaced from the cathodic member of the backplate and the conducting means are at the periphery of the backplate.

This may be accomplished by passing the electrical current from the cathodes of the first cell of a pair of cells through the backplate toward the conductor means in a direction lateral to the overall flow of current through the electrolyzer, thereafter passing the electrical current through the conductor means, and then passing the current through the backplate to the anodes laterally to the direction of the overall flow of current.

Figure 4.1: Bipolar Electrolyzer

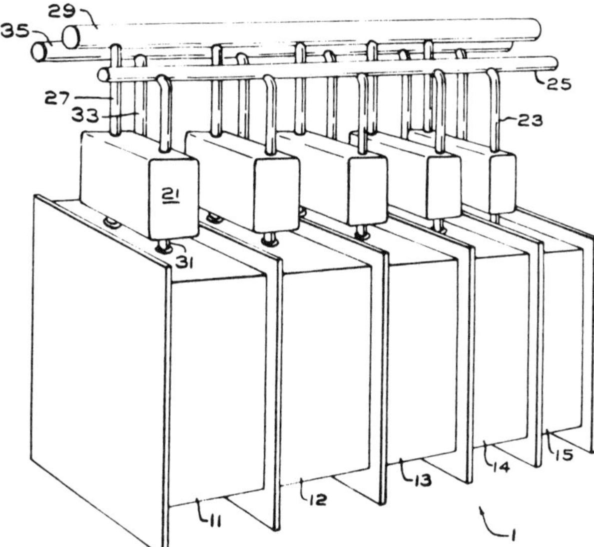

Source: U.S. Patent 4,152,239

A bipolar electrolyzer **1** is shown in Figure 4.1. The bipolar electrolyzer **1** has a plurality of individual electrolytic cells **11** through **15** electrically and mechanically in series, with an anodic end cell **11** at one end of the electrolyzer and a cathodic end cell **15** at the opposite end of the electrolyzer **1**. Intermediate cells **12** through **14** are between the anodic end cell **11** and the cathodic cell **15** of the electrolyzer.

On top of the electrolyzer are the brine tanks **21**. Brine is fed from a brine header **25** through brine lines **23** to the brine tanks **21** and from the brine tanks to the individual electrolytic cells **11** through **15**. The brine tanks also receive chlorine gas from the individual cells **11** through **15** through lines **31** to the brine tank and discharge chlorine from the brine tank through chlorine lines **27** to the chlorine header **29**.

Hydrogen is recovered from the individual cells **11** through **15** through hydrogen lines **33** that lead to the hydrogen header **35**. Liquid catholyte product, for example, a cell liquor of potassium chloride and potassium hydroxide in a diaphragm cell having a potassium chloride feed, or a cell liquor of sodium chloride and sodium hydroxide in a diaphragm cell having sodium chloride feed, or sodium hydroxide in a permionic membrane equipped cell having sodium chloride feed, is recovered from the cells through catholyte recovery means, i.e., cell liquor perc pipes. The effluent of the cell liquor perc pipes is collected in a cell liquor trough.

In the operation of a bipolar electrolyzer an electrical current passes from the anodes of the first electrolytic cell through electrolyte to cathodes of the first electrolytic cell, evolving chlorine on the anodes, hydrogen on the cathodes, and alkali metal hydroxide in the catholyte liquor. The electrical current then passes from the cathode of one cell to the anodes of the next adjacent cell in the electrolyzer.

According to this method, the electrical current typically will undergo four changes of direction. First, the current will change direction from the direction of the overall resultant flow of current from the one cell to the next, i.e., the vector flow of current, to a direction lateral thereto. Second, when the electrical current encounters a conductor means, the direction of flow of the current will generally be in the parallel to the vector flow. Third, as the current passes from the conductor through the anodic element of the backplate, the current will again change direction to a direction lateral to the vector flow of current. Fourth, as the current enters the anode of the next adjacent cell in the electrolyzer, it will return to the direction of the vector flow of current. In this way an indirect path is provided for the electrical current.

Titanium Plate Explosion-Bonded to Iron Plate

In the bipolar system electrolytic cell developed by *M. Seko, S. Ogawa, N. Ajiki and M. Yoshida; U.S. Patent 4,111,779; September 5, 1978; assigned to Asahi Kasei Kogyo KK, Japan*, having a partition wall made of explosion-bonded titanium plate and iron plate which is electrically connected to an anode of titanium substrate at its titanium side and to a cathode of iron at its iron side, space is preferably given between anode and the partition wall and also between cathode and the partition wall. An assembly having a number of such unit cells arranged in series is useful for electrolysis of sodium chloride which can be performed under a low voltage per unit cell.

The explosion-bonded titanium plate and iron plate herein used refers to titanium plate and iron plate which are pressure bonded to each other by utilizing the explosive force of explosive powder.

Example 1: In the electrolytic cell as shown in Figure 4.2, the partition wall **3** which is 1.2 m long and 2.4 m wide is prepared by explosion-bonding iron plates with titanium plate, followed by hot rolling.

Figure 4.2: Electrolytic Cell with Titanium/Iron Plate

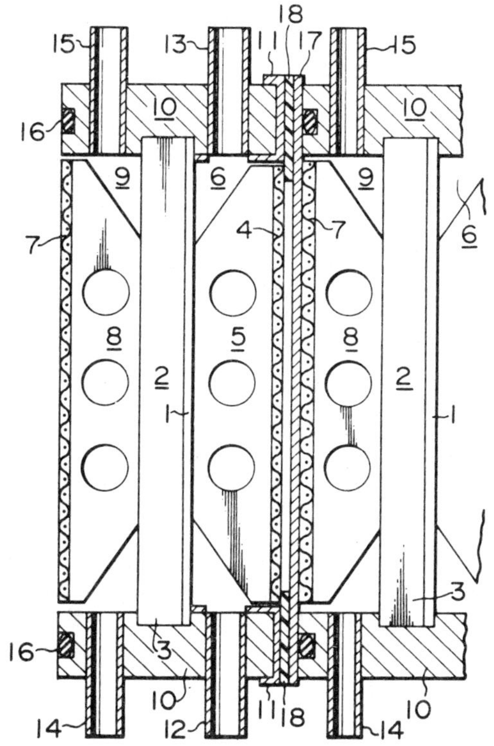

Source: U.S. Patent 4,111,779

The titanium plate **1** is 1 mm and the iron plate 9 mm in thickness. A porous titanium plate which is prepared by expanding the titanium plate with a thickness of 1.5 mm, having an opening ratio of 60%, is coated, 5 μ thick, with a eutectic mixture comprising 60 mol % ruthenium oxide, 30 mol % titanium oxide and 10 mol % zirconium oxide to provide the anode **4**.

In order to provide 25 mm of the space **6** for the anode chamber between the anode **4** and the titanium of the partition wall, the titanium plate **5** which is 4 mm thick, 25 mm wide and 1.2 m long is arranged at the interval of 10 cm.

The titanium plate is arranged vertically so as not to disturb the stirring effect by gas and provided with 10 holes of about 10 mm in diameter so as to permit horizontal mixing of the liquid.

The titanium plate **5**, the titanium **1** of the partition wall and the anode **4** are connected to one another by means of welding so as to reduce the electric resistance as much as possible. As the cathode **7**, an expanded porous plate with an opening ratio of 60%, which is prepared from 1.6 mm iron plate, is used.

In order to provide 45 mm of the space **9** for the cathode chamber between the cathode **7** and the partition wall, there is vertically arranged the iron plate **8** which is 6 mm thick, 45 mm wide and about 1.2 m long and provided with 10 holes of about 10 mm in diameter. The cathode **7**, the iron plate **8** and the iron **2** of the partition wall are connected to one another by means of welding so as to reduce the electric resistance as much as possible.

There is the iron frame with thickness of 16 mm around the partition wall **3** and it is lined with titanium plate **11** with thickness of 2 mm at the surface in contact with the anolyte. The interval between the cathode **7** and the anode **4** is maintained at about 2 mm by the packing of ethylene-propylene rubber with a thickness of 2 mm. As the cation exchange resin **17**, a sulfonic acid type resin made from fluoro-resin substrate reinforced with fluoro-fiber cloth is used. A bipolar system electrolytic cell assembly is made by arrangement of 80 unit electrolytic cells as described above.

To the liquid supply nozzle **12** of each anode chamber is supplied an aqueous sodium chloride solution through the pipes arranged parallel to each other from the anolyte tank. From the liquid discharge nozzle **13** is discharged the anolyte comprising sodium chloride solution and chlorine gas through the pipes arranged similarly parallel to each other, which is then returned to the anolyte tank.

To the liquid supply nozzle **14** of each cathode chamber is supplied an aqueous caustic soda solution through the pipes arranged parallel to each other from the catholyte tank and from the liquid discharge nozzle **15** is discharged 20 wt % aqueous caustic soda solution and hydrogen gas, which is then returned to the catholyte tank.

When direct current of 14,000 A is passed through the bipolar system electrolytic cell as described above at an electrolysis temperature of 92°C, the voltage per unit cell is only 3.6 V. The voltage drop between the cathode **7** and the anode **4** through the partition wall **3** is merely several millivolts, which clearly shows the advantage of the structure having an explosion-bonded partition wall.

Example 2: *Reference* – In this reference example, heat-resistant polyvinyl chloride resin plate is used as the partition wall. The same anode and cathode as in Example 1 are used. The corresponding titanium plates **5** are arranged at the intervals of 10 cm and titanium plates with thickness of 10 mm and width of 15 cm are arranged for distribution of current between the titanium plates. Rod of 10 cm in diameter which is welded with the aforementioned titanium plate, is penetrated through the partition wall of heat-resistant polyvinyl chloride. The cathode side is also provided with the same structure as the anode side and both sides are connected by setscrew on the penetrated portion of the heat-resistant polyvinyl chloride.

Although the size of the cathode chamber, the size of the anode chamber, the cation exchange membrane, the anolyte concentration and the catholyte concentration are the same as in Example 1, the voltage drop between the cathode and the anode through the partition wall is as much as about 200 mV when direct current of 14,000 A is passed. The heat-resistant polyvinyl chloride is observed to be molten at the penetrated portion by the heat evolved at an electrolysis temperature of 70°C. Therefore, electrolysis can no longer be continued. Furthermore, the electrolysis voltage is as high as 4.7 volts per unit cell. Thus, no large scale electrolytic cell can be produced by using heat-resistant polyvinyl chloride as the partition wall.

Sinuous Flow Path Preventing Stagnant Areas

D.G. Stevenson; U.S. Patent 4,124,480; November 7, 1978; assigned to Paterson Candy International, Ltd., England has developed a bipolar cell comprising a plurality of spaced electrode plates defining a sinuous flow path therethrough, wherein the flow path sweeps out the entire space between adjacent plates so as substantially to prevent the formation of stagnant areas and thus minimize the growth of precipitate between the plates.

In Figure 4.3 a bipolar cell comprises eleven spaced plates contained between insulating end plates **1,1'** and separated by gaskets **2**. Two outer plates **3** are connected to a common positive source and hence act as anodes, and the central plate **4** is connected to a negative source and hence acts as a cathode.

Figure 4.3: Bipolar Cell with Sinuous Flow Path

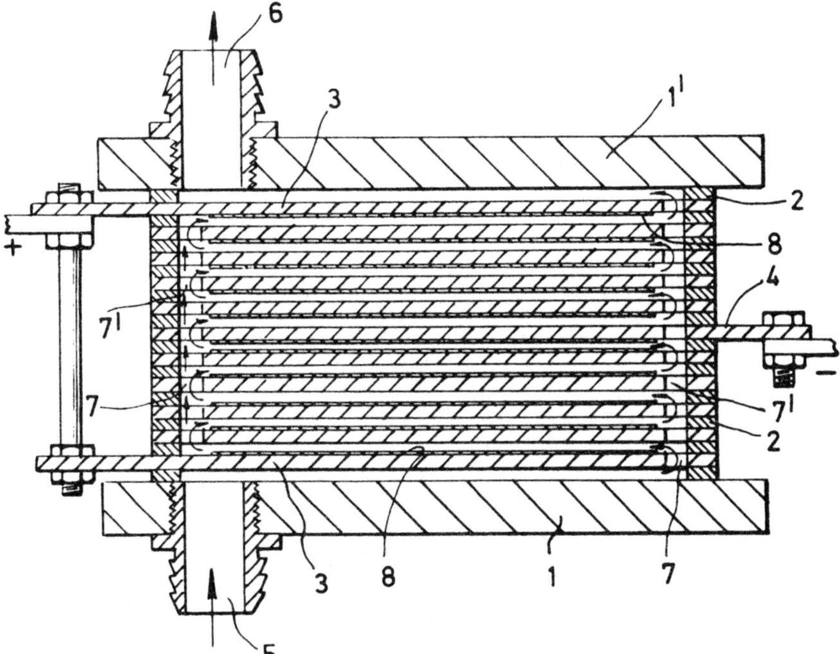

Source: U.S. Patent 4,124,480

Liquid enters through an inlet **5** in one of the insulating plates **1** and leaves through an outlet **6** in the other insulating plate **1'**.

Each electrode plate has a slot **7** cut along one end so that liquid flows along each plate over the whole width thereof, through the slot **7** and back along the adjacent plate and also over the whole width thereof. Since the slot **7** also extends over the whole width of the plate the whole flow path between the plates is swept out by the liquid, which follows an unrestricted sinuous path through the plates. Thus the risk of liquid stagnation is substantially reduced.

Vertical Anodes and Cathodes

O. De Nora and V. De Nora; U.S. Patents 4,130,468; December 19, 1978; and 4,073,715; Febraury 14, 1978; both assigned to Oronzio De Nora Impianti Elettrochimici SpA, Italy describe an electrolysis cell and method of operation in which metal anodes (preferably titanium) provided with an electrically conducting electrocatalytic coating, in an anode compartment, face metal cathodes (preferably diaphragm covered) in a cathode compartment, in which the anodes are spaced from an imperforate valve metal separating partition by a separating wall behind which the anolyte can recirculate downward.

The anodic gases rising in the anode compartment discharge into a brine box above the anode compartment near the center thereof and the anolyte recirculates downward near at least one end of the anode compartment. The method of operation provides circulation from front to back of the anode compartment and from center to the sides of the anode compartment.

Figure 4.4 shows a single bipolar cell, **1a**, which is part of an assembly of cell units **1a**, **1b**, **1c**, etc. As shown for the single cell, the brine feed funnels or dip tubes **9d** extend through the brine box approximately to the bottom of each of the cell units **1a**, **1b**, etc., so that the fresh brine is delivered to or below the bottom of the anode fingers **5**. The anode fingers **5** and the cathode fingers **6** extend approximately from the top to the bottom of the anode and cathode compartments, but are spaced a short distance from the bottom and top of these compartments to permit circulation and recirculation of the electrolyte within the cell units and permit escape of anodic and cathodic gases.

Each of the brine boxes **10a**, **10b**, **10c**, etc., is connected with its corresponding cell unit **1a**, **1b**, **1c**, etc., by brine box connections **11a** and **11b** at each end and a gas riser **11** at the center. The bottom of the gas riser **11** is flush with the top of each anode compartment and the top extends close to the normal operating brine level or just above it so that most of the chlorine (or other anodic gas) flows into the brine boxes through the center connection. The connections **11a** and **11b** extend a short distance below the top of the anode compartments and into the brine in the flooded anolyte compartment.

Fresh brine flows downward through feed funnel **9d** and connections **11a**, and recirculated brine flows downward through connections **11a** and **11b** into the anode compartments, so that, in operation, there is a constant recirculation of brine or anolyte downward at each end of the cell units **1a**, **1b**, **1c**, etc., and upward, together with chlorine through gas riser **11** in the center of each brine box **10a**, **10b**, etc., as indicated by the arrows.

Figure 4.4: Electrolysis Cell with Vertical Anodes and Cathodes

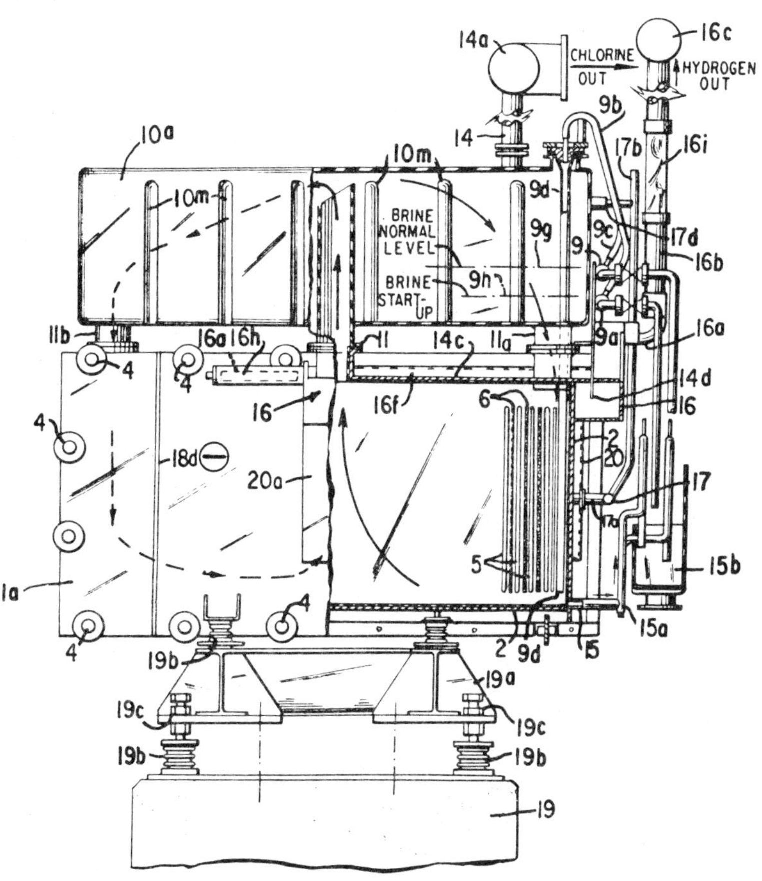

Source: U.S. Patent 4,130,468

Recirculation of the anolyte from the front to the back of each anode compartment of cell unit **1a**, **1b**, **1c**, etc. is provided at the same time by the chlorine gas bubbles rising in the interelectrodic gaps between the anodes **5** and the diaphragm covered cathodes **6** and in the spaces inside the hollow anode fingers **5**.

Most of the chlorine gas bubbles, upon reaching the top of the anode compartment, escape through the gas riser **11** imparting an upward thrust to the electrolyte present inside the gas riser. The electrolyte rises with the gas to the top of the riser **11** and overflows into the electrolyte pool in the brine boxes **10a**, **10b**, etc., whereby the chlorine gas separates from the brine. Simultaneously an equivalent volume of electrolyte moves down through the connections **11a** and **11b** and enters the anode compartment near the side walls of the compartment.

In this way, a circulation motion of the electrolyte to and from the anode compartment and the cooperating brine box is created. This motion, in cooperation with the recirculation motion of the anolyte within each anode compartment resulting from the division of the anode compartment by the separating wall formed by anode support bars into a front portion, affected by the upward flow of the gas and of the anolyte, and rear portion wherein anolyte brought to the top of the anode compartment is recirculated to the bottom of the compartment, produces an intense recirculation of the anolyte from near the side walls towards the center of the anode compartment and through the brine box riser **11** and downcomers **11a** and **11b** and from the front to the back or from the top to the bottom of the anode compartment.

Bipolar Cell with Bipolar Intermediate Electrode

S.M. Wheatley, G.R. Sherfield and D.A. Burton; U.S. Patent 4,140,616; Feb. 20, 1979; assigned to A. Johnson & Co. (London) Ltd., England describe a bipolar electrolytic cell comprising a row of spaced-apart electrodes which include at least one bipolar intermediate electrode. There is an inlet for the supply of electrolyte liquid and an outlet for the discharge of treated liquid, the cell being such that liquid can flow from the inlet to the outlet via a path in which it passes in succession through all the spaces between the electrodes in the row, in each case across the faces of the two electrodes on opposite sides of the space.

There is a coating of electrically insulating material on that external surface of each of the electrodes which is on the outside of the row, the insulating material, for example, being a plastics material such as nylon.

The cell shown in Figure 4.5 has a vertical row of twenty-one wedge-shaped graphite electrodes, with an anode **1** at the lower end of the row, a cathode **2** at the upper end of the row and nineteen bipolar electrodes **3** each of which acts as a cathode at its lower face and as an anode at its upper face. Each bipolar electrode has an anode and a cathode integral with one another.

Electrodes **1** and **2** are similar to one another, having one flat face which is perpendicular to the longitudinal axis of the row of electrodes and an opposite face which is inclined to that axis by an angle other than 90°. Electrodes **3** are similar to one another; each has two opposite faces which are inclined in opposite senses to the aforementioned axis. Each electrode **3** has its thickest part lying between the thinnest parts of the two adjacent electrodes. Each electrode **3** has a hole **4** through it from one flat face to the opposite one and near the thinnest part of the electrode.

Each of the electrodes **1** and **2** has a hole through it from one flat face to the opposite one and near the thinnest part of the electrode, these holes communicating with an inlet **5** and an outlet **6** respectively for the supply and discharge of liquid to and from the cell respectively.

The vertical row of electrodes is between a pair of registration plates **7** and **8** serving on the one hand for the registration of the inlet **5** with the hole through the anode and of the outlet **6** with the hole through the cathode; and on the other hand for the registration of an electrical connector **9** with the anode and of an electrical connector **10** with the cathode. The connectors **9** and **10** are electrically connected to the anode **1** and the cathode **2** respectively via tips **11** and **12** screwed into the anode and cathode respectively.

Figure 4.5: Electrolytic Cell

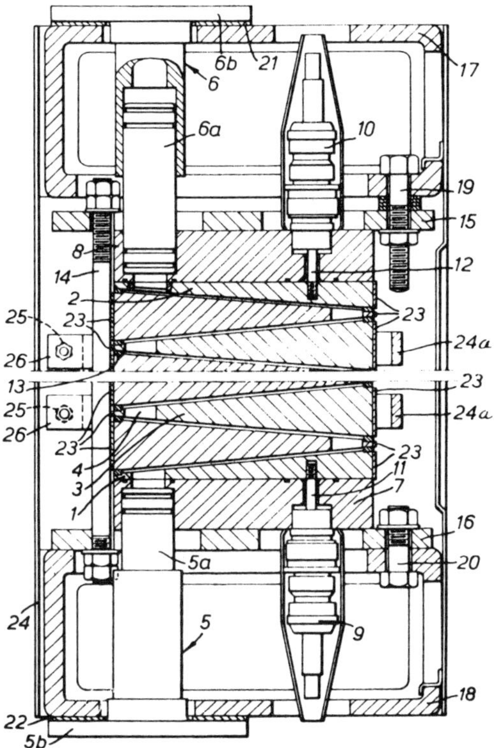

Source: U.S. Patent 4,140,616

The stack comprising the electrodes and the registration plates **7** and **8** is clamped between two end plates **15** and **16** by means of six nut and bolt arrangements **14** (of which only one can be seen in the figure) there being gaskets **13** of electrically insulating material around the edges of the electrodes and between them, these being the only means spacing apart the electrodes and providing fluid-tight seals to the exterior.

On that external surface of each of the electrodes which is on the outside of the row, there is a thin coating of electrically insulating material, for example, plastics material such as nylon. In the figure, for the sake of clarity, the coatings are indicated by reference numerals **23** and drawn to an exaggerated scale so that they can be seen. The electrodes are provided with the coatings **23** individually by powder deposition, for example, before they are stacked together.

The cell described above is disposed in a cabinet **24** open to the atmosphere, the row of electrodes therefore not being in a housing the interior of which is closed off from the exterior. The coatings **23** serve to prevent voltage breakdowns as a result of this.

Electrolyte flowing through the cell from the inlet to the outlet flows in an ever-descending path, passing in succession through all the spaces between the electrodes in the row, in each case across the faces of the two electrodes on opposite sides of the space.

Metal Anode Connected to Metal Cathode

O. De Nora and V. De Nora; U.S. Patent 4,161,438; July 17, 1979; assigned to Oronzio De Nora Impianti Elettrochimici SpA, Italy describe an electrolysis cell having metal anodes (preferably titanium) and metal cathodes connected together back to back by a metal to metal contact forming a bimetallic partition. The anodes and cathodes are in wave form with their active surfaces intermeshed together and the cell may be unipolar or bipolar with terminal positive and negative end unit cells and a plurality of intermediate cell units.

Figure 4.6 illustrates a three unit bipolar cell having a terminal positive end unit **A**, an intermediate unit **B** and a terminal negative end unit **C**. Only one intermediate unit **B** has been illustrated, but it will be understood that any number of intermediate units **B**, **B**, etc. may be used.

The unit **A** consists of a positive (anode) end plate **1**, preferably of steel, to which the positive electrical connections **2** are secured. The plate **1** is provided with a titanium, tantalum or other valve metal lining **3** which is resistant to the electrolyte and the electrolysis conditions encountered in the cell and the anode waves or fingers **4** are connected to the titanium lining by titanium connectors **5**, which space the anodes from the lining **3** and insure good electrical connections between the end plate **1** and the anode waves or fingers **4**. The interior of the anode waves are hollow.

The titanium or other valve metal lining **3** is secured to the end plate **1** by sandwich welding, using intermediate sandwich metals if necessary, or by bolting or any other connection which insures a good metal-to-metal electrical contact between the end plates **1** and the electrolyte resistant lining **3**. Titanium, tantalum or other valve metals or alloys of these metals may be used for the lining and the anode waves or fingers **4**.

The end anode plate **1** is spaced from a steel cathode supporting end plate **1a**, from which the steel screen cathode waves or fingers are supported by welded strips or projections **7** which space the cathode from the end plates **1a** and form the electrical connection between the cathode fingers and the steel plate **1a**.

A rectangular spacer frame **8** forming the side walls of each cell unit, extends between the lining **3** and a squared pipe **9** which surrounds the catholyte compartment **10** formed between the inside of the cathode fingers **6** and the plate **1a**. The spacers are lined with a titanium lining **8a** or with a polyester or other lining which is resistant to the anolyte and the corrosive conditions encountered in an electrolytic cell.

The rectangular spacer frames **8** are provided with outwardly extending flanges **11a** which form the joints between the spacers **8** and the end plates **1**, **1a**, etc.

and rubber gaskets **11** seal the joints between the plates **1** and **1a** and the spacers **8** so that a fluid-tight box-like structure housing the anode waves **4** and the cathode waves **6** is formed between the plates **1** and **1a** in each of units **A**, **B** and **C** of the bipolar cell.

Figure 4.6: Plan View of Three-Unit Bipolar Cell with Metal-to-Metal Connection Between Electrodes

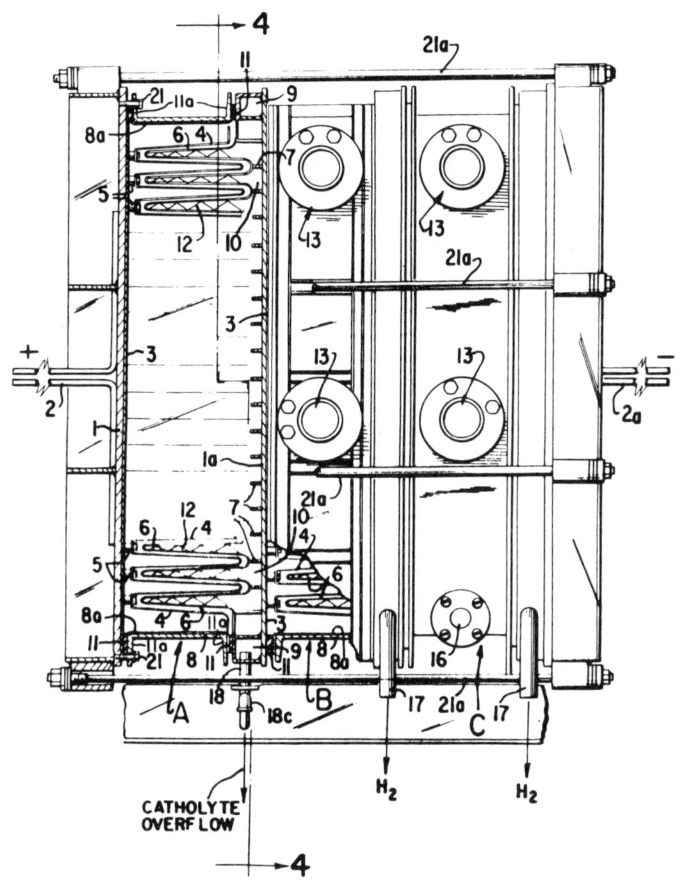

Source: U.S. Patent 4,161,438

Inside each cathode finger **6**, zigzag bent steel reinforcements **12** are welded at spaced intervals inside the cathode fingers to prevent collapse of the screen cathode waves or fingers **6** when an asbestos or other diaphragm material is deposited on the screen cathode fingers under vacuum. The steel screen cathode waves or fingers are closed at the top and bottom and are covered with a diaphragm material, usually either woven asbestos fiber or asbestos flock applied under vacuum.

The diaphragm material covers the side walls as well as the top and bottom of cathode waves or fingers **6**. The diaphragms separate the anolyte compartment from the catholyte compartment and keep the gases formed in each of these compartments separate.

When the cell is used for the electrolysis of sodium chloride brine to produce chlorine, caustic soda and hydrogen, the electrolyzing current flows from the anode waves **4** to the cathode waves **6**. Chlorine is released at the anode waves or fingers, the brine flows through the diaphragms surrounding the cathode waves **6** and caustic soda and hydrogen are formed at the cathode surfaces inside the diaphragms.

Chlorine (or other anodic gases) released at the anodes **4** rises along both the front and the back of the anodes through the electrolyte and escapes through the chlorine passages **13** into brine containers (not shown) on the top of each cell unit **A, B, C** and flows out of the chlorine outlets to the chlorine recovery system. A pipe connection **16** feeds brine from each of the brine containers to the spaces between the anode and cathode fingers of the cell units **A, B** and **C** and a sight glass (not shown) indicates the level of the brine in the brine containers.

Sodium hydroxide and hydrogen released at the cathode fingers flows into the catholyte space between diaphragms surrounding the cathode fingers **6** and the end plates **1a** and into a squared pipe **9** which surrounds the catholyte space. The hydrogen flows upward through the holes at the top of the squared pipe **9** and out through the hydrogen outlets **17** and the depleted brine containing the sodium hydroxide (about 11 to 12%) flows to the catholyte outlet **18**.

Gas-Lifting of Catholyte

B.E. Kurtz and R.H. Fitch; U.S. Patent 4,198,277; April 15, 1980; assigned to Allied Chemical Corp. describe a method for accomplishing series catholyte flow in a multicompartment bipolar permselective membrane electrolyzer which utilizes the hydrogen gas evolved in the cathode compartment for an autogenous gas-lift in order to transport the catholyte from the cathode compartment of one cell to the cathode compartment of the next cell.

The process establishes a high electrical resistance in the stream of catholyte between the exit of the preceding cell and the entrance of the succeeding cell so as to electrically isolate the cathode compartment of the preceding cell from the cathode compartment of the succeeding cell.

Figure 4.7 shows two cells **100** and **200** connected for series catholyte flow. Each cell has a cathode compartment **104** and **204**, and an anode compartment **110** and **210** separated by permselective membranes **108** and **208**, respectively. Each cathode compartment has a cathode **106** and **206**. Lines **112** and **212** feed into the cathode compartments containing catholyte **102** and **202**, respectively.

The upper portion of each catholyte compartment is equipped with risers **114** and **214** through which the catholyte flows by means of a gas-lift generated by hydrogen gas bubbles **116** and **216**.

Figure 4.7: Electrolysis of Aqueous Salt Solutions

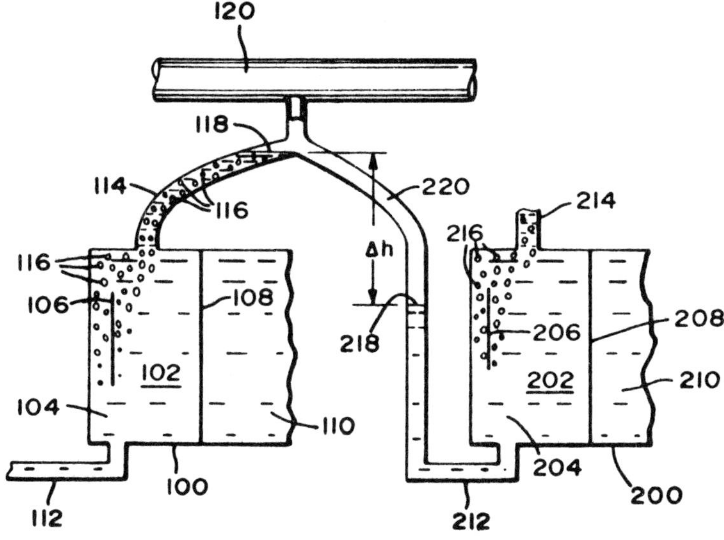

Source: U.S. Patent 4,198,277

The gas and liquid caustic soda separate at separation point **118** from which the gas flows upward into header **120** and the liquid drops through void space **220** to collection point **218.**

The horizontal cross-sectional area of the cathode compartment should be confined within certain limits. The cathode compartment horizontal cross-sectional area required for satisfactory gas-lift effect will depend on the volumetric rate of hydrogen evolution. More specifically, it will be proportional to the product of the cathode area and the current density according to the equation

$$DW = KHWC_d$$

wherein D = depth of cathode compartment,

W = width of cathode compartment,

H = height of cathode compartment,

C_d = current density on cathode,

K = proportionality constant.

Accordingly, the required current density on the cathode is given by the equation $KC_d = D/H$. It has been found that for operation at or near atmospheric pressure, producing caustic soda in concentrations of 7 to 20 wt % at a temperature in the range of 30° to 90°C, the proportionality constant K should be in the range of 0.01 to 2.0, preferably, 0.05 to 1.0. At K values less than about 0.01 the cell will operate at an undesirably high voltage due to the large fraction of the catholyte compartment occupied by the hydrogen. At K values more than about 2.0, the rate of hydrogen evolution will be insufficient to create the necessary gas-lift effect.

Examples 1 through 3: In the following examples, data were obtained from a three cell electrolyzer in which the individual cells were assembled together so that common end plates served to separate the cathode compartment of one cell from the anode compartment of the adjacent cell. The cathode of each individual cell is electrically connected externally to the anode of the adjacent cell. As illustrated in Figure 4.7, the catholyte and hydrogen from the first cell exits from the top of the cell body into an external disengaging point from which the catholyte flows downward through a confined void space and thence into the bottom of the adjacent second cell and so on from the second cell to the third cell. Caustic soda at the final concentration is withdrawn from the top of the third cell.

The anodes employed were constructed of titanium coated with rare earth metal oxides, DSA. The cathode was mild steel. The membrane was a cationic permselective membrane, Nafion. The current density employed was 0.25 A/cm^2. The pressure within the anode compartment was maintained at about 7" of water, that within the cathode compartment at about 1" of water.

Example 1 comprises the average result of two runs at approximately the same final caustic concentration, Example 2 is the average of 3 runs at approximately the same final caustic concentration, and Example 3 is the average of 2 runs at approximately the same final caustic concentration. The duration of each run was approximately 1 hour. The results are shown below.

	Ex. 1	Ex. 2	Ex. 3
Operating Conditions			
Feed brine concentration (g/ℓ)	322	317	312
Brine depletion (%)	11.4	10.8	10.6
Average cell temperature (°C)	82	82	84
EMF Applied (V)			
Cell #1	4.2	4.2	4.0
Cell #2	3.8	3.9	3.7
Cell #3	4.2	4.4	4.3
Average	4.1	4.2	4.0
Cell Efficiencies (% based on NaOH produced)			
Voltage efficiency	55.0	53.5	55.7
Current efficiency	91.7	90.4	83.2
Power efficiency	50.0	48.4	46.3
NaOH Concentration (weight %)			
Cell #1	6.0	4.6	8.5
Cell #2	10.1	7.8	13.8
Cell #3	13.5	10.9	17.8
Power Consumption (kWh/ton NaOH)	2,695	2,800	2,930

For all of these runs, the value of K discussed above is calculated to be about 0.5 on the basis of H = 10 cm, C_d = 0.25 A/m^2, and D = 1.2 cm. In all the runs, the voltage was satisfactorily low. The difference in level between the separation point and the collection point was, in all cases, in the range of 3 to 6 cm, quite satisfactory for electrical isolation between adjacent cathode compartments.

For periods of operation at current densities less than 0.25 A/cm^2, it was observed that the difference in catholyte levels between adjacent cathode compartments was diminished, but was still adequate (2 to 3 cm), at current densities in the range of 0.12 A/cm^2, corresponding to a K value of about 1.0.

As compared to the alternative of a mechanical pump for transporting the catholyte, the autogenous gas-lift method of this process avoids increased complexity and cost, and decreased reliability of the electrolyzer. It also utilizes the energy generated by the buoyancy of the hydrogen bubbles which would otherwise be wasted.

As compared to the alternative of gravity flow for transporting the catholyte, the gas-lift method avoids the need for having adjacent cathode compartments at successively lower positions which would seriously complicate the design and increase the cost of the electrolyzer. Regarding the necessity for a high electrical resistance in the stream of catholyte for a series catholyte flow, the gas-lift method accomplishes this by creating a discontinuity in the catholyte stream where the stream falls freely through a confined void space created by the difference in head.

Vertical Multiunit Electrolyzers Utilizing Gas Lift

O. De Nora; V. De Nora and P.M. Spaziante; U.S. Patent 4,108,756; August 22, 1978; assigned to Oronzio De Nora Impianti Elettrochimici SpA, Italy provide a vertical bipolar electrolyzer which takes up a minimum of space and has increased efficiency due to the utilization of a gas-lift effect occurring in the electrolyte during electrolysis to sweep the electrolysis product and precipitates out of the apparatus to prevent fouling thereof.

Due to the vertical construction of the electrolyzer, the flow of the electrolyte is essentially straight and turbulence in the cell units is avoided. This straight flow of the electrolyte through the space between adjacent anodes and cathodes forming the electrolysis gap reduces the accumulation of insoluble particles which may precipitate in the cells, particularly a problem when seawater is being electrolyzed. The solids are carried out of the cell by this electrolyte flow and the speed of the electrolyte is increased from the inlet to the outlet by utilizing the gas-lifting effect of the gases formed in the electrolysis.

The hydrogen bubbles formed during the electrolysis do not have the opportunity to stagnate in the cell units and to form gas pockets because of the straight forward flow of the electrolyte. Instead, the hydrogen bubbles are dispersed throughout the electrolyte and increase in concentration as the electrolyte rises through the vertical cell, thus increasing the speed of the electrolyte in the upper sections of the electrolyzer.

The uniformity of the flow across the entire section of the electrolyzer, the progressive increase of the speed of the electrolyte gas dispersion rising through the electrolyzer, and the absence of turbulence induced internal recirculation paths or stagnant zones prevent the settling of solid particles such as precipitates of calcium and magnesium, organic matter etc. inside the electrolyzer. These solid particles are maintained in suspension and are effectively swept away by the flow of the electrolyte gas dispersion.

A typical set of operating data of the electrolyzers used to generate active chlorine in the form of hypochlorite to chlorinate seawater to be used as a cooling medium in large industrial complexes is as follows:

Electrolyte	Untreated seawater
NaCl concentration, g/ℓ as chlorine	20
Electrolyte inlet temperature, °C	24
Electrolyte outlet temperature, °C	26
Electrolyte retention time, sec	15
Average electrolyte speed, cm/sec	11
Electrode spacing, mm	3.75
Current density, A/m^2	1,600
Active chlorine concentration in the effluent, g/ℓ	2
Current efficiency, %	95
Chlorate content in the effluent	Undetected

Production of Chlorine Under Pressure

The process developed by *G.M. Kamarian, L.A. Kostandov and V.M. Zimin; U.S. Patent 4,136,004; January 23, 1979* permits production of chlorine under pressure.

An electrolyzer for electrolysis of aqueous solutions of chlorites of alkali metals comprises a horizontally placed casing **1** (Figure 4.8a) filled with an electrolyte **2** which is an aqueous solution of sodium chloride and two end monopolar electrodes: an anode **3** and a cathode **4**.

The anode **3** is provided with a base **5** with feeding busbars **6** connected to the outer side of the base and anode members **7** secured on the inner side of the base **5** and immersed into the electrolyte **2**. The base is made of a titanium or steel sheet with a protective coating against the action of the electrolysis products. The anode members are made of a titanium grid of a perforated titanium sheet coated by an anode-active film, e.g., ruthenium dioxide.

The cathode **4** is provided with a base **8** with feeding busbars **9** connected to the outer surface of the base **8**. Cathode members **10** are secured on the inside surface of the base so that a common cathode space **11** is formed between the cathode members and the base. The base **8** is made of a steel sheet, whereas the cathode members **10** are made of a steel grid.

The electrolyzer also comprises an electrolyte supplying device **12** which is a connecting pipe placed on the casing **1**, a chlorine tapping device **13** which is a connecting pipe located on the base **5** of the anode **3** over the level of the electrolyte **2**, an alkali tapping device **14** and a hydrogen tapping device **15** which are also connecting pipes located on the base **8** of the cathode **4**. Arrows **A, B, C, D** indicate directions for supply of electrolyte and tapping of chlorine, alkali and hydrogen, respectively.

The electrolyzer casing **1** is a cylinder of a corrosion-resistant nonconductive material, e.g., fiber glass plastic, the bases of the cylinder being attached to the bases **5** and **8** of the anode **3** and the cathode **4** forming a tight cylindrical chamber. In order to ensure pressurization, sealing gaskets **16** and **17** are placed between the casing **1** and the bases **5** and **8**, respectively. The casing **1**, the anode and the cathode are tightened by means of anchor bolts **18** and nuts **19**.

Figure 4.8: Lateral Views of Electrolyzers

a.

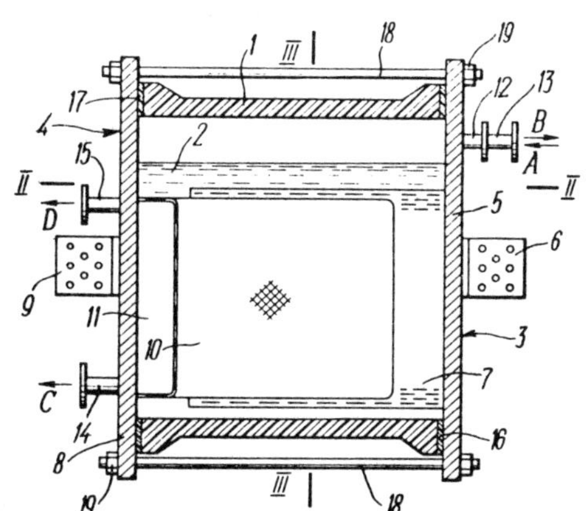

b.

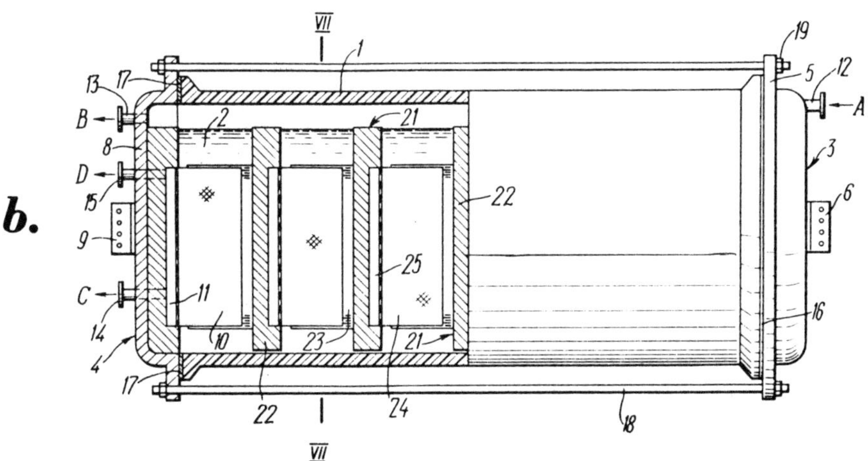

Source: U.S. Patent 4,136,004

An electrolyzer shown in Figure 4.8b comprises bipolar electrodes **21** located between the anode **3** and the cathode **4**. Each bipolar electrode **21** has a current distributing base **22**, anode members **23** being secured on one side and cathode members **24** being secured on the other side.

The anode members **23** of the bipolar electrode **21** are similar to the anode members **7** of the anode **3**, (Figure 4.8a) whereas the cathode members **24** are similar to the cathode members **10** of the cathode **4** (Figure 4.8a). A common cathode space **25** is formed between the cathode members **24** and the base **22**.

G.E.D. Nasser; U.S. Patent 4,077,863; March 7, 1978; assigned to Linde AG, Germany provides apparatus suitable for substantially larger throughput quantities than the devices known previously, while maintaining the initial investment costs at a low level even though the arrangement is designed for higher pressures.

As shown in Figures 4.9a and 4.9b, a square cell block **1** is disposed in a pressure vessel **2**. Coolers **3** for cooling the electrolyte as well as the hydrogen gas are mounted on both sides of the cell block **1**. The cells (Figures 4.9c and 4.9d) are constructed so that the openings **4a** on the anode side of the diaphragm (openings for oxygen) are disposed at the part of the cell block **1** located toward the middle, and the openings **4b** on the cathode side are located at the part of the cell block **1** on the wall side.

Figure 4.9: Pressure Electrolyzer

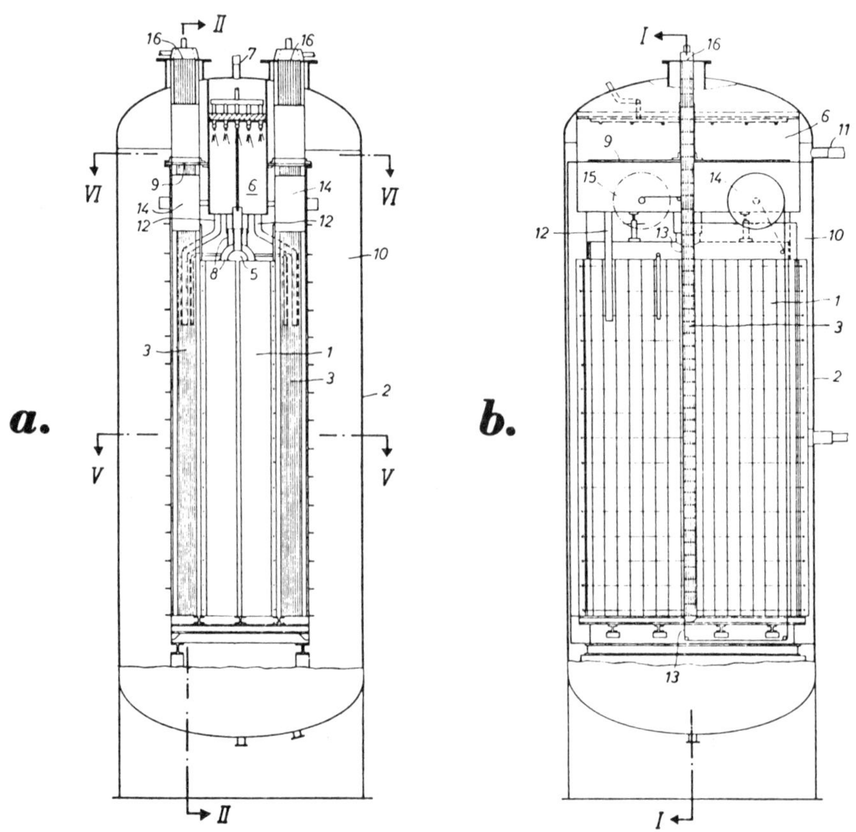

(continued)

Figure 4.9: (continued)

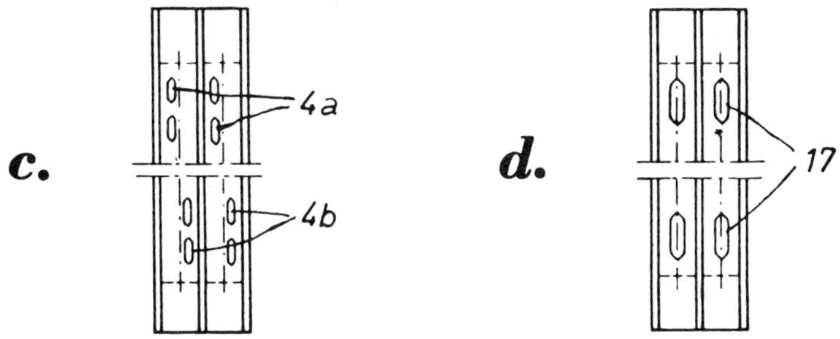

(a)(b) Longitudinal sectional views
(c)(d) Parts of cells

Source: U.S. Patent 4,077,863

The product oxygen is passed through the chamber **5**, cooled in chamber **6** countercurrently to entering feed water, and discharged from the pressure jacket via the pipe connection **7**. The path along which the product oxygen is conducted is surrounded by a hollow shell **8** in which the high pressure of the feed water is ambient, in order to avoid at any event an intermixing of the oxygen with the hydrogen gas in case of a possible leakage.

The hydrogen gas is forced, with the aid of a separator plate **9**, to pass the upper portion of the coolers **3** before entering the outer chamber **10**; via the pipe connection **11**, the hydrogen gas is discharged from the plant. The electrolytic fluid is automatically circulated by the bubbles of product gases rising within the cells. The liquid leaving the cell block through the oxygen ports is combined in chamber **6** with fresh water and conducted through the two pipes **12**. Care is taken, by the provision of corresponding baffles, that the electrolytic fluid must pass the two coolers before reentering the cell block from below via openings **17** (Figure 4.9d).

Upstream of the coolers **3**, filter mats **13** are arranged to clean the electrolyte; these filter mats are continuously renewed, in that they are unwound from the reel **14** and wound onto reel **15**. The coolers **3** can be removed through two hatches **16**, thus providing a manhole for servicing work. The cell block **1**, consisting of 32 cell parcels can be lifted out of the pressure vessel parcel by parcel for servicing purposes.

Liquor Emission Control Process

H. Cunningham; U.S. Patent 4,085,015; April 18, 1978; assigned to PPG Industries, Inc. describes a bipolar electrolyzer having a plurality of individual electrolytic cells electrically and mechanically in series. Each of the cells has an anolyte chamber, a catholyte chamber, and catholyte liquor withdrawal means.

A catholyte liquor trough is disposed alongside the electrolyzer and beneath the catholyte liquor withdrawal means. The bipolar electrolyzer is characterized by apparatus for limiting the emission of catholyte liquor laden moisture from the cell liquor trough while simultaneously being capable of indicating a low catholyte liquor level in an individual electrolyte cell of the electrolyzer. Additionally, the apparatus may interrupt the flow of catholyte liquor between the catholyte chambers of the individual electrolytic cells and the catholyte liquor trough, thereby reducing corrosion of the perc pipes.

Example: As shown in Figure 4.10, a cell liquor trough **61** is provided that is 8" wide by 4½" high serving a bipolar electrolyzer of 11 individual diaphragm electrolytic cells.

Figure 4.10: Electrolyzer with Cell Liquor Recovery Means

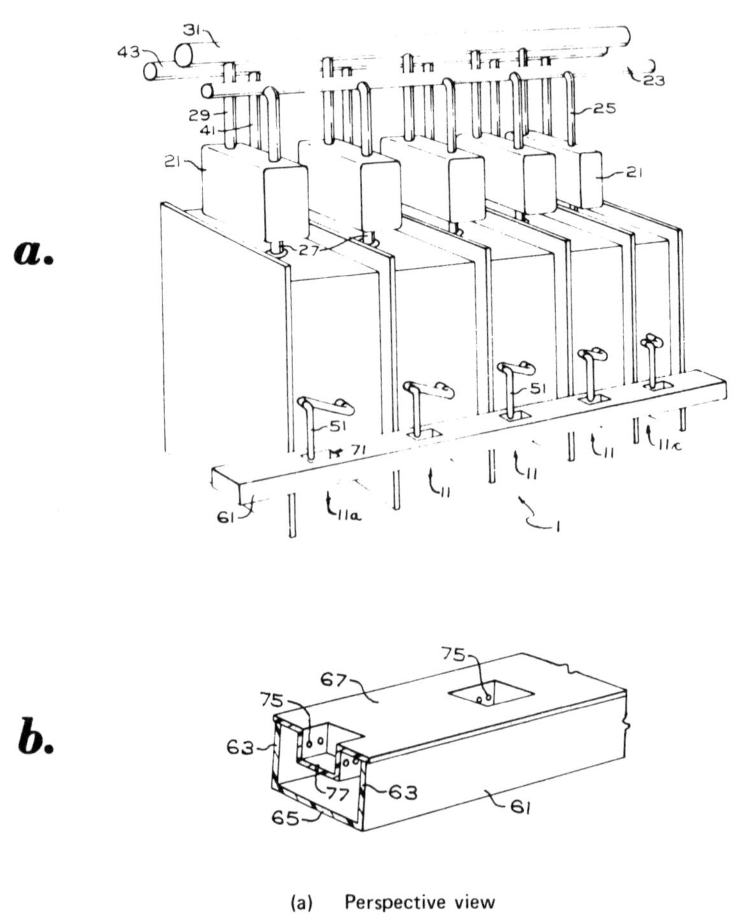

(a) Perspective view
(b) Cell liquor trough

Source: U.S. Patent 4,085,015

The individual perc pipes **51** having a diameter of about 1/12" are inserted in recesses **71** that are 3" deep by 5" in interior diameter. Each individual recess **71** has twelve 3/16" diameter first apertures for a total area of first apertures **75** of 11.5 square inches. The first apertures **75** have their center line approximately 1" above the bottom of the recess **71** and approximately ½" below the intended level of catholyte cell liquor in the recess **71**. The second aperture means **77** are provided by a single drain of 1/32" diameter whereby the second aperture means provides approximately 0.01% of the total aperture area.

According to the method, catholyte cell liquor passes from the catholyte liquor through the perc pipe **51** to the apertured recess **71** for a residence time of about 2 to 4 seconds. In the event of interruption of flow from a catholyte perc pipe the catholyte liquor flows out of the recess **71** first through the first aperture means **75** until the first aperture means are exposed, as well as slowly through the second aperture means **77**.

When the first aperture means **75** are no longer covered by catholyte cell liquor, it is possible for water vapor within the cell liquor trough **61** to pass out of the cell liquor trough through the first aperture means. The water vapor condenses upon contact with the cooler atmosphere providing the cell operator with a signal that the individual electrolytic cell corresponding thereto is not operating normally.

ELECTRODES

Conductive Pins Connecting Anode to Cathode Through Composite Partition

T. Ichisaka and T. Ikegami; U.S. Patent 4,116,805; September 26, 1978; assigned to Chlorine Engineers Corp., Ltd., Japan describe a bipolar electrode comprising:

(1) An anode member comprising a substrate made of an anticorrosive metal or metal alloy and an electrically conductive coating formed on the surface thereof;

(2) A cathode member comprising a metal or a metal alloy;

(3) A partition wall for separating the anode member from the cathode member, the partition wall comprising an anode-side sheet made of the same type of anticorrosive metal or metal alloy used as the substrate of the anode member and a cathode-side sheet made of the same type of metal or metal alloy used as the cathode member; and

(4) A composite member for electrically and structurally connecting the anode member and the cathode member to each other.

One embodiment is shown in Figure 4.11 which is an enlarged view of the bonded part of the partition wall and the composite member in the bipolar electrode. **2** is an anode-side sheet, **3** is a cathode-side sheet, **6** is a spacer and **8** is a spacer.

The composite member is shown by reference numeral **12**, and is a mutally bonded structure comprising an anode-side portion **13** made of a metal or a metal alloy such as titanium or a titanium alloy used as the substrate of the anode member, a cathode-side portion **15** made of a metal such as mild steel or a metal alloy used as the cathode member, and an interlayer portion **14** made of copper or an alloy of copper disposed between the portions **13** and **15**.

Figure 4.11: Cross-Sectional View of Bipolar Electrode

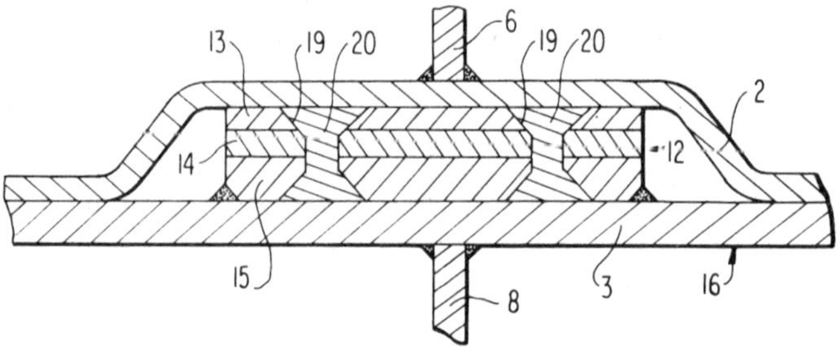

Source: U.S. Patent 4,116,805

The portions **13, 14** and **15** may be formed into plates of various shapes such as a circular shape, an elliptical shape or a rectangular shape, and are bonded by an explosive welding method or a friction welding method.

The composite member **12** includes through-holes **19** which diverge toward both of the surfaces like a funnel. A pin **20** made of an electrically conductive metal or an alloy of an electrically conductive metal, such as copper or a copper alloy, (e.g., brass) which is resistant to the migration of hydrogen and is substantially impermeable to atomic hydrogen is fitted by caulking in each of the through-holes **19**.

The anode-side sheet of the partition wall is welded by resistance welding onto the top surface of the anode-side portion of the composite member. Thus, even if the surface of the pin **20** is exposed on the surface of the anode-side sheet of the partition wall, resistance welding can be easily performed without being affected by the pin because the pin has good electric conductivity.

In addition, the composite member is not connected in the through-hole of the partition wall by mere insertion, but the cathode-side portion of the composite member is welded to the cathode-side sheet which has no through-hole for insertion, of the partition wall in the superimposed state. Thus, tolerance exists at the welded portion between the cathode-side portion of the composite member and the cathode-side sheet of the partition wall. Hence, cracks do not easily form due to stress during welding.

Even if cracks should occur, there is no likelihood of the permeation of the catholyte solution since the welded portion between the cathode-side portion of the composite member and the cathode-side sheet of the partition wall is not exposed to the catholyte solution. For this reason, the interlayer of the composite member is not corroded, and the bonded portions of the composite member are not destroyed. The electrode can, therefore, be operated in a stable manner for long periods of time.

A variation of such a bipolar electrode is described by *T. Ichisaka and T. Ikegami; U.S. Patent 4,141,815; February 27, 1979; assigned to Chlorine Engineers Corp., Ltd., Japan.*

In Figure 4.12, reference numeral **17** is a picture frame-like electrode frame which is made of, for example, mild steel. Titanium can also be used as the electrode frame.

Figure 4.12: Partial Cross-Sectional View of Bipolar Electrode

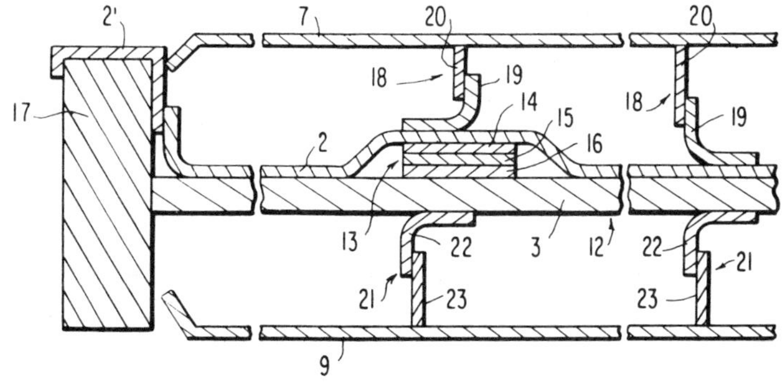

Source: U.S. Patent 4,141,815

A partition wall **12** comprises a composite structure of an anode-side sheet **2** and a cathode-side sheet **3**. The partition wall **12** is welded to the electrode frame **17**. At a portion **2'**, the anode-side sheet **2** is fixed to the electrode frame **17**. The members **2** and **2'** may be formed as a single continuous sheet.

A composite member **13** is a triple clad material composed of a portion **14** made of the same type of metal or metal alloy used as the anode-side sheet, e.g., a metal such as titanium or a titanium alloy; a portion **15** made of an electrically conductive material resistant to atomic hydrogen migration, such as copper, gold, tin, lead, nickel, cobalt, chromium, tungsten, molybdenum and cadmium, and alloys of these metals; and a portion **16** made of the same type of metal or metal alloy used as the cathode-side sheet, e.g., mild steel or the like.

Elongated Metal Anode and Cathode

T.W. Boulton; U.S. Patent 4,124,479; November 7, 1978; assigned to Imperial Chemical Industries Ltd., England has devised a bipolar unit for use in bipolar electrolytic cells which allows very small or even zero anode/cathode gaps without damage to the diaphragms or membranes, and which can be manufactured without resorting to the considerable accuracy which is required in bipolar units comprising plate anodes.

The bipolar unit for an electrolytic cell comprises

(A) An anode comprising a group of elongated members of a film-forming metal carrying on at least part of their surfaces an electrocatalytically active coating, the members being electrically conductively mounted on and projecting from a sheet of a film-forming metal so that a part of the members lie in a plane laterally spaced from the sheet; and

(B) A cathode comprising a group of elongated metal members electrically conductively mounted on and projecting from a metal sheet so that a part of the members lie in a plane laterally spaced from the sheet, the elongated members in at least one of the groups being flexible and the sheets of film-forming anode metal and of cathode metal being electrically conductively bonded to each other.

Example: A titanium anode of the same construction as the anode of the bipolar unit shown in Figure 4.13a comprised 6 rows of titanium wires **2**, with each row containing 32 wires and each wire having a 154 mm long and 3 mm diameter straight portion **5**. The wires **2** were capacitor discharge stud welded to the titanium sheet **1** which had dimensions of 300 mm x 970.5 mm. The titanium wires **2** were coated with a mixture of ruthenium oxide and titanium dioxide.

Figure 4.13: Elevational Cross-Sectional Views

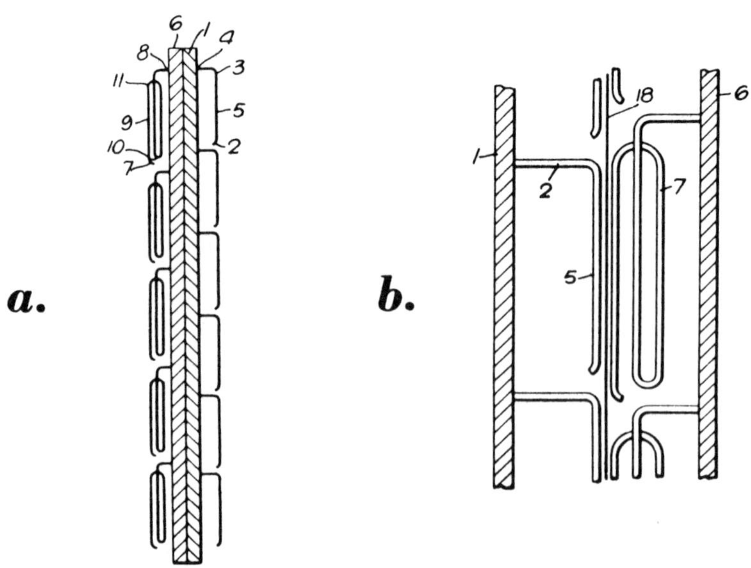

(a) Bipolar unit
(b) Electrolytic cell

Source: U.S. Patent 4,124,479

The cathode of the same construction as the cathode shown in Figure 4.13a comprised 5 rows of looped mild steel wires **7** with each row containing 32 wires which were capacitor discharge stud welded to the mild steel sheet **6**.

The anode and cathode were assembled into a vertical laboratory membrane cell as shown in Figure 4.13b to reproduce under monopolar conditions the performance of the bipolar unit. The distances between the titanium sheet **1** and the membrane **18**, i.e., the width of the anolyte compartment, and between the mild steel sheet **6** and the membrane **18**, i.e., the width of the catholyte compartment, were each 28 mm.

The membrane was a perfluorosulfonic acid membrane based on copolymers of tetrafluoroethylene and fluorinated vinyl ethers, Nafion (DuPont). The membrane was adjacent to both the cathode and the anode, i.e., the anode/cathode gap was zero.

Sodium chloride brine (conc 300 g/ℓ of NaCl) was fed to the anolyte compartment at a rate of 6 ℓ/hr. Deionized water was added to the catholyte compartment. The temperature of the cell was maintained at 85°C.

A current of 300 A (equivalent to a current density of 1.8 kA/m^2) was passed through the cell. The cell operating voltage was 2.9 V. The chlorine produced contained 94% by weight of Cl_2 and less than 0.1% by weight of H_2. The sodium hydroxide produced contained 10% by weight of caustic soda. The cell operated at a sodium hydroxide current efficiency of 86%. The membrane was undamaged by the wires of the anode and cathode.

Electrode Assembly

An improved assembly of bipolar electrode elements for use in the electrochemical preparation of hypochlorites or chlorine from dilute water solutions of chlorides, particularly mildly saline swimming pool water, is described by *J. McCallum; U.S. Patent 4,142,959; March 6, 1979; assigned to Electro-Chlor Corp.*

The end of each bipolar electrode element is enclosed in a nonconductive plastic strip, and guard baffles of corresponding thickness and height are inserted on each side of the bipolar electrode. A desired number of these elements are grouped and cemented together by solvent welding. Special adhesives enhance the sealing of the bipolar electrodes. With this electrode assembly, the deposition of mineral deposits on the electrode surfaces can be minimized by reversing the flow of direct current as often as twice a day or as seldom as once every two weeks.

Example: 1/32" thick titanium was plated on both sides with 50 microinches of platinum, (Englehard Industries). This material was cut into pieces 2" by 2" for the double-coated, bipolar electrodes. 1/8" thick PVC, Type 1, gray plastic sheet, (DeFabCo, Incorporated), was cut into 0.25" wide strips. These strips were then cut into 6" long pieces for the electrode spacers.

Two, 1.5" o.d., 0.015" thick Gyros brand circular saw blades were mounted together on one spindle and placed in the arbor of a 23,000 rpm router. A horizontal guide was mounted under the Gyros saw blades so that by adjusting this horizontal guide with simultaneous adjustments of the arbor protrusion, grooves about 0.030" thick were cut exactly in the center of one cut edge of each electrode spacer.

The platinum coated bipolar electrodes were then forced into these grooves. Shields were made from 1/32" thick PVC, Type 1, gray plastic sheet (Plastic Piping

Systems Company). They were cut into sections 1" wide by 2" long. Next, pieces of 3/32" o.d. PVC welding rod were cut into lengths about 1" for spacers, and fastened thereto with diluted PVC cement.

These shields with their attached supports on one side were slid into the electrode spacer grooves against each side of the bipolar electrodes. They were cemented in that position with VC-2 Vinyl Cement (Schwartz Chemical Co.). This cement has been diluted about 1:1 with Purple Primer (Celanese Piping Systems, Inc.). The same diluted cement was fed with an eye dropper around the entire periphery of one side of the bipolar electrode.

Nine bipolar electrodes were made this way and these nine were cemented together with the smooth, rolled sides of each electrode spacer against one another. After about 20 minutes of drying time for the diluted cement, the exposed cut edges of the spacers were passed through a table-bench saw to make a uniform, smooth surface above and below the assembled bipolar electrodes.

End electrodes were then made with shields and protruding tabs on titanium electrodes with the 50 microinch platinum coating on one side only. These end electrodes were cemented to the previously cemented group of nine bipolar electrodes with additional 1/32" thick by 0.2" wide by 6" long spacers at top and bottom of the assembly. 1/16" by 5/16" holes in the cover for the extreme electrode tabs, were made in a drill press fitted with a 1/16" router bit.

Structural adhesive 3M No. 2216B/A was spread all over the top of the spacers and end electrodes just prior to placing the cover over the tabs. Excess adhesive at edges was wiped away and some of it was placed over the cover around the protruding tabs. After 24 hours of set-up time for the adhesive, the electrode assembly with its cover was welded into a preassembled cell. Integral tabs on ancilliary grounding electrodes were also sealed through the plastic plumbing with structural adhesive No. 2216B/A.

Using a control unit and with a manual, double pole, double throw switch to reverse the direction of current flow about once-a-day, this unit operated satisfactorily on a 13,600 gallon pool for 53 days with no perceptable loss of performance for making the desired chlorine.

At this time, this unit was taken off line for examination. There were no hardness deposits between electrodes nor on the grounding electrodes. All internal parts seemed almost new. After winter storage, the unit was again put on line and continued to operate satisfactorily for another consecutive 182 days using a manual reversal of current every 12 to 48 hours, thereby indicating the bipolar electrode connection problems, the electrode sealing problem, the bottom sealing problem, and the current reversal problem had all been solved.

Hybrid Bipolar Electrode

K.A. Poush and J.E. Reynolds; U.S. Patent 4,085,027; April 18, 1978; assigned to Kerr-McGee Chemical Corp. describe an improved bipolar electrode which includes an anodic member and a cathodic member secured in a spaced apart relationship via at least one fastener assembly.

The anodic member and the cathodic member are electrically connected for conducting current therebetween, and the space between the anodic member and the cathodic member is sealed to substantially inhibit the flow of electrolyte into such space during the operation of the hybrid bipolar electrode in an electrolytic cell. In one form, the bipolar electrode includes a barrier member interposed in the space between the anodic member and the cathodic member and constructed of a material inhibiting the contact of hydrogen with the anodic member in an electrolysis application.

As shown in Figure 4.14, the first spacers **82** operate to space the second face **22** of the anodic member **12** a distance from the first face **60** of the barrier member **18**, and the second spacers **84** operate to space the second face **36** of the cathodic member **14** and the ends **54** portion of the second face **36** of the cathodic member **14** a distance from the second face **62** of the barrier member **18**.

Figure 4.14: Fragmentary, Partial Sectional, Partial Elevational View of Hybrid Bipolar Electrode

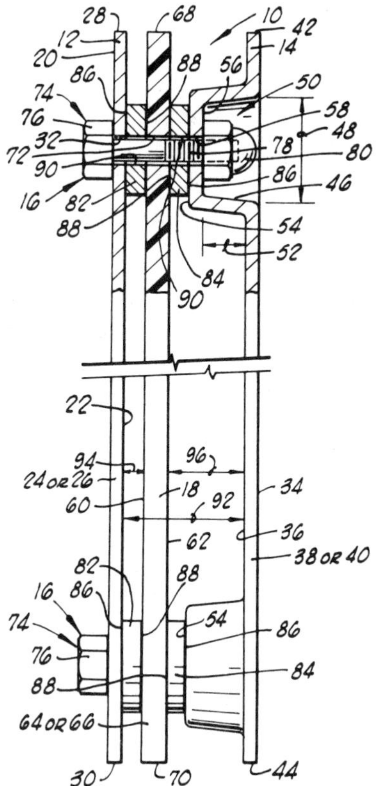

Source: U.S. Patent 4,085,027

The first and second spacers **82** and **84** cooperate with the barrier member **18** to space the second face **22** of the anodic member **12** the distance **92** from the second face **36** of the cathodic member **14**. Thus, the elements of the hybrid bipolar electrode **10** which are constructed of a metal comprising titanium are isolated and spaced from the cathodic member **14**, i.e., the hydrogen producing member or elements as in the case of the second spacer **84**.

The spacing of the elements constructed of a metal comprising titanium from the hydrogen producing elements (the cathodic member **14**) and the disposition and construction of the barrier member **18** cooperate to substantially reduce the possibility of hydrogen attacking the elements constructed of titanium and provide a metal bipolar electrode having an anodic member constructed of titanium which is serviceable in an alkali metal chlorate or chlorine electrolytic cell application.

Hollow Cathode

H. Cunningham; U.S. Patent 4,165,272; August 21, 1979; assigned to PPG Industries, Inc. describes an electrolytic cell cathode having a hollow cathode finger with fins extending outwardly therefrom. A synthetic separator surrounds the cathode and rests upon the fin-like extensions.

The relationship of the cathode structure to an electrolytic cell, e.g., an electrolytic cell of a bipolar electrolyzer, is shown in Figure 4.15a. The bipolar electrolyzer **1** has a plurality of individual electrolytic cells **11**, e.g., from 2 to 100 or more. Each individual electrolytic cell **11** has an anodic unit **51** of one bipolar unit **21** and a cathodic unit **71** of an adjacent bipolar unit **21**.

Each cell contains an anodic unit **51** containing brine feed means **31**, brine recovery means **33**, and a chlorine outlet **35**. Additionally, the anodic unit **51** has an anolyte-resistant sheet **24** on the backplate **22** and an anolyte-resistant lining **28** on the interior surfaces of the unit **51** such as the cell walls **25**, the top of the cell **26**, and the bottom of the cell **27**. Extending outwardly from the anodic unit **51** are anode blades **53**.

The cathodic unit **71** of the cell **11** has water feed means **37**, cell liquor recovery **39**, hydrogen recovery **41**, a cathode element **81**, and a back screen **73** spaced from the backplate **22**.

The basic structural element of the electrolyzer **1** is the bipolar unit **21** shown generally in Figure 4.15a and in detail in Figure 4.15b. Backplate **22** separates the anodic unit **51** of the bipolar unit from the cathodic unit **71** of the bipolar unit **21**. The backplate **22** has two members, a heavy catholyte-resistant plate **23** and a thin anolyte-resistant sheet **24**. A thin anolyte-resistant sheathing, layer, lining, or sheeting **28** lines the top **26**, bottom **27**, and walls **25** of the bipolar unit **21** within the anolyte compartment. Anodes **53** extend outwardly from the backplate while cathode elements **81** extend outwardly from the opposite side **23** of the bipolar unit **21**.

The cathode unit **71** of the bipolar unit **21** shown in detail in Figure 4.15c, includes a plurality of individual cathode elements **81** having cathode fingers **83** and a synthetic separator **101** spaced from the cathode fingers by extensions **91**.

The cathode unit **71** also includes a back screen **73** spaced from the backplate **22** and substantially parallel to the backplate. The back screen can be fabricated of the same material as the cathodic electrodes **83**, and it may or may not have extensions **91**.

Figure 4.15: Isometric Views, in Partial Cutaway

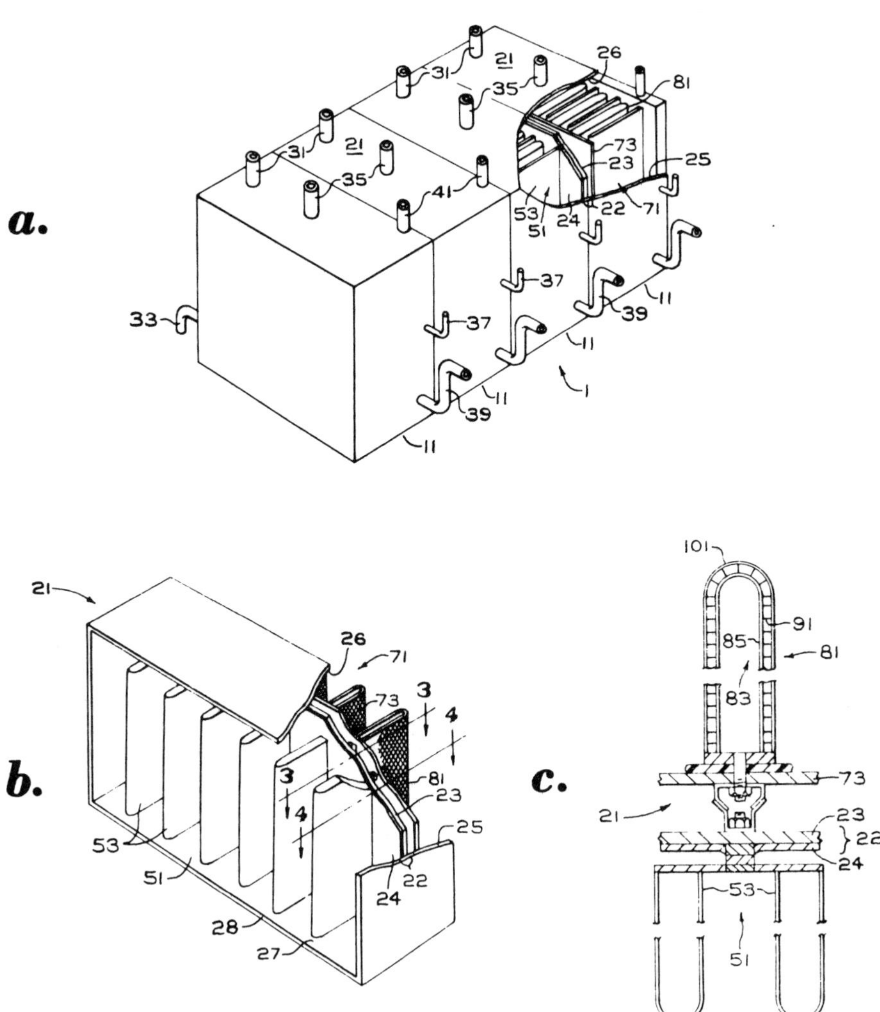

(a) Bipolar electrolyzer
(b) Bipolar electrode
(c) Cathode element

Source: U.S. Patent 4,165,272

When the back screen **73** is made of the same material as the cathodic electrodes **83**, it can serve as an auxiliary electrode area. It can, however, be fabricated of a nonactive but permeable metal to allow the synthetic separator **101** to function.

The back screen **73** may also be fabricated of material that is substantially impermeable to the flow of either electrolyte or ion.

The electrolyte impermeable side walls **85** of the individual cathode element **81** are parallel to each other and spaced from each other. They may be fabricated of materials of construction conventionally used to fabricate cathodes such as iron, alloys of iron with cobalt, nickel, manganese, carbon, and the like. Alternatively, the electrolyte permeable side walls **85** may be fabricated of stainless steel.

The side walls **85** of the hollow cathode fingers **83** may be electrolyte permeable as perforate or foraminous mesh, plate, or sheet. Alternatively, they may be louvered whereby to allow electrolyte to pass through. The cathode element **81** can either be open or closed at the top as appropriate to allow for bonding, joining, and sealing of the synthetic separator **101**, for example, by masked compressive means or the like.

Extensions **91** are joined to the side walls **85** of the cathode fingers **83**. The purpose of the extensions is to space the synthetic separator **101** from the cathode side walls **85**.

Titanium Alloy

D.W. Du Bois and W.B. Darlington; U.S. Patent 4,133,730; January 9, 1979; assigned to PPG Industries, Inc. disclose an improved method of electrolysis utilizing an electrode fabricated from an alloy of titanium and a rare earth metal. The electrode may be a cathode, or, when having a suitable electrocatalytic coating, an anode, or even a bipolar electrode with anodic and cathodic regions.

One surface of the bipolar electrode, which may or may not be coated, faces the anode of a prior bipolar electrode and functions as the cathode of the bipolar electrode.

The opposite surface of the electrode, coated with an electrocatalytic material, faces the cathode of a subsequent electrode, thereby functioning as the anode of the bipolar electrode.

Example 1: Three titanium coupons were tested as cathodes in a 10 wt % aqueous Na_2SO_4 solution. One coupon was prepared from an alloy containing 0.2 wt % palladium and the balance titanium. The second coupon was prepared from commercial Ti-38A titanium alloy (containing 0.3 wt % molybdenum and 0.8 wt % nickel). The third coupon was prepared from a titanium-yttrium alloy containing 0.02 wt % yttrium, 0.07 wt % iron, 0.061 wt % oxygen, 0.008 wt % nitrogen, 0.03 wt % carbon, and 25 ppm hydrogen.

The coupons were cleaned in an aqueous solution prepared from 3 vol % HF, 30 vol % HNO_3, balance water. Thereafter, each coupon was taped so that only a 1" by 1" segment was exposed to the electrolyte.

Each coupon was then placed in a separate container of 10 wt % Na_2SO_4 and tested as a cathode at a current density of 232 A/ft^2. The weight increases shown in the following table were obtained.

Cumulative Percentage Weight Increases of Titanium Coupons

Days Under Test	Ti-0.3% Mo-0.8% Ni Alloy (19.0678 g)	Ti-0.2% Pd Alloy (15.2014 g)	Ti-0.02% Y Alloy (20.0745 g)
		(%)	
7	0.059	0.024	-
11	-	-	0.012
14	0.093	0.030	-
16	-	-	0.014
20	-	-	0.016
21	0.114	0.034	-
27	-	-	0.018
28	0.124	0.030	-
34	-	-	0.018
35	0.111	0.025	-
41	-	-	0.020
46	0.077	0.023	-
48	-	-	0.020
51	0.088	0.020	-
91	-0.062	0.016	0.020

Note: Actual weight losses indicated physical separation of the titanium hydride.

Example 2: The hydrogen evolution voltages on the Ti-0.2 wt % palladium alloy coupon and on the Ti-0.02 wt % yttrium alloy coupon were tested at 50°C and 232 A/in^2 versus a silver-silver chloride electrode in saturated potassium chloride. The measured hydrogen evolution voltages were 1.64 volts for the titanium-palladium alloy coupon and 1.56 volts for the titanium-yttrium alloy.

OTHER CELL COMPONENTS

Separating Web

B.S. Wallace; U.S. Patents 4,139,448; February 13, 1979; and 4,137,145; Jan. 30, 1979; both assigned to Hooker Chemicals & Plastics Corp. describes an assembly of electrolyte compartment frames and a separating web for an electrolytic apparatus which comprises an anolyte compartment frame, a catholyte compartment frame and a separating web between them to prevent passage of electrolyte between such compartments. The separating web is capable of being held by pressure in liquid-tight relationship to the anolyte compartment and catholyte compartment frames while being movable between them to compensate for expansion and contraction of the web without distortion of the web and without loss of liquid-tight separation of the anolyte and catholyte compartments by the web.

In preferred embodiments the electrodes are bipolar; the apparatus is for the electrolysis of brine or hydrochloric acid, preferably brine; the electrolyte compartment frames are of synthetic organic polymeric material, preferably asbestos or glass fiber-filled polypropylene; the separating web is of similar material, although metal webs may also be employed; the separating web is held in liquid-tight and gas-tight relationship with the electrolyte compartment frames, as in a sandwich, with gasketing between it and the frames, allowing longitudinal movement of the web; and recesses are provided at the web ends to allow expansion of the web into such recesses upon heating of the cell and thereby to prevent strains that could be caused by having such expansion occur against an abutting or restraining portion of an electrolyte compartment frame.

In Figure 4.16a, anolyte compartment frame **11** and catholyte compartment frame **13** are shown, for clarity of representation, separated from intermediate web assembly **15**. Anolyte compartment frame **11** includes framing walls **14** and **16**, bottom framing portion **17** and top framing portion **19**, all of which combine to form the generally rectangularly shaped framing structure of member **11**.

Figure 4.16: Electrolytic Compartment Frames and Intermediate Web Assembly

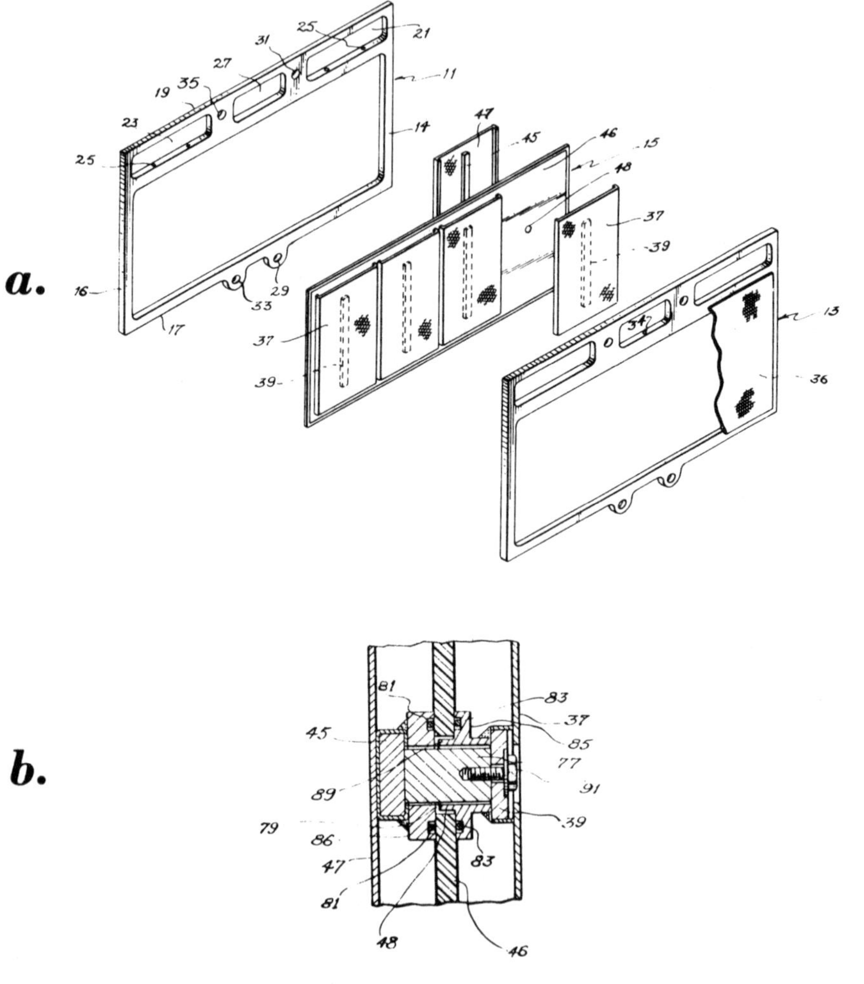

(a) Partially disassembled view
(b) Horizontal section

Source: U.S. Patent 4,139,448

Internal header openings **21** and **23** are for gaseous products of electrolysis generated at the anode and such gas passes through openings **25** from the anolyte compartment section below to the headers, which are formed by the walls about such openings. Header opening **27** is for gas which may be generated at the cathode. Thus, in a cell for the electrolysis of brine, headers **21** and **23** are for chlorine and header **27** is for hydrogen. Additionally, headers **21** and **23** are for anolyte liquid and header **27** is for caustic solution. Drains **29** and **33** are for brine and caustic, respectively and inlets **31** and **35** are for brine and caustic solution feeds.

Compressive forces are applied to the frames and intermediate webs, by tightening of threaded nuts (not shown) onto threaded ends of tightening rods or bolts (not shown), so as to tightly press a series of electrolyte compartment framing members together. Catholyte compartment framing member **13**, as illustrated, is clearly of a structure similar to that of anolyte compartment frame **11** and accordingly the parts thereof need not be further described except for mention of passage **34** for the removal of gas generated at the cathode from the cathode area to header passageway or opening **27**.

Ion exchange membrane **36** is held to the catholyte compartment frame **13** by suitable means (not shown), but may be held to the far side (that away from the viewer) of the anolyte compartment frame **11** instead. It is continuous and a single and unitary membrane, such as one of fluorinated polymeric plastic of the ion exchange type, e.g., Nafion, (E.I. DuPont de Nemours & Co., Inc.), covers the frame and overlies all four (or other number, e.g., 1 to 8) cathodes. The ion exchange membrane may sometimes be replaced with a suitable semipermeable membrane or diaphragm.

Web assembly **15** includes web **46** which is mountable in sandwiched relationship between frames **11** and **13**. The cathode screens **37**, as illustrated, being four separate screens, each fastened to a cathode conductor bar **39**, are each essentially rectangular in shape with sides being formed or bent so as to extend from the major surface of the cathode screen toward the web (although such parts do not normally contact it). The tops and bottoms of such screens may be similarly formed.

Cathode conductor bars **39** are joined to cathode screens **37** by suitable conductive welding or other conductive joinders and to intermediate conductive sections, not illustrated in Figure 4.16a but shown in Figure 4.16b, by bolt or other fastening means. Anode conductor bar **45** and anode screen **47** are viewable when moved away from the web, as is passageway **48** through web **46** when cathode screen **37** and conductor bar **39** are similarly moved.

In Figure 4.16b, cathode **37**, suitably fastened, as by welding, to cathode conductor bar **39**, has electrical current communicated from it through the conductor bar **39** and intermediate conductive material **77**, which may be joined to cathode conductor bar **39** or to anode conductor bar **45**, to anode **47**.

Although it is highly desirable for the conduction of the electrical current to be essentially perfect, with little or essentially no resistance, it is also important that there should be no transfer of liquid or gas from one electrolyte compartment to another. For this reason opening **48** in web **46** (which allows passage of the conductor from cathode to anode) is sealed off by ring gaskets **81** and **83**, contained in accomodating recesses in collars **86** and **85**, the former being joined to anode conductor bar **45**.

It will be seen that the clearance space **79** between such collar-shaped members in passage **48** allows for tightening of the parts thereof by bolt **91** without causing strains due to movements of one part against another and yet ledge portion **89** helps to position collar **85** with respect to web **46**.

The apparatus illustrated in Figure 4.16a, with an internal conductor similar to that shown in Figure 4.16b being employed to transmit electricity between electrode pairs, has been manufactured and tested. A two-cell unit equipped with steel cathodes and dimensionally stable anodes has been operated over a four-month period, utilizing a 25% brine solution to generate chlorine, hydrogen and cell liquor containing 200 g/ℓ of sodium hydroxide. The semipermeable membrane employed is DuPont Nafion and the unit operated at an efficiency of 75 to 80%, at 8,000 A and 8 V.

Conduction of electricity from an upstream cathode to a downstream anode was with a minimal voltage drop and the neoprene and EPDM elastomeric gasketing utilized satisfactorily maintains sealing contact with the asbestos-filled polypropylene web and frames.

Intercell Connector

B.B. Smura; U.S. Patent 4,115,236; September 19, 1978; assigned to Allied Chemical Corp. provides an intercell connector for a bipolar permselective membrane electrolyzer which maximizes both mechanical connection and electrical communication between the cells while substantially precluding fluid and/or gaseous flow.

As shown in Figure 4.17, the anode boss **14** is formed with a blind threaded bore **38**. The cathode boss has a corresponding through bore **40**, while the electrically conductive insert **36** has a bore **42**. Preferably, the insert **36** is a copper tube or bushing. A fastener **44** is inserted through the bores in the cathode boss, tubular insert, and into mating engagement with the threaded bore in the anode boss. The fastener **44** is, most advantageously, a standard steel or ferrous alloy bolt having a head **46** and shoulder **48**.

Where the anode boss **14** meets the face of insert **36**, there is defined an anode interface **50** peripherally about bolt **44**. Likewise, a cathode interface **52** is formed where cathode boss **24** mates with the insert **36**. Because each of the anode and cathode bosses has a transverse dimension greater than that of the insert **36**, there are also formed an anode/web interface **54** and a cathode/web interface **56**, respectively.

To preclude fluid and gaseous flow across the connector, gaskets **58** are provided at the electrode/web interfaces **54**, **56**. These gaskets may be fabricated from various chemically resistant materials, among which might be mentioned rubber, chlorinated plastics, polypropylene, polymers and copolymers of trifluorochloroethylene, tetrachloroethylene, tetrafluoroethylene, polyvinyl acetate, polyesters, etc., with or without fillers such as, e.g., asbestos.

Figure 4.17: Intercell Connector

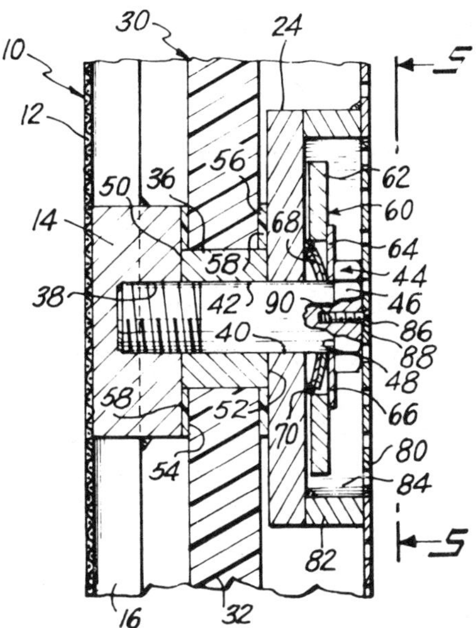

Source: U.S. Patent 4,115,236

When the bolt **44** is tightened within the threaded bore **38**, an axial compressive force is exerted which compresses the gaskets **58** at the interfaces **54**, **56**, to insure a fluid and gas tight connection.

In order to assure the maintenance of a low resistance electrical path, it has been found essential to maintain a constant compressive force on the electrode interfaces **50** and **52**. Thus, in conjunction with the axial force applied by bolt **44**, there is provided a biasing force in opposition thereto. This opposing force is achieved by a biasing device, designated generally as **60**.

The biasing member **60** includes a bolt head skirt **62**, which, in combination with a washer **64** resting against the shoulder **48** of bolt **44**, defines an annular channel **66**. Disposed within this channel is a biasing spring member **68**, which might be simply a spring washer. In order to effectuate a fluid and gas tight seal, an O-ring **70** is included within the annular channel **66** about the circumferential periphery of spring **68**. This O-ring may be of a material selected from the same group of materials for the gaskets **58**.

From the foregoing, it is evident that both the mechanical connection and electrical communication either between cells (i.e., intercell) or at the terminal cells (i.e., end cell) are maximized. Fluid and gaseous integrity are maintained by virtue of the O-ring seals and elastomeric gaskets at all points at which fluid or gas might otherwise penetrate the connector.

Mechanical connection is positive by virtue of the design of the bolt **44** in combination with the electrode bosses **14** and **24**, along with the insert **36**. Due to materials' selection and the effect of the biasing member **60**, electrical conductivity across the connector is maintained, whereby a low resistance electrical path is established.

Spring-Loaded Intercell Connectors

An electrolytic cell bank **A** is generally depicted in Figure 4.18a comprised of a plurality of individual cell units **B** which are electrically connected in series to form a bipolar cell bank by conductive strips **C** which are sandwiched between adjacent cell units **B**.

G.R. Pohto and M.J. Kubrin; U.S. Patent 4,108,752; August 22, 1978; assigned to Diamond Shamrock Corp. utilize only conductive strips **C** and compressive contact thereof between adjacent cell units **B** to establish electrical connection.

A linear form conductive strip **C** is shown in Figure 4.18b. A plurality of louvers **50** extend outwardly of the planar faces of the strip **C** in alternating pairs outwardly of one face or the other of the conductive strip. As can be seen from the illustration in Figure 4.18b, louvers **52** establish contact between parallel surfaces abutting thereto represented by dashed lines **54**, which dashed lines may represent planar pans **10** and **12**. In order to assure positive electrical connection, there is compressive contact of louver edge portions **52** with the adjacent abutting surfaces **54** so that an elastic bending or spring-like force is developed in louvers **50** to assure positive electrical connection.

Figure 4.18: Electrolytic Cell Bank Having Spring-Loaded Intercell Connectors

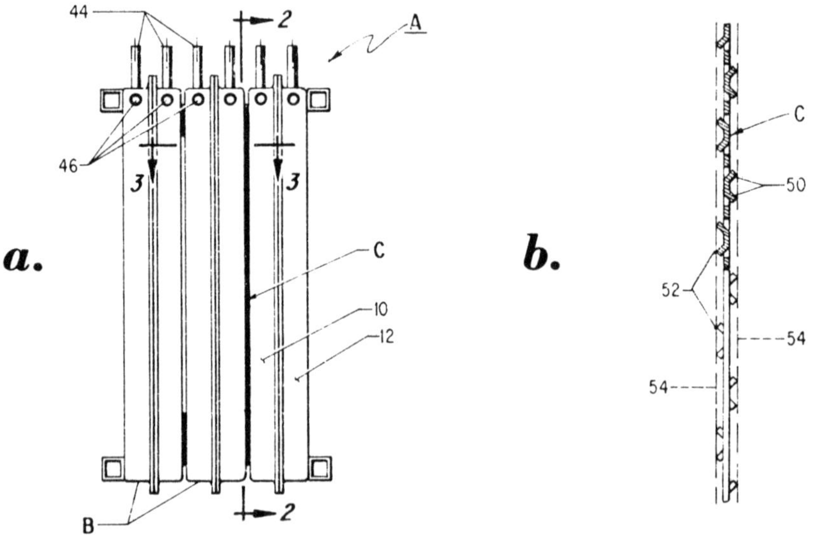

(continued)

Figure 4.18: (continued)

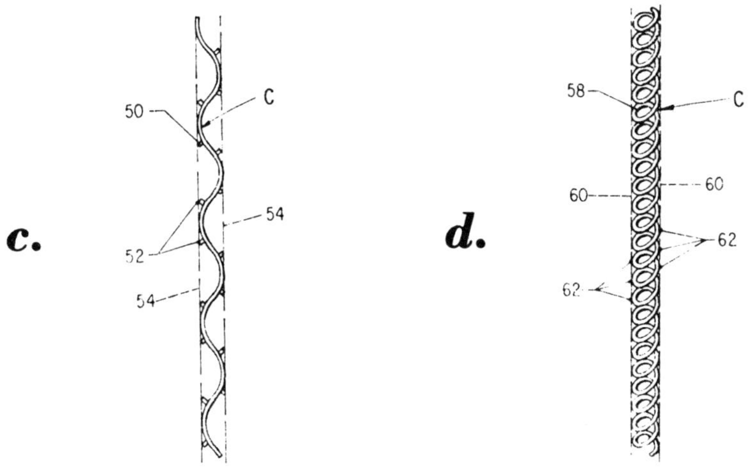

(a) Side elevational view of a cell bank with spring-loaded connectors
(b)-(d) Side elevational views, in partial section, of conductive strips

Source: U.S. Patent 4,108,752

By employing the undulate form of Figure 4.18c, the action of adjacent planes such as pans **10** and **12** in compressive contact against louver edge portions **52** causes an additional springing force to be realized over that provided by the bending flexure of louvers **50** alone, the additional springing force being provided by the undulate form. These two independently springing forces further assure positive electrical connection between the parallel facing surfaces **54**.

An askew helical form may also be employed to establish electrical connection through a plurality of contacts such as is illustrated in Figure 4.18d. A helical spring, such as of beryllium copper, is formed having a helical axis passing therethrough. The laterally opposite sides of the helix are then forced in opposite directions parallel to the helical axis so that the helix is askew from its prior form. This askew form causes a spring force to be developed within the helix tending to return the helix to its original form because the helix has a smaller lateral width than that of its original form.

With the interposition of an askew helic **58** such as shown in Figure 4.18d between adjacent planar surfaces **60** a plurality of contacts **62** are made at the lateral edges of each loop of the helix. The normal springing force of the helix combined with the forces developed because of its askew configuration causes compressive contact between the contact points **62** and the parallel surfaces **60** to establish electrical connection between the parallel surfaces at a plurality of points.

Through the use of identical cell units **B** and conductive strips **C**, it can be clearly seen that replacement of defective or worn out cell bank components may be easily effected by merely withdrawing one cell unit and plugging in a new or remanufactured cell unit.

Bipolar Electrode Backplates

L.M. Meyer; U.S. Patent 4,138,324; February 6, 1979; assigned to Diamond Shamrock Corp. discloses a method for electrically and mechanically connecting the backplates of a bipolar electrode to be used in a filter press electrolytic cell for electrochemical production.

This method employs the use of a metal laminate strip having surfaces of metallic substances identical and corresponding to the metallic makeup of the given backplates which can be welded between the anode and cathode backplates using standard weldment procedures. The metal laminate strips are placed in a spaced series such that the anode and cathode backplates present two parallel planes in spaced relation to each other thereby leaving a space for the escape of hydrogen gas, preventing hydrogen embrittlement of the titanium anode backplate.

In a related patent, *E.J. Peters; U.S. Patent 4,116,807; September 26, 1978; assigned to Diamond Shamrock Corp.* employs the use of explosion bonded solid metallic strips between the two backplates of a bipolar electrode to provide the essential electrical and mechanical connection.

This process can be generally achieved by supporting a layer of one material parallel to the surface of the other material, the inside surfaces being spaced apart slightly and placing on the other surface of one layer a detonating explosive having a velocity of detonation <120% of the velocity of sound in that metal in the system having the highest sonic velocity and thereafter initiating the explosive layer. Usually it is desirable to use an explosive having a detonation velocity not greater than the velocity of sound in that metal with the higher sonic velocity.

The metal layers must be separated from each other a distance at least sufficient for the explosive propelled layer to achieve an adequate velocity before impact with the stationary layer; a spacing of 0.001" between the facing surfaces of the two layers represents the minimum spacing to produce consistently adequate results. The maximum separation will depend almost entirely upon the reduction of the velocity of the propelled layer caused by the air layer between the two metals. By increasing the explosive loading or evacuating the space between the layers, spacings much greater than 0.001" are feasible. In general, however, separation of more than 0.5" is not convenient or necessary.

A 0.0625" layer of copper can be clad onto a 0.5" plate of mild steel in the following manner. The copper sheet was covered on one side with a 1" thick layer of polystyrene foam and the polystyrene layer was covered with a layer of an explosive composition having a weight distribution of 10 grams per square inch.

The explosive employed in this example was a thin uniform sheet of flexible explosive composition comprising 20% very fine pentaerythritol tetranitrate (PETN), 70% red lead, and as a binder, 10% of a 50–50 mixture of butyl rubber and a thermoplastic terpene resin mixture of polymers of β-pinene of the formula

$(C_{10}H_{16})_n$, Piccolyte S-10, (Pennsylvania Industrial Chemical Corp.). Complete details of this composition and a suitable method for its manufacture are disclosed in the U.S. Patent 3,093,521.

The composition is readily rolled into sheets and detonates at a velocity of about 4,100 m/sec. The edges of the copper-polystyrene-explosive sandwich were sealed with waterproof tape, and it was placed on the mild steel with a spacing between the copper layer and the steel layer of 0.0138" provided by uniform particles of iron powder. These are particles which have been screened to pass through a number 45 mesh and held on a 100 mesh. The edges of the completed assembly were sealed with tape and an electrical initiator was attached to one corner on the explosive layer. The assembly was then immersed in water and the explosion initiated. Excellent bonding of the copper onto the steel resulted.

The next portion of the procedure employed a duplication of the prior process to prepare a titanium on copper cladding. The titanium layer was 0.05" thick and the copper layer was the same as the preceding. The spacing, which in this case was provided by particles of titanium powder, was 0.0138" and the weight of the explosive was 10 g/in^2. Following the detonation of the explosive, the titanium and copper sheets were firmly and uniformly bonded. This then formed a sandwich of the cathode backplate made of mild steel to the copper metallic electrical conductor to the anode backplate made of titanium.

A one step process can be accomplished by sandwiching all three components with the same amount of spacing as in the prior procedure, using iron particles and titanium particles as described above, placing the explosive to a weight distribution of approximately 15 g/in^2 on top of the titanium anode backplate, and the entire structure being sealed inside a box such that the structure may be submerged in water and the initiator detonated. A solid bonding between all three components results from such a process.

IMPROVED ELECTRODES

Electrodes for the electrolysis of alkali metal brines, whether for the production of chlorine or alkali metal chlorate, generally have been bulk graphite slab or plates. The problems associated with the use of graphite electrodes in such applications are well-known to practitioners in the art. The products of the electrolysis reaction tend to chemically and physically attack the surface of the graphite. As the graphite wears away, the distance between electrodes increases, thus causing an increase in energy consumption of the electrolytic cell. Furthermore, the chemical and physical deterioration of the graphite produces undesirable carbon products within the cell.

There have been numerous attempts to improve the dimensional stability of graphite electrodes. For this purpose, the graphite has been coated with various substances in an attempt to improve its resistance to chemical, physical and electrochemical attack. Catalytically effective coatings have also been applied to the graphite with the intention of increasing power efficiency. Hower, such coatings themselves tend to erode over extended periods of time.

Another approach that has been adopted in the art to achieve dimensional stability is to replace the graphite with metallic electrodes. In their common form, such metallic electrodes have a film-forming metal base coated with an electrocatalytically active material, usually a noble metal or oxide of a noble metal.

Film-forming metal bases that have been suggested include titanium, tantalum, tungsten, niobium, zirconium, and alloys thereof. The most frequently used film-forming metal base has been titanium. However, in many applications solid metallic electrodes may not be economically attractive for commercial use due to the large amounts of high cost film-forming metals, noble metals, and noble metal oxides that are required.

Another approach that has been adopted in the art to achieve dimensional stability is to coat a graphite substrate with a film-forming metal, and then apply an electrocatalytically active material to the film-forming metal, e.g., U.S. Patent 3,770,613. However, the known methods of joining the graphite substrate to the film-forming metal create an undesirable voltage drop across the interface between the graphite and the film-forming metal.

Consequently, there is a need for an electrode which is relatively inexpensive, energy efficient, resistant to chemical, physical, or electrochemical attack, dimensionally stable, and resistant to voltage drops across its surface.

METAL ANODES

Layered Electrode

An electrode is produced by *R.J. Pangborn; U.S. Patent 4,181,585; January 1, 1980; assigned to The Dow Chemical Company* by interposing a nonferrous filler layer between the solderable metal surfaces of the electroconductive substrate and the film-forming metal.

A portion of the outer surface of the film-forming metal is electrocatalytically active or is capable of being rendered electrocatalytically active and corresponds to at least a portion of the working area of the electrode. The nonferrous filler layer is heated to sufficient temperature to effect an electrically conductive bond between the solderable metal surfaces.

Example 1: *Preparation of the Electrode* – Two titanium sheets approximately 6 x 16 x 0.003 inches, were cleaned by grit blastings with 46 mesh (U.S. Standard Sieve Series) alumina (Al_2O_3) grit. The sheets were then coated with a solid solution of ruthenium and titanium oxide in a manner substantially as described in Example 1 of U.S. Patent 4,112,140.

The sides of the titanium sheets not coated with ruthenium and titanium oxide were cleaned with an emery cloth to remove any titanium dioxide formed by air oxidation of the exposed titanium metal surface. A coating of copper metal was applied to the thus cleaned surfaces by metal spraying.

In this method, a ⅛ inch diameter copper wire was fed into an oxyacetylene flame at a speed just sufficient to permit melting of the copper. The molten copper was atomized and projected against the surface of the titanium. The impingement of the copper particles on the surface of the titanium formed a continuous coating of copper approximately 400 microns thick.

Both faces of a piece of graphite approximately 5 x 15 x 1.2 inches were copper sprayed by the method described above. The copper coating on the graphite faces was approximately 400 microns thick.

A solder containing 50 parts by weight tin and 50 parts by weight lead was placed on the copper-coated faces of the graphite. One of the titanium sheets was placed against each of the faces of the graphite, so that the copper-coated surface of each titanium sheet was also in contact with the solder. That portion of each titanium sheet extending beyond the edges of the graphite was bent inwardly.

The titanium sheets, solder, and graphite were then pressed together between heated plates at about 450°F. The cavity formed between the graphite and the inwardly extending edges of the titanium was filled with a vinyl ester resin. The resin was cured at 70°C for about 72 hours. The resulting composite electrode was allowed to cool slowly while still under pressure, and the pressure was then released.

Example 2: *Use of the Electrode in an Electrolytic Cell* – The electrode produced in Example 1 is employed as an anode in a laboratory electrolytic cell to produce gaseous chlorine from an aqueous solution containing about 300 grams per liter sodium chloride. The anode is suitably spaced apart from a steel screen cathode by an asbestos containing diaphragm. The cell is operated at an anode current density of 1.0 amp per square inch and a voltage of 3.5 volts. The sodium hydroxide concentration in the catholyte is about 140 grams per liter and the pH of the anolyte is from about 2.0 to 4.5. The cell is operated at a temperature of from about 75° to 85°C. Substantially no carbon dioxide from chemical attack of the graphite anode is formed during operation.

Expandable Anode

K.A. Poush and O.C. Taylor; U.S. Patent 4,080,279; March 21, 1978; assigned to The Dow Chemical Company have developed an electrolytic cell expandable anode assembly particularly useful as anodes in commercial chlor-alkali production cells, the assembly having two parallel, foraminous working faces and a simple spreading means.

Referring to Figure 5.1a, the anode assembly can be seen. The anode working faces **10** have been spread apart by two spreading bars, **14a** and **14b**. The working faces **10** are physically and electrically connected to an intermediate electrical conducting means **18** by means such as welding.

Figure 5.1: Expandable Anode

a.

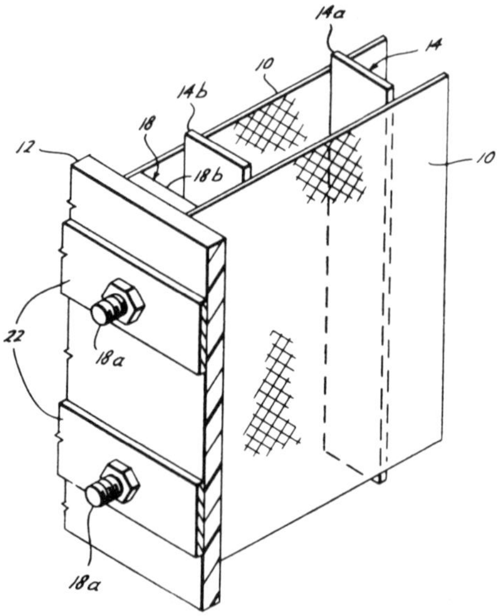

(continued)

Figure 5.1 (continued):

b.

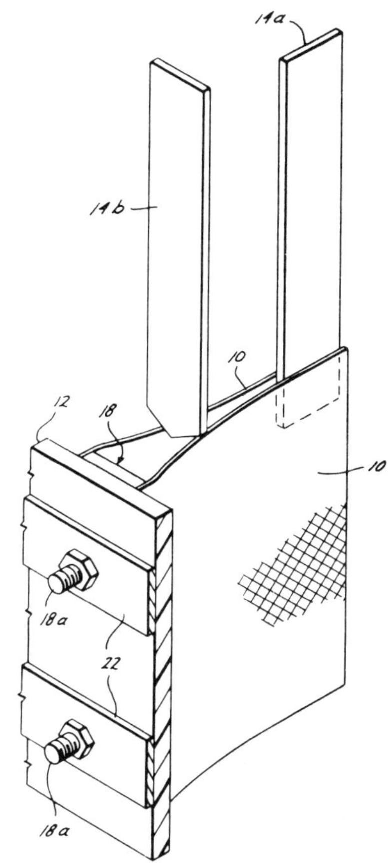

(a) Perspective view of anode assembly with base **12** shown broken away from its remainder

(b) Perspective view of anode assembly shown with electrode faces **10** in collapsed positions

Source: U.S. Patent 4,080,279

In Figure 5.1b, the anode working faces **10** are shown in their collapsed position and the spreading means **14a** and **14b** are shown just before they are inserted between the working faces **10**. Working faces **10** are seen to be close together before the spreading means **14** (**14a** and **14b** in Figure 5.1b) are positioned in place. This is due to the resilience of the dimensionally stable metals needed to be used for the construction of these working faces, and the fact that these are originally constructed with a spring memory which causes them to tend to remain or return to this collapsed position, close to one another. Their position, when the spreading means **14** has spread them apart from this collapsed position, is referred to as their expanded position or spread position.

The spreading means **14** can be constructed from a variety of materials. Electrical conductivity of these materials, unlike the prior art, is of no consequence in this assembly inasmuch as the spreading means plays no part in the conduction of electricity to the working faces **10** from the external dc power connecting means **22**.

The only limitation on the type of material used is that it be rigid enough to force and maintain the working faces **10** apart. Also, these materials must be made of materials which can withstand attack of the environment of the type electrolytic cells in which it is used. Thus when used as an anode in a chlor-alkali cell, the spreading means must be able to withstand the chemically corrosive attack of an aqueous mixture of brine, hydrochloric acid and chlorine gas. Many materials such as glass, wood, many plastics, and some metals meet these criteria for chlor-alkali electrolytic cells.

The working faces **10** of the electrode assembly must, of course, be electrically conductive, resilient with a spring memory inward toward each other, foraminous to allow the passage of electrolyte therethrough, and resistive to the corrosive attack of its environment.

Generally the working faces **10** are made foraminous by expanding them (in the expanded metal sense) or punching holes in sheets of suitable metals such as titanium, tungsten, and tantalum.

Converting Box Anodes to Expandable Anodes

G.R. Pohto and R.O. Olson; U.S. Patent 4,154,667; May 15, 1979; assigned to Diamond Shamrock Corporation describe a method for converting box form dimentionally stable anodes to the more useful and energy-saving expandable type. The method utilizes the old style box anode as a basic structure which could be converted, thus allowing chlorine and caustic producers utilizing such box anodes to obtain the economic benefits of expandable type dimensionally stable anodes without the high capital cost of disposal of old box anodes and purchase of all-new expandable types.

Figure 5.2a shows a typical box-style anode assembly **10** comprising an anode riser member **12** to which anode plates **14** and **16** are welded as at welds **18**. Anode riser **12** comprises generally a cylindrical member having a conductive core **20** which has a titanium cover **22** clad or otherwise applied to the core **20** as is shown in Figure 5.2b.

Anode plates **14** and **16** may take any form but are generally of a foraminous structure having a plurality of openings therethrough so that the evolution of gas at the surface of the anode is facilitated and a large surface area is presented. The anode plates **14** and **16** are usually made of a valve metal such as titanium and are preferably coated with an electrocatalytic metal or metal oxide as is common in the art. Anode plates **14** and **16** have a configuration so that they may be joined to each other at their ends by butt or lap joints.

Figure 5.2: Converting Box Anodes to Expandable Anodes

a.

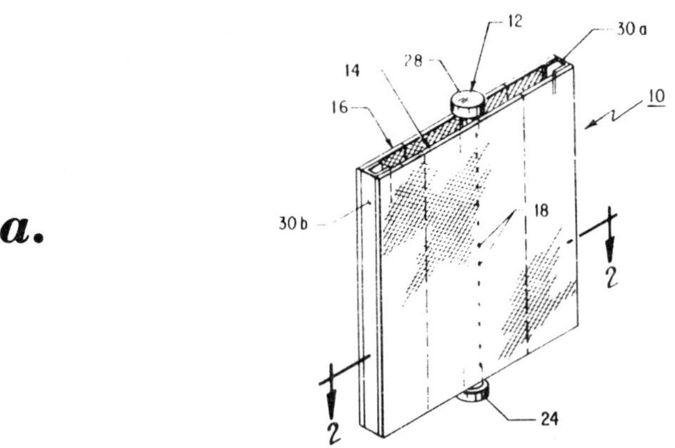

b.

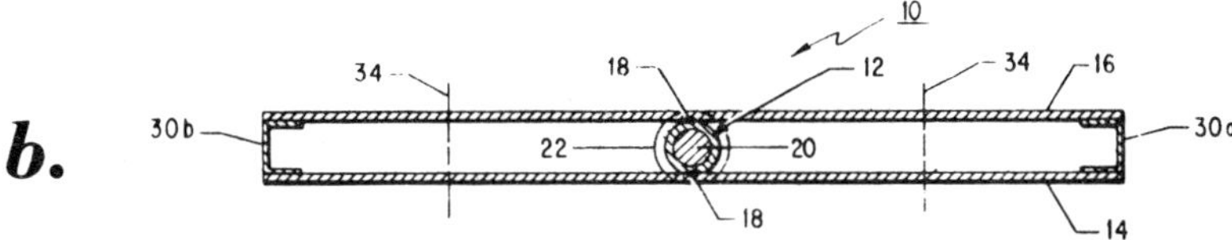

c.

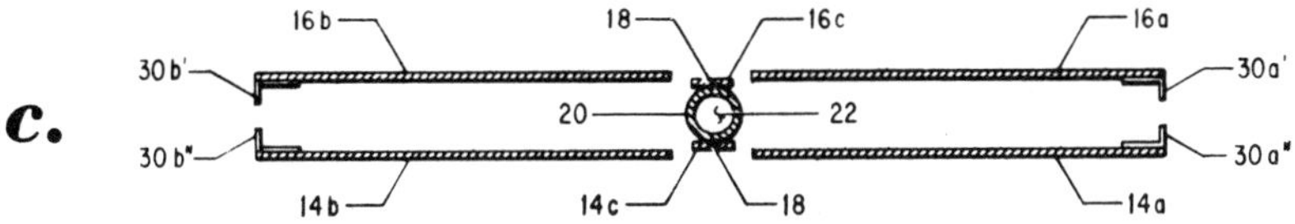

d.

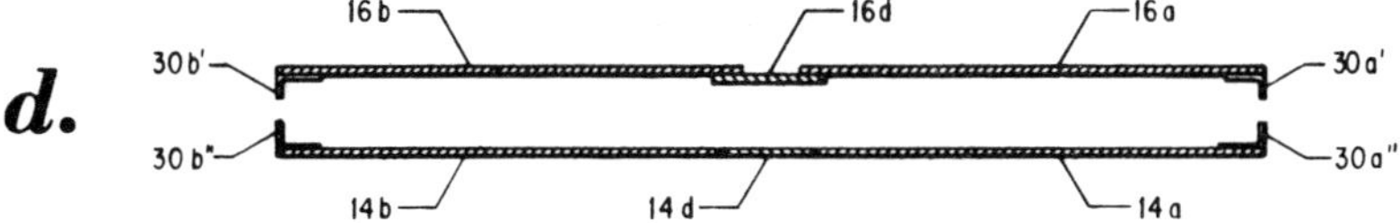

(continued)

Figure 5.2 (continued):

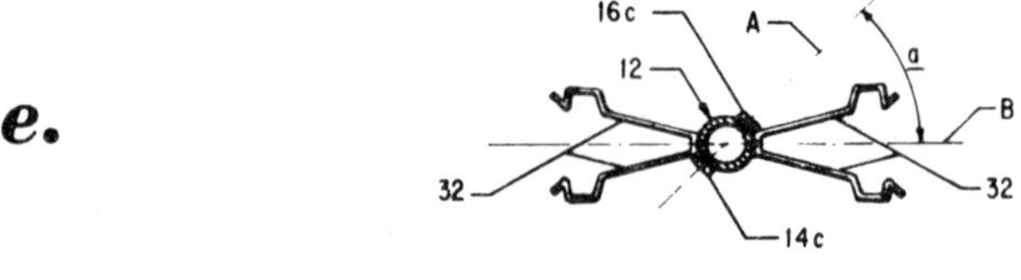

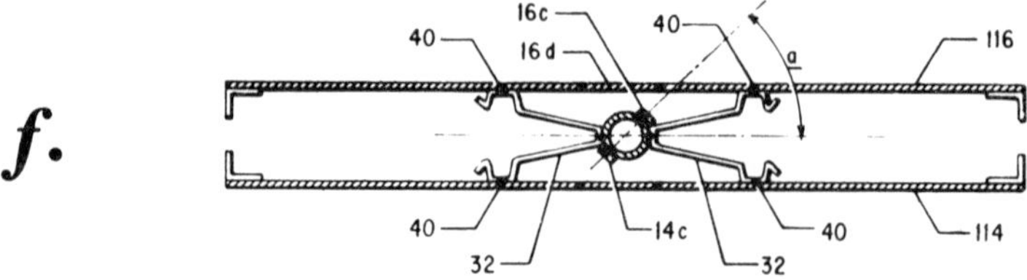

(a) Perspective view of a box-style anode
(b) Cross-sectional view of the box anode shown in Figure 5.2a taken along line **2–2** thereof
(c) Cross-sectional view of the box-style anode shown in Figure 5.2a in which portions of the anode plates have been cut off
(d) Cross-sectional view illustrating the reconstruction of anode plates
(e) Cross-sectional view illustrating the attachment of expander members to the anode riser
(f) Illustration of final assembly of the anode riser having expander portions and repaired mesh portion forming an expandable anode

Source: U.S. Patent 4,154,667

Flange member **24** is located at the base portion of the anode riser **12** and constitutes a mounting flange for attaching the anode riser to the anode base of an electrolytic cell. Titanium cap member **28** is positioned over the top end of the anode riser **12** so as to totally enclose and seal the copper core **20** at the upper end of the anode riser **12**.

As seen in Figure 5.2b, the anode plates **14** and **16** and end members **30a** and **30b** form a four-sided box enclosure which is substantially rigid and, when installed in an electrolytic cell, is in a position relative to the adjacent cathode tubes which cannot be varied.

Box anode **10** may be converted from a rigid structure to an expandable-type anode having all of the economic advantages inherent thereto through the process illustrated in Figures 5.2b through 5.2f.

The first step is to remove plate members **14** and **16** from anode riser member **12** at a point adjacent the attaching welds **18**. This may be accomplished by any of several cutting processes such as grinding, shearing, sawing, laser cutting or the like. Anode plate quarter sections **14a** and **16a** are connected by transverse member **30a** and form a U-shaped member upon separation from anode riser **12**.

Similarly, anode quarter sections **14b** and **16b** have a U-shape through their interconnection by end member **30b**. A small portion **14c** and **16c** of each anode plate member **14** and **16** is left attached to anode riser **12** by welds **18**. End portions **30a** and **30b** are then cut by any of the abovementioned processes to separate plate members **14a, 16a, 14b** and **16b.**

Following the removal of anode plate quarter sections **14a, 14b, 16a** and **16b,** these portions are rejoined as illustrated in Figure 5.2d by welding a new center section **14d** and **16d** to the corresponding anode quarter sections. Resistance welding is preferred for this splicing procedure since the parts are fused during the process resulting in a flat, planar repaired screen. It will be understood, however, that resistance welding is merely preferred and any other welding process may be utilized to accomplish this result.

Center members **14d** and **16d** are of anode material similar to that forming the original anode plates **14** and **16**. The anode plate members are now in a condition to be straightened, flattened and coated with electrocatalytic material, if necessary, prior to attachment to an anode riser.

Anode riser **12** having the attached residual pieces **14c** and **16c** is utilized as a mounting side for resilient anode expanders **32** which are common in the art of expandable anodes. Any common form of anode expander may be utilized, one anode expander being mounted at a point diametrically opposite the other anode expander **32** on anode riser **12**.

In accordance with a preferred embodiment (Figure 5.2e and Figure 5.2f) the anode riser **12** is oriented so that a plane **A** passing through welds **18** is disposed at an acute angle **a** relative to a plane **B** defined by the center axes of anode expanders **32**. This avoids interference between residual center portions **14c** and **16c** and an installed plate member when the anode is in a compressed condition. It will be understood that this is only a preferred step and that anode riser **12** may be oriented in any manner with respect to anode expanders **32**.

Finally, new and/or reassembled anode plate members **114** and **116** (Figure 5.2f) are attached as at welds **40** to the anode expanders to complete the expandable anode structure.

Valve Metals Coated with Doped Mixed Metal Oxides

G. Bianchi, V. de Nora, P. Gallone and A. Nidola; U.S. Patent 4,070,504; January 24, 1978; assigned to Diamond Shamrock Technologies, SA, Switzerland, describe chlorine-resistant metal electrodes, preferably of valve metals such as titanium and tantalum, having coatings of mixed metal oxides, preferably valve metal oxides and platinum group metal oxides, which have been doped to provide semiconducting surfaces on the electrodes. The coatings have the capacity to catalyze chlorine discharge from the electrodes and to resist corrosive conditions in a chlorine cell.

Example: Titanium trichloride in HCl solution is dissolved in methanol; the $TiCl_3$ is converted to the pertitanate by the addition of H_2O_2. This conversion is indicated by a change in color from $TiCl_3$ (purple) to Ti_2O_5 (orange). An excess of H_2O_2 is used to insure complete conversion to the pertitanate. Sufficient $RuCl_3 \cdot 3H_2O$ is dissolved in methanol to give the desired final ratio of TiO_2 to RuO_2.

The solution of pertitanic acid and ruthenium trichloride are mixed and the resulting solution is applied to both sides of a cleaned titanium anode surface and to the intermediate surfaces by brushing. The coating is applied as a series of coats with baking at about 350°C for 5 minutes between each coat. After a coating of the desired thickness or weight per unit of area has been applied, the deposit is given a final heat treatment at about 450°C for 15 minutes to 1 hour.

The molar ratio of TiO_2 to RuO_2 may vary from 1:1 $TiO_2:RuO_2$ to 10:1 $TiO_2:RuO_2$. The molar values correspond to 22.3:47 weight percent Ti:Ru and 51:10.8 weight percent Ti:Ru.

Base of Passivatable Metal with Covering Layer of Activating Substance

F. Brandmair, O. Rubisch and D. Honig; U.S. Patent 4,078,988; March 14, 1978; assigned to Sigri Elektrographit GmbH, Germany provide an electrode for electrochemical processes wherein the adhesion of the covering layer to the electrode base is so improved that reductions in the electrochemical activity of the electrode due to partial loosening or detachment of the covering layer are completely avoided.

The electrode comprises a base formed of passivatable material, and a covering layer of activating substance at least partly covering the base, the material of the base consisting of titanium oxide TiO_x, wherein x = 0.25 to 1.50.

The following steps are performed: mixing titanium powder and titanium dioxide powder in a ratio of 7:1 to 1:3, adding a binding agent thereto, compressing the resulting mixture and sintering it at temperature of 1200° to 1400°C in an argon atmosphere, and coating the thus compressed and sintered body with a covering layer containing an activating substance.

The covering layer contains at least one metal of the group platinum, palladium, iridium, ruthenium, osmium, rhodium, gold and silver or of a compound of these metals, such as an oxide, nitride or sulfide thereof. Suitable methods of applying the covering layer are, for example, precipitation from solutions, the spreading on of a suspension, galvanic deposition, plasma-spraying, flame-spraying or pyrolytic deposition from the gas phase. The covering layer which is baked or burned on by heating to about 300° to 600°C, should cover at least 5% of the surface of the electric base and should have a thickness of about 0.5 to 10 μm.

Example: Titanium powder with a grain size <0.06 μm and rutile TiO_2 powder with a grain size <0.01 μm were premixed in a high-speed blade mixer, 5 pbw of a 2% aqueous polyvinyl alcohol solution were added thereto, and the mixture was then mixed for an additional 10 minutes. The ratio of Ti powder to TiO_2 powder was 7:1 to 1:3.

The resultant mixture was compressed in a forging press at a pressure of 2 Mp/cm^2 (megaponds per square centimeter) into cylindrical members having a diameter of 100 mm and a height of 50 mm, which were initially dried at a temperature of 105°C and then heated and sintered in argon at 1250°C.

The cylinders were then provided by flame-spraying with a platinum layer having a mean thickness of about 5 μm, the adhesive strength of which was tested by quenching the cylinders that had been heated to 200°C in water of about 18°C.

In comparison, coated cylinders of oxygen-free titanium, after quenching only three to five times, already exhibited local cracks or ruptures in the covering layer; with cylinders having the composition TiO_x, wherein $0.25 < x < 0.42$ and wherein $0.60 < x < 1.50$, the first very small defects were able to be observed after quenching more than ten times; and the covering layer of cylindrical members of the composition TiO_x, wherein x = 0.42 to 0.60 remained free of defects even after being quenched twenty times.

A further advantage of members having an oxygen content of from 0.42 to 0.60 is the relatively low specific electrical resistance thereof, whereas members having an oxygen content $x > 1.50$ are little suited for electrodes because of their high electrical resistance.

Polymer-Laminated Surface

K. Motani, T. Sata and M. Nishimura; U.S. Patent 4,090,931; May 23, 1978; assigned to Tokuyama Soda KK, Japan describe an anode-structure for liquid-phase electrolysis comprising an anode and a polymer containing a cation exchange group, the polymer being laminated in the form of a film on one surface of the anode.

Example: A plain weave fabric of polytetrafluoroethylene was interposed between two 5-mil thick films of a copolymer of tetrafluoroethylene and perfluoro-(3,6-dioxa-4-methyl-7-octenesulfonyl fluoride) which has an ion exchange capacity corresponding to 0.91 meq/g of dry membrane (H^+ type, 1100 equivalent weight) in the hydrolyzed state, and by melt-adhesion under heat, made into a single film structure.

Furthermore, a 1.5 mil thick film of the same copolymer having an ion exchange capacity corresponding to 0.67 meq/g of dry weight (H^+ type, 1500 equivalent weight) in the hydrolyzed state was superimposed on the resulting structure and melt-adhered to form a single polymeric membranous product.

One surface of a titanium lath material was mechanically roughened, and the other surface was coated with 2 $RuO_2 \cdot TiO_2$ in such a way that the coating did not adhere to the roughened surface. A dispersion of polytetrafluoroethylene was coated on the roughened surface, dried, and heated to 370°C to coat the uncoated surface with the polytetrafluoroethylene.

That side of the resulting polymeric membranous product which had a higher exchange capacity was pressed into the anode by heating, whereby the titanium lath electrode entered the polymeric membranous product. The assembly was immersed in an 8% methanol solution of potassium hydroxide at room temperature for 48 hours to convert the sulfonyl fluoride group to a potassium sulfonate group to obtain a unitary structure of the cation exchange membrane and the insoluble anode.

A two-compartment electrolytic cell was built by combining this structure with a nickel-plated iron mesh cathode. In this case, that side of the anode-structure on which the anode was not exposed was placed facing the cathode and the distance between the cathode and the anode was adjusted to 3 mm.

Using this electrolytic cell, a saturated aqueous solution of sodium chloride was electrolyzed. Specifically, the saturated solution of sodium chloride was fed into the anode compartment at such a speed that the rate of decomposition would become 45%. Pure water was fed into the cathode compartment so that a 6.0 N aqueous solution of sodium hydroxide could be steadily obtained from the cathode compartment. The electrolyzing temperature was 85°C and the current density was 40 A/dm^2.

For comparison, the following experiment was conducted to illustrate the case where the anode was not laminated to the ion exchange membrane.

A woven fabric of polytetrafluoroethylene was interposed between two polymeric films, 5 mils thick and the same as those used hereinabove, having an ion exchange capacity of 0.91 meq/g of dry membrane (H^+ type), and melt-adhered under heat.

A 1.5 mils thick polymeric film having an ion exchange capacity of 0.67 meq/g of dry membrane (H^+ type) was melt-adhered under heat to the resulting assembly to form a single polymeric membranous product. The resulting product was immersed in an 8% methanol solution of potassium hydroxide to hydrolyze it and obtain a sulfonic acid-type cation exchange membrane.

Separately, an insoluble anode was produced by coating the entire surfaces of the same titanium lath material with $2RuO_2 \cdot TiO_2$. A filter-press type two-compartment electrolytic cell was built by interposing the ion exchange membrane between the insoluble anode and the same cathode as used hereinabove. The distance between the anode and the cathode was maintained at 3 mm, and the membrane was brought into contact with the anode by applying a water pressure of 100 mm.

The electrolyzing temperature, the current density, the decomposition ratio of saturated solution of sodium chloride, and the sodium hydroxide concentration in the cathode compartment were maintained the same as in the case of using the above laminated anode-structure.

This electrolysis was continued for 3 months by maintaining the conditions as constant as possible. When the electrolysis was performed for another one year by maintaining the conditions as constant as possible, the coating of the anode at the part of contact with the cation exchange membrane somewhat stripped off owing to friction with the membrane. The effective electricity-flowing area was 1 dm^2 in either case.

The cell voltage, the current efficiency, and sodium chloride concentration were measured, and the results are shown as follows.

	Current Efficiency (%)	NaCl in NaOH (ppm)*	Cell Voltage (V)
Disclosed anode structure			
At outset	84	40	3.85
3 months later	83	45	3.90
Filter-press type cell			
At outset	82	75	4.00
3 months later	80	80	4.25

*Calculated on 48% NaOH.

Anode with Membrane and Noble Metal or Oxide Coating

E.H. Cook, Jr. and G.R. Marks; U.S. Patent 4,100,050; July 11, 1978; assigned to Hooker Chemicals & Plastics Corp. disclose an anode structure for a chlor-alkali type diaphragm cell wherein the anode is a porous, valve metal, e.g., titanium, anode having on one side a coating of a noble metal or noble metal oxide and on the uncoated side a membrane composed essentially of a cation-active permselective material which is substantially impervious to liquids and gases. The membrane covered surface is disposed spaced from but facing the cathode member. Anodes of this structure give minimum cell voltage and a surprisingly reduced consumption rate of the noble metal or oxide coating.

Figure 5.3 shows a schematic view of the electrolysis cell **1**, which comprises an anode **2** and a cathode **3** separated by a cation-active permselective membrane **4** which is positioned on the face of anode **2** facing the cathode **3** and forming an anolyte compartment **13** and a catholyte compartment **14**.

The anode consists essentially of a porous section **15** coated on the back face, i.e., the face remote from the cathode, with a conductive coating **16** of a noble metal, noble metal alloy, or oxide thereof. Thereby the combination of the membrane **4**, porous anode section **15**, and noble metal or oxide coating **16**, forms a unitary element.

The cell **1** has an inlet **5** in the anode compartment **13** for the electrolyte, an outlet **10** for spent electrolyte, and an outlet **6** for gaseous product, e.g., chlorine, formed at the coated surface or face **16** of the anode.

There is also provided an inlet **7** for charging liquid such as water or dilute aqueous caustic soda, to the cathode compartment, an outlet **8** for discharging concentrated cathode liquor, such as concentrated caustic soda, from the cathode compartment and an outlet **9** for hydrogen gas.

Saturated electrolyte, e.g., brine, is continuously circulated in the anolyte compartment, by introducing the solution through inlet **5** and withdrawing the spent electrolyte through outlet **10** and directing it to replenishing zone **11** wherein the depleted solution is resaturated with, e.g., sodium chloride and acidified with acid if desired. The replenished electrolyte flows, via line **12**, to reenter cell **1** at inlet **5**.

The unitary structure of the porous anode and membrane wherein the anode is coated on the back face only functions to limit the formation of gaseous products on this coated surface.

Figure 5.3: Schematic View of a Two-Compartment Cell

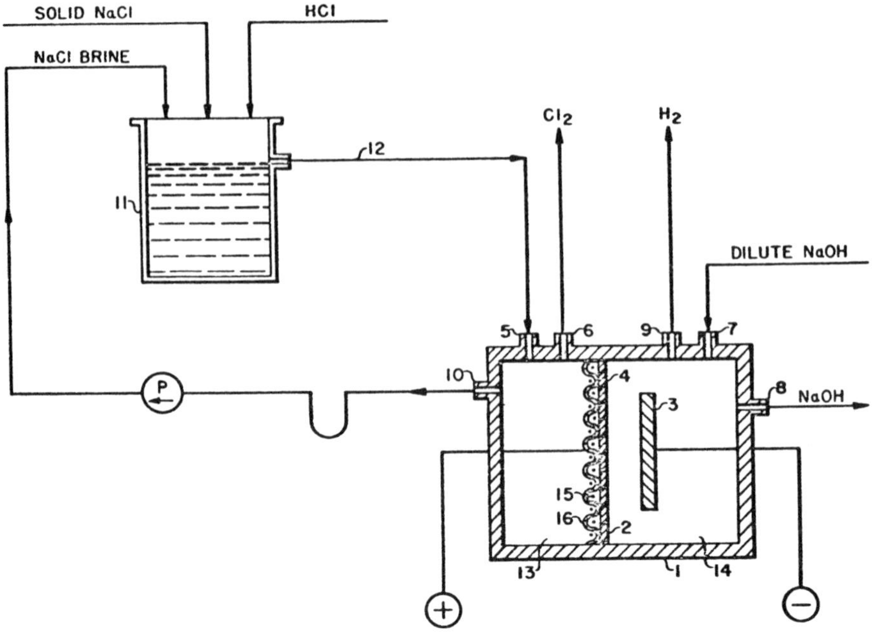

Source: U.S. Patent 4,100,050

Electrolysis does not occur on the uncoated surface. Gas bubbles are rapidly released from the back face of the anode and travel upwardly to the gas outlet. No gas is formed on the front uncoated surface of the porous anode and is prevented from collecting between the anode and membrane by the placement of the latter on the front face of the former. Accordingly, voltage increases, due to such entrapped gas do not occur.

Moreover, by positioning the noble metal coating on the back face of the porous anode section, relatively remote from the highly alkaline catholyte, i.e., remote from or outside of the diffusion layer of caustic liquor at the membrane surface the coating consumption is reduced to and maintained at a surprisingly low rate.

Example: A saturated solution of sodium chloride was continuously introduced into the anode compartment of a two-compartment electrolysis cell as illustrated in Figure 5.3. The anode member was a titanium anode of flat diamond mesh coated on the back face only with ruthenium oxide.

The front face of the anode was covered with a cation-active permselective membrane of 2.14 in^2 effective area composed of a hydrolyzed copolymer of tetrafluoroethylene and fluorosulfonated perfluorovinyl ether of equivalent weight of about 1100, prepared according to U.S. Patent 3,282,875. The membrane had been conditioned to the free acid form by soaking in boiling water prior to emplacement upon the uncoated surface of the porous titanium mesh anode.

The brine was circulated continuously within the anode compartment through a conduit in communication with the brine inlet and outlet. The cathode compartment was initially filled with dilute aqueous sodium hydroxide containing 50 g/ℓ NaOH. An overflow pipe for removal of concentrated caustic soda liquor was located in the cathode compartment, and the level of catholyte was maintained by the continuous addition of water to the cathode compartment.

Chlorine gas discharged at the anode was taken-off from the anode compartment through the gas vent pipe provided therein and hydrogen discharged at the cathode was similarly vented from the cathode compartment. A cell temperature of about 90°C was maintained in the cell which was operated at a current density of 2 A/in^2 of diaphragm. The cell was operated continuously under these conditions for four months.

For comparison a similarly constructed cell but containing a titanium mesh anode having a ruthenium oxide coating over all surfaces of the titanium was operated for a similar period under substantially the same conditions.

The coating consumption rate of the ruthenium oxide coating in the cell containing the coating on the back face only was 0.22 g of ruthenium per ton of chlorine produced. In the comparison cell containing the coating on both faces of the titanium mesh the consumption rate of ruthenium was found to be 2.0 g of ruthenium per ton of chlorine produced.

When an anode constructed of unflattened diamond shape titanium mesh and coated on the back face only, was used under similar operating conditions, the voltage increase was substantially that observed in the comparison cell.

Porous Anode Separator

Electrolysis of alkali metal chloride solutions to produce chlorine and alkali metal hydroxides is accomplished by *I.V. Kadija; U.S. Patent 4,120,772; Oct. 17, 1978; assigned to Olin Corporation* in a cell comprising an anode compartment, a cathode compartment and an anode separator which divides the anode compartment from the cathode compartment.

The anode separator is comprised of a porous plate of a valve metal having an electrochemically active coating on the face, and an electrochemically nonactive coating on the back and a portion of the interior. The electrochemically nonactive barrier layer may include resinous materials, for example, polyarylene compounds or polyolefin compounds.

The anode separator provides improved gas separation properties, eliminates the need for a separate diaphragm or membrane and enables the cell to operate with reduced power requirements.

Example: A commercially available porous titanium plate 1/16 inch thick and having a porosity of 60% and an average pore size of 25 μ was coated on one side with a thin protective coat of silicone rubber (RTV-102, General Electric

Co.). The silicone rubber penetrated the interior of the porous plate, but was prevented from coating the face of the plate. The rubber coated side was cured at room temperature over a 2 hour period. The face or uncoated side of the porous titanium plate was then painted with a 10% solution of $RuCl_4$ in 0.1 N HCl. The plate was then baked in an oven at 400°C for 5 minutes.

Following cooling, the face was recoated with the $RuCl_4$ solution and the porous plate then heated in an oven having an air atmosphere for about 6 hours at 400°C. During this heating, the silicone rubber coated titanium was oxidized and a mixture of silicon dioxide and titanium dioxide formed on the back and through the porous structure of the plate. An electrochemically active coating of ruthenium dioxide formed on the front of the plate.

Photomicrographs obtained using a scanning electron microscope established that the silicon dioxide was evenly distributed throughout the barrier layer as a mixture with titanium dioxide containing about 30% by volume of SiO_2. The barrier layer mixture covered about 50% of the interior structure of the porous plate.

The overpotential characteristics of the anode of the above example were determined by connecting the anode in an electrolytic cell containing a cathode, a reference electrode and sodium chloride as the electrolyte. The anode-cathode gap was about 1 mm. Electrolysis of the sodium chloride was conducted at the following current densities and the overpotential determined.

Current Density (kA/m^2)	Overpotential of Anode Separator of Example (millivolts)
0.1	35
1.0	55
3.0	75
5.0	95
10.0	125

The anode separator was thus shown to function as an anode in the electrolysis of sodium chloride.

In a related patent, electrolysis of alkali metal chloride solutions to produce chlorine and alkali metal hydroxides is accomplished by *I.V. Kadija and B.K. Ahn; U.S. Patent 4,066,519; January 3, 1978; assigned to Olin Corporation* in a cell comprising an anode compartment, a cathode compartment, a cation permeable divider separating the anode compartment from the cathode compartment, and the anode compartment contains a porous metal separator.

The porous metal separator is comprised of a porous plate of, for example, a valve metal having a porosity of from about 30 to 75% and an air flow value of from about 0.1 to 60 cfm. The anode separator is positioned in the anode compartment so it is spaced apart from the cation permeable divider and from the anode.

During electrolysis, an alkaline brine zone is formed between the porous metal separator and the cation permeable divider which increases the service life of the cation permeable divider. In addition, the porous metal separator provides improved chlorine gas separation properties.

Example: An electrolytic cell was assembled. The anode was a titanium mesh having an electroactive coating of ruthenium dioxide on the outside layer. A porous diaphragm of a perfluorosulfonic acid resin supported by a polytetrafluoroethylene fabric (Nafion Diaphragm 701, Du Pont) separated the anode compartment from the cathode compartment, which housed a foraminous steel cathode.

A porous titanium metal separator was installed in the cell between the mesh anode and the porous diaphragm. The Ti separator having a thickness of 1/16 inch, a porosity of 60 to 65% and an air flow value of about 5 cfm/ft^2 of porous plate, was spaced apart from the diaphragm 1/8 inch, and the space between the anode and the diaphragm was ½ inch.

Sodium chloride brine at a temperature of 80°C and having a concentration of 300 g/ℓ of NaCl was fed to the anode compartment to provide a head level of about 12 inches over the level of the catholyte in the cathode compartment. Electric current was supplied to the anode to provide a current density of 1.5 kA/m^2.

During operation of the cell, the pH of the brine in the area between the Ti separator and the diaphragm was measured periodically and found to be in the pH range of 14 where the NaOH concentration in the catholyte was 264 g/ℓ. Similarly, the pH of the brine in front of the anode was measured periodically and found to average about 4.7.

During operation of the cell over a period of 430 hours, favorable flow properties of the diaphragm were maintained without plugging of the diaphragm occurring. Current efficiency during the period of operation averaged 75% based on the production of caustic soda. The caustic liquor has an average sodium chloride concentration of 133 g/ℓ.

Comparative Example: The procedure of the above Example was repeated using the same cell apparatus with the exception that the porous titanium separator was removed. The cell was operated using brine at a concentration of 292 g/ℓ of NaCl in the anode compartment and employing a current density of 1.5 kA/m^2. Caustic liquor having an NaOH concentration of 93 to 137 g/ℓ was produced at a current efficiency of 64% over a period of 210 hours. After that period, the NaOH concentration rose rapidly to concentrations of 288 to 337 g/ℓ of NaOH indicating the diaphragm had become badly plugged. After 340 hours from start up, cell operation was discontinued.

Comparing the results from the Example with the Comparative Example shows that use of the porous Ti separator in the anode compartment provides an alkaline brine area adjacent to the porous diaphragm which results in increased cathode current efficiencies of 11%.

Louvered Anode

An anode developed by *T.W. Boulton; U.S. Patent 4,121,990; October 24, 1978; assigned to Imperial Chemical Industries Limited, England* is comprised of a pair of parallel slotted plates (e.g., louvered plates) mounted on a support, which are maintained at a fixed distance apart by means of a removable spacer adapted to engage the plates at their unsupported ends in combination with spacer studs located between the plates.

Referring to Figure 5.4a through Figure 5.4c, the anode comprises a pair of anode plates fabricated of titanium and each having three rows of vertical louvres **2**, one above the other. The louvres **2** are formed by pressing out with a slitting and forming tool from a single sheet of titanium (of a size corresponding to the overall dimensions of the anodes). The louvered anode plates are coated with an electrocatalytically active material, for example, a coating comprising ruthenium oxide and titanium oxide.

Each pair of anode plates **1** is resistance seam welded at their lower ends to a titanium bridge piece **3**. The titanium bridge piece **3** is argon-arc welded to titanium studs **4** previously capacitor discharge stud welded to a titanium sheet **5** which serves as the base plate of the cell.

The titanium base plate **5** is, in turn, conductively bonded to a mild steel slab (not shown) which serves as a conductor providing a low-resistance electrical flow path between the anodes **1** and copper connectors (not shown) bolted to a side edge of the mild steel slab.

The anode plates **1** are held apart in their correct position during use in the cell by means of a spacer **6** in the form of a removable comb or clip, preferably of plastics material such as polyvinylidene chloride, which fits over the upper ends of the plates, and by means of spacing studs **7**, for example, of titanium which are capacitor discharge stud welded to the back surface of one of the plates **1**.

Figure 5.4: Louvered Anode

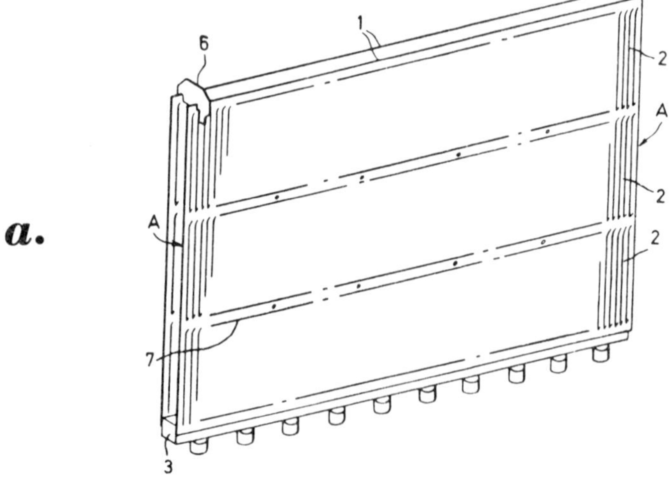

(continued)

Figure 5.4 (continued):

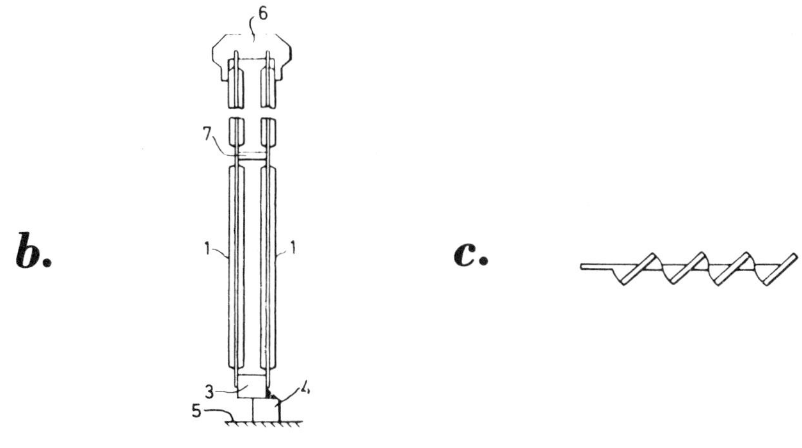

(a) Diagrammatic view
(b) End elevation
(c) Section along **A–A** of Figure 5.4a

Source: U.S. Patent 4,121,990

The spacer **6** may also be in the form of a clip comprising a loop of titanium wire. While assembling, the plates may, if desired, be protected at their upper edges by any convenient clamping means, for example, a cover strip, preferably of plastics material, in order to reduce the risk of damage to the diaphragms.

Example: A diaphragm cell was provided with one pair of louvered anodes according to the process, a mild steel gauze cathode and a polytetrafluoroethylene diaphragm and an anode/cathode gap of 6 mm. The polytetrafluoroethylene diaphragm was prepared by calendering a mixture of an aqueous polytetrafluoroethylene dispersion, titanium dioxide and starch, and subsequently removing the starch by electrolytic extraction in situ in the cell.

The cell was fed with sodium chloride brine (at a concentration of 310 g/ℓ). A current of 400 A was passed through the cell which corresponded to a current density of 2.0 kA/m^2 when compared with the effective area of the diaphragm. The cell operating voltage was 3.02 V. The chlorine produced contained 98.0% chlorine and less than 2.0% oxygen. The aqueous sodium hydroxide produced contained 10% by weight of NaOH. The cell operated at current efficiency of 96.08%.

Inclined Louvered Anode

As a modification of the preceding patent, *T.W. Boulton; U.S. Patent 4,141,814; February 27, 1979; assigned to Imperial Chemical Industries Limited, England* has developed an anode comprised of two groups of coated film-forming metal parallel elongated members lying in separate diverging planes which are connected to each other.

The plates **1** are inclined (as shown in Figure 5.5) so that the separation of the plates increases from the bottom of the anode to the top. The cell comprises a plurality of such anodes in parallel array (not shown) so that the anode/cathode gaps of adjacent anodes and cathodes decrease from the bottom of the anodes to the top of the anodes. This advantageously reduces the electrolytic resistance in the vicinity of the upper half of the anodes, thereby leading to a more even distribution of current.

A diaphragm cell using a pair of these anodes operated at a current efficiency of 96.0% under the conditions described in the preceding patent.

Figure 5.5: Inclined Louvered Anode

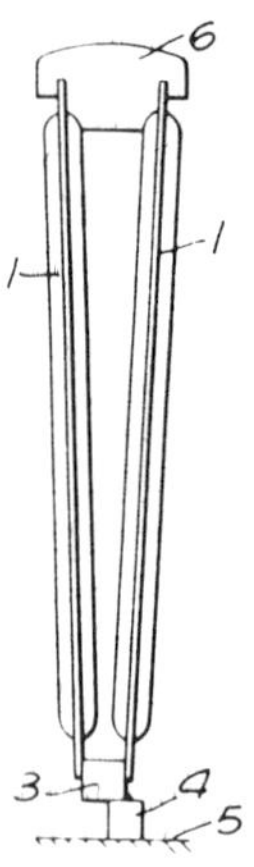

Source: U.S. Patent 4,141,814

Closed-Loop Flexible Wire Anode

An anode comprised of one or two groups of coated film-forming metal parallel elongated members lying in a plane, or in separate parallel planes connected to one another, is described by *T.W. Boulton; U.S. Patent 4,154,665; May 15, 1979; assigned to Imperial Chemical Industries Limited, England.* The members (e.g., wires) extend lengthwise from the point of connection and are resiliently mounted to give them flexibility.

Referring to Figure 5.6a and 5.6b, the anode assembly comprises a plurality of pairs of wires **1**, each of which is provided at its lower end with a loop **2** and wherein each wire is joined at its upper end to form a closed loop **2'**, the loops **2** and **2'** imparting flexibility.

The wires **1** are arranged one behind another to form effectively two parallel rows of anodes, each row being adjacent to a diaphragm **3** and a cathode **4**.

Figure 5.6: Closed-Loop Flexible Wire Anode

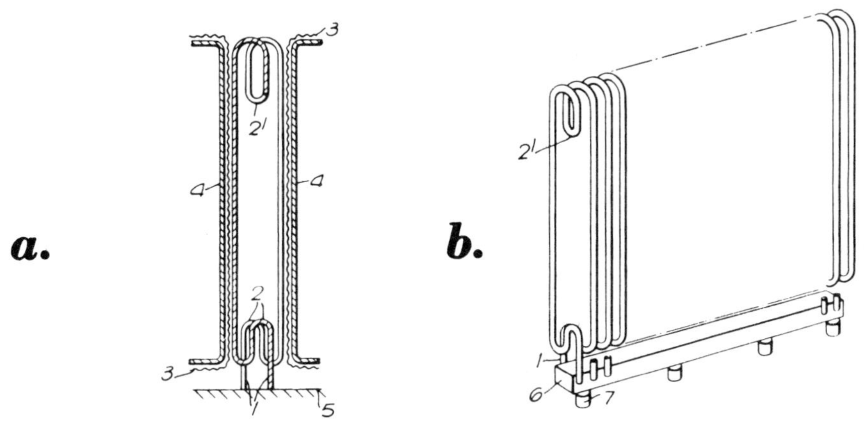

(a) Sectional end elevation
(b) Diagrammatic view

Source: U.S. Patent 4,154,665

Example: A titanium wire (3 mm diameter) anode, coated with a mixture of ruthenium oxide and titanium dioxide, was assembled in a vertical laboratory diaphragm cell. The cell was provided with a mild steel gauze cathode and a polytetrafluoroethylene diaphragm. The anode/cathode gap was zero and the diaphragm was in contact with the cathode.

The cell was fed with sodium chloride brine (300 g/ℓ NaCl), at a rate of 6 ℓ/hr and a current of 395 A (equivalent to a current density of 2.0 kA/m^2) was passed through the cell. The cell operating voltage was 2.8 V. The chlorine produced contained 98% by weight of Cl_2 and less than 0.1% by weight of H_2. The sodium hydroxide produced contained 9% by weight of NaOH. The cell operated at a current efficiency of 96%.

All-Welded Anode Assembly

F. Smith; U.S. Patent 4,078,986; March 14, 1978; assigned to Imperial Chemical Industries Limited, England provides an anode assembly of an electrically-conductive metal which is resistant to the electrolyte used in the cell, a plurality of metal anodes mounted on and attached in electrical contact with the upper surface of the base plate and a current lead-in member or members electrically and mechanically-bonded to the undersurface of the base plate by means of a plurality of electrically-conducting stud members each of which is connected by welding at one end to the undersurface of the base plate and is connected at the other end by welding or other means to the current lead-in member or members.

Referring to Figure 5.7, a plurality of sheet anodes **1** fabricated of titanium and provided with an electrocatalytically-active coating are fillet-welded to a series of parallel vertically disposed spaced apart titanium anode support ribs **2**.

Figure 5.7: All-Welded Anode Assembly

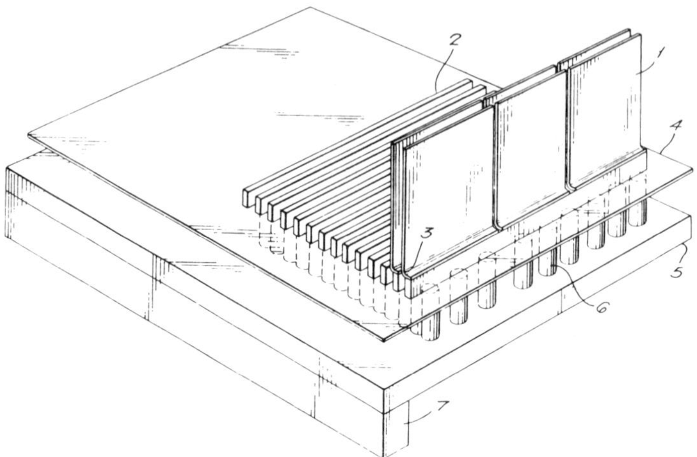

Source: U.S. Patent 4,078,986

The anodes **1** are folded over at their lower ends **3** so that the welding of the anodes **1** to the support ribs **2** will not distort the anode sheets. The support ribs **2** are argon-arc welded along their lengths to a titanium sheet **4** which serves as the base plate of the cell. The titanium base plate **4** is connected to a current lead-in member in the form of a slotted aluminum plate **5** by a plurality of aluminum studs **6**.

The aluminum studs **6** are friction-welded at their top ends to the underside of the titanium base plate **4**. At their lower ends the aluminum studs **6** are sunk into and hand-welded to the aluminum plate **5**. The aluminum studs **6** and the aluminum plate **5** serve as the means for providing a low-resistance electrical path between the anodes **1** and a source of electricity. Aluminum conductor bars **7** are welded to the underside of the aluminum plate **5**. Connector tapes or flexibles (not shown) are, in turn, connected to the aluminum bars **7** and the source of electricity.

In order to prevent distortion of the anodes **1** during welding they may be provided with ribbing below the coated area. The titanium base plate **4** can be provided with drop-edges in order to stiffen the plate. The base plate **4** is provided with a drainage hole or holes (not shown). The anode assembly is conveniently supported on a mild steel frame (not shown).

An important advantage of the abovedescribed all-welded anode assembly is that single anodes can be removed from the assembly with ease simply by cutting the base of the anode **1** free from the support rib **2**. The technique of welding the anode to the support member is particularly advantageous in the case of non-planar anodes.

Bonding of Base Plate to Steel Slab

F. Smith; U.S. Patent 4,116,802; September 26, 1978; assigned to Imperial Chemical Industries Limited, England has found a particularly suitable method of achieving an efficient mechanical and electrical bond between a base plate of an electrolytic cell and a metallic conductor.

Referring to Figure 5.8a, the assembly comprises a plurality of pairs of sheet anodes **1**, e.g., louvered sheet anodes, fabricated of titanium and provided with an electrocatalytically active coating, for example, ruthenium oxide and titanium dioxide, which are resistance seam welded at their lower ends to titanium bridge pieces **2**. The bridge pieces **2** are argon-arc welded to titanium studs **3** which are, in turn, capacitor discharge stud welded to a titanium sheet **4** which serves as the base plate of the assembly.

The titanium base plate **4** is bonded to a mild steel slab **5**, using the apparatus shown in Figure 5.8b (as described below), and preferably using a silver/copper alloy as the metallic interlayer. The mild steel slab serves as a conductor providing a low-resistance electrical flow path between the anodes **1** and copper connectors (not shown) bolted to a side edge of the mild steel slab **5**. The titanium base plate is provided with a drainage hole or holes (not shown).

Figure 5.8: Bonding of Base Plate to Steel Slab

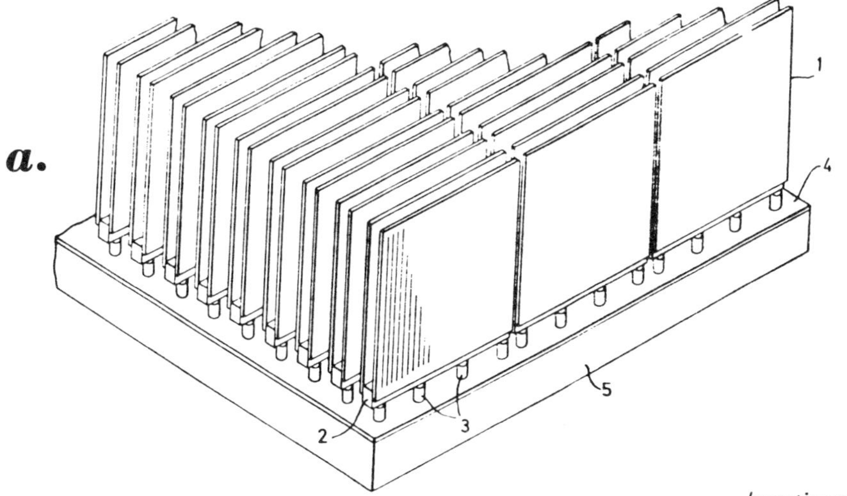

(continued)

Figure 5.8 (continued):

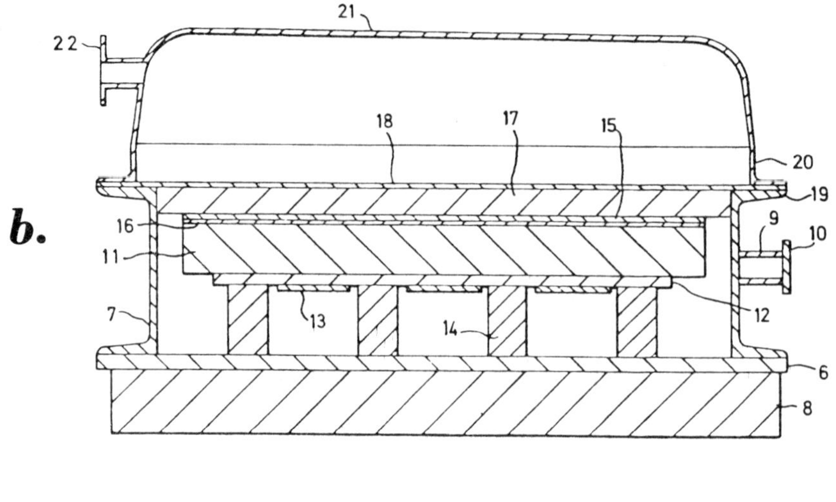

(a) Anode assembly
(b) Bonding apparatus

Source: U.S. Patent 4,116,802

The bonding apparatus (shown in Figure 5.8b) comprises a mild steel vessel having a base **6** and sidewalls **7**. The vessel is supported on a mild steel frame **8**. The sidewalls **7** are provided with an exit pipe **9** fitted with a flange **10** for connection to a vacuum pump (not shown). A mild steel slab **11** (part of the cell base) is placed on a mild steel platen **12** provided with heating elements **13**, and the slab **11** and platen **12** are supported on legs **14** of insulating material.

A titanium plate **15**, forming the cell base plate, is placed on the upper surface of the mild steel slab **11**, and a metallic interlayer **16**, for example, a silver alloy, is interposed between the titanium plate **15** and the mild steel slab **11**. The titanium plate **15** is covered with a layer of insulating material **17**, which also acts as a uniform load spreader, and the vessel is covered with a flexible metal diaphragm **18** which is sealed between a flange **19** on the sidewalls **7** and a flanged cover **20**. If desired, the flanged cover **20** may be provided with a dome-shaped top **21** having a flanged exit pipe **22**.

The vessel was evacuated to a pressure in the range 10^{-1} to 10^{-5} mm, e.g., 10^{-3} mm, of mercury, corresponding to a loading of about 15 psi on the upper surface of the titanium plate **15** (as exerted through the membrane **18** and the insulating material **17**). The mild steel slab **11** was heated to a temperature in the range 20° to 100°C, e.g., about 50°C, above the melting point of the metallic interlayer **16**. Typical temperatures were as follows:

Temperature, °C	Metallic Interlayer
950°	93% Ag, 7% Cu
830°	72% Ag, 28% Cu
910°	85% Ag, 15% In
870°	95% Ag, 5% Al

The temperature and vacuum were maintained for a period of 60 minutes and the titanium base plate bonded to the mild steel slab was then allowed to cool and was then removed from the vessel. In each case the titanium base plate was found to be firmly bonded to the mild steel slab. Examination of the bond showed that liquid metal accelerated diffusion bonding had taken place.

If desired, extra pressure may be exerted on the membrane **18** and thence via the insulating material **17** onto the titanium plate **15**, by applying fluid pressure through the pipe **22**, for example, by introducing compressed air through the pipe **22**.

Anode Between Membranes

An anode unit is provided by *G.O. Westerlund; U.S. Patent 4,069,128; Jan. 17, 1978; assigned to Gow Enterprises Limited, Canada* which comprises a generally rectangular parallelepiped framework including a pair of spaced-apart gridlike structurally rigid peripheral outer sidewalls; an ionically conductive gas impermeable membrane secured against the inner face of each such outer peripheral walls; an anode sealingly disposed between the membranes to provide a pair of fluid-tight anode compartments; means for feeding anolyte to the anode compartments; and means for withdrawing spent anolyte and entrained and/or occluded gaseous products of electrolysis therein from the anode unit.

As shown in Figures 5.9a and 5.9b, the anode/membrane/cathode unit **10** is a hollow rectangular parallelepiped including two outer sidewalls comprising generally rectangular structural supports **11** which are generally of the same rectangular shape as the unit **10**. It is essential that the structural supports **11** be porous. The structural sheeting **11** employed has openings, or perforations, **12** or is provided as expanded sheeting or porous fused powder sheeting. Plastic sheeting is possible, but titanium is preferred because of its structural advantages and best performance.

Disposed against each of the inside faces of each of the structural sheetings **11** is an internal membrane **13**. The membrane provides an ionically conductive impermeable barrier. One particularly effective material for the membrane **13** is Nafion film (DuPont), 1 to 20 mils thick.

Sealingly held centrally within the hollow rectangular parallelepiped is an anode **14** providing a pair of anode compartments **15**. One end of the anode **14** is set within the spaced-apart flanges **16** of a vertically positioned anolyte recirculation channel member **17**.

While not shown, the flanges **16** at the top and bottom of the channel member **17** are each provided with perforations to enable anolyte flow. The channel member **17** is held against, but maintained in liquid-tight relationship to, membrane **13**, by means of gasket **18**.

Figure 5.9: Anode/Membrane/Cathode Unit

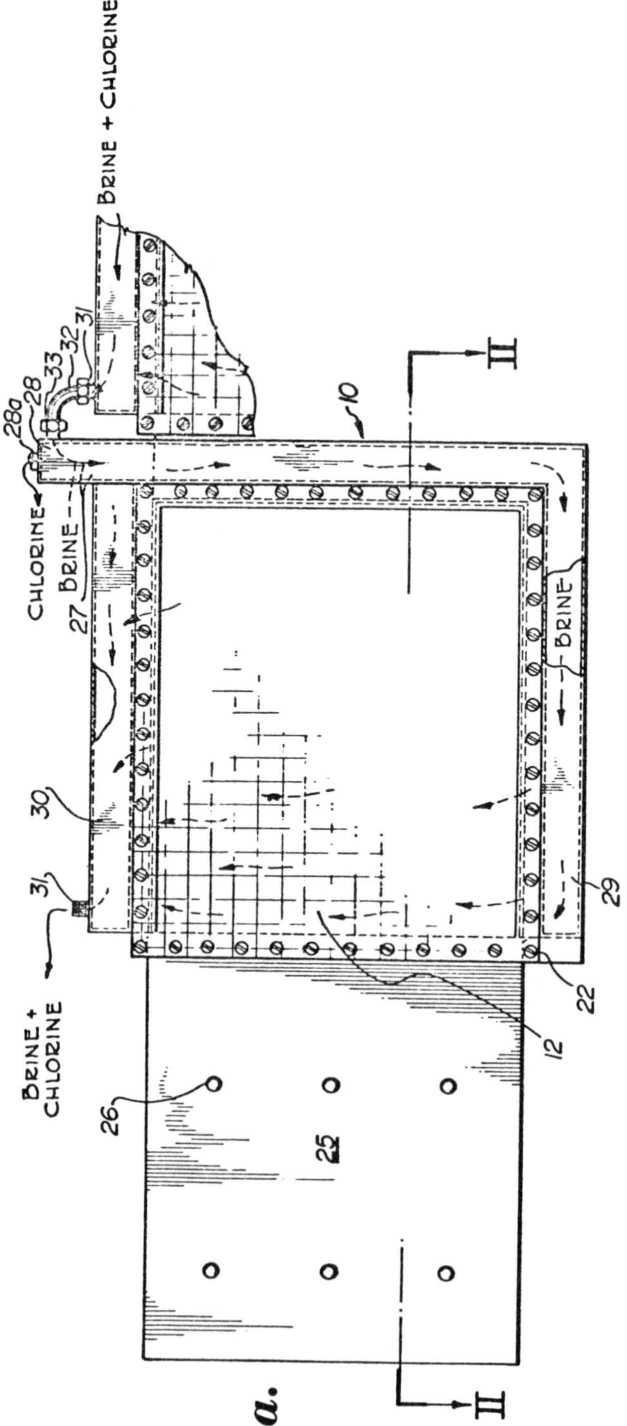

(continued)

Figure 5.9 (continued):

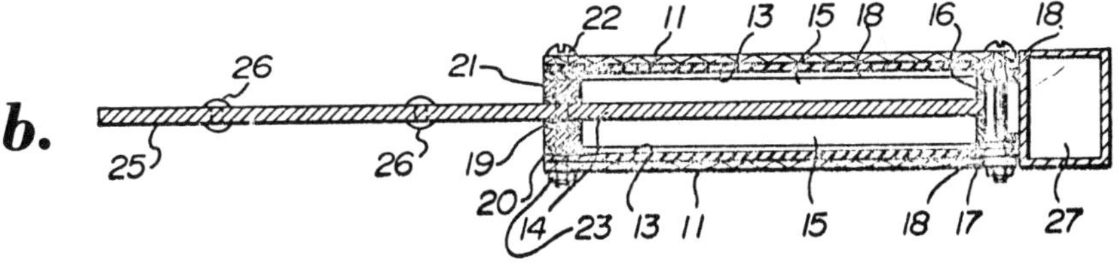

(a) Side elevational view
(b) Section along line II–II of Figure 5.9a

Source: U.S. Patent 4,069,128

The other end of the anode **14** projects beyond the anolyte chambers **15** and is held in liquid-tight contact by means of gaskets **19** in contact with the anode **14**, and **20** in contact with the membrane **13**, as well as suitable spacers **21**. The anode molecule section is held together by a plurality of peripherally spaced bolt **22** and nut **23** units.

The portion of the anode **14** which protrudes from the anode compartments **15** of the module **10** is the cathode **25**. The cathode **25** preferably is titanium but may also be other cathodic material, e.g., steel; in such case, there should be an overlap at the bolt joint. The cathodes **25** are equipped with plastic buttons or rivets **26**. Suitable materials include Teflon, polyvinyl chloride (PVC), etc. The purpose of these buttons is to space the cathode sheet from an adjacent module to ensure that they do not short circuit.

Provided on the upstream side of the anode unit **10** is a downcomer channel support anolyte inlet **27**. Anolyte inlet **27** has a height greater than the height of the anode unit **10**, and so provides a space **28** for gas/liquid separation. The gas is drawn off via anolyte gas outlet **28a**, while the liquor is led to a horizontal distributor base trough **29** leading freely to the anode compartments **15**. The anolyte rises by gas lift to an upper header **30** which is provided with a common liquor and gas outlet pipe **31**. This may be connected to an elbow **32** to an inlet **33** to anolyte inlet **27**.

In one embodiment, the porous structural support was a perforated 0.3 mm thick titanium sheet, 2 mm holes about 45% open area. This gave a differential voltage increase of 0.1 V at 1500 A/m^2. A plastic sheet would have to be thicker (up to 1 mm) and voltage increase would become significant (0.5 and higher). 2 mm thick sheeting is about the maximum to achieve reasonable voltage readings; 1 mm is preferred and 0.2 mm would be the extreme minimum due to the low mechanical strength.

For thicker sheets it is important electrically to short circuit this membrane with the cathode/anode, otherwise the cell voltage could significantly increase (probably gas build-up inside the plate if partially working as a cathode).

The membrane may be of different types. The first type is a film about 0.03 to 0.3 mm thick; this film is the membrane without additives. This film is lower in voltage but has less mechanical strength and readily swells with very large width and length linear dimensional increase at the normal cell conditions up to 20%. Thus, when used in a dimensionally stable frame the film, after a few days, appears as a corrugated mesh (i.e., expansion occurs in both directions) which could seriously upset good flow conditions as well as seal itself against the electrodes in some areas which also contributes to increased voltage. This uncontrolled dimensional swelling also lowers the life cycle of the film.

The second type is a film as above but reinforced with Teflon fabric. This does mechanically improve the film as well as lower its linear dimensional increase to less than 5%; however, even this is large. Furthermore, the cell voltage is significantly increased by the resistance of Teflon fabric (up to 1 V at 1600 A/m^2).

In a further embodiment in producing the membrane, Nafion powder, 0.2 to 0.5 mm in size, was ground under liquid nitrogen freezing conditions. Porous powder titanium sheeting (70% porosity, 10 to 30 μ pore size) was plugged with this powder and successfully operated as a dimensionally stable membrane, i.e., in this case the module would not require any structural member.

Fuel-Cell-Type Spaced Porous Catalytic Anode Providing pH Control

An improved process and apparatus for pH control and energy savings in chlor-alkali electrolysis cells is disclosed by *W.A. McRae; U.S. Patent 4,217,186; August 12, 1980; assigned to Ionics Inc.*

A fuel-cell-type spaced porous catalytic anode is utilized to chemically oxidize a controlled, substoichiometric amount of hydrogen to provide hydrogen ions to a recirculating anolyte. The pH is monitored and the flow of hydrogen fuel adjusted to provide a resultant desired pH in the range of about 2 to 4. Optionally, hydrogen gas produced at the cell cathode may comprise the fuel supply and a spaced porous catalytic cathode may be employed for hydrogen supply control and depolarization.

Figure 5.10 shows a schematic representation of an electrolysis cell **10** comprising an anolyte compartment **12** and a catholyte compartment **14** separated by a cation permselective membrane **16**. Anode **18** is comprised of a spaced porous material such as graphite or titanium having a catalyst such as platinum or ruthenium oxide deposited thereon. Cathode **20** may be a conventional steel or nickel cathode or optionally a spaced porous type such as porous carbon having a silver oxide or colloidal platinum catalyst.

The membrane may be composed of a conventional cation exchange membrane such as is well-known in the art or preferably a perfluorinated carboxylic or sulfonic acid type such as Nafion (DuPont). A voltage is impressed on the electrodes through lines **22** and **24** from a source not shown.

The anolyte (a concentrated substantially saturated brine solution) may be constantly recirculated and replenished by means **26** shown schematically and composed of apparatus as would be obvious to those skilled in the art or passed through the anolyte compartment on a once-through basis.

Figure 5.10: Electrolysis Cell Providing pH Control

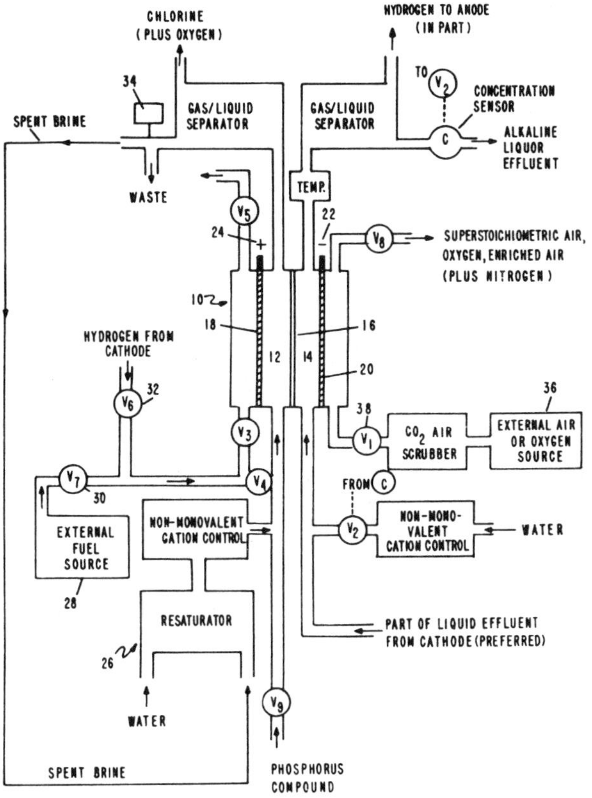

Source: U.S. Patent 4,217,186

Example 1: This example illustrates a preferred operation but without pH control of the anolyte. An electrolyte cell is constructed in accordance with Figure 5.10.

The membrane is a perfluorosulfonic acid type (Nafion, DuPont) and consists of a thin skin having an equivalent weight of about 1350 laminated to a substrate having an equivalent weight of about 1100. The membrane is reinforced with a woven polyperfluorocarbon fabric (Teflon, DuPont).

The effective area of the membrane is about 1 dm^2. A perfluorocarboxylic acid membrane, such as that manufactured by the Asahi Chemical Industry Co. of Tokyo may also be used.

The cathode is woven nickel wire mesh; the anode is a woven titanium wire mesh which has been coated on the face adjacent to the membrane with several layers of finely divided ruthenium oxide powder, baked at an elevated temperature to promote adhesion to the mesh as is well-known in the art.

The electrodes also have apparent areas of about 1 dm^2. The electrodes are spaced from the membrane to permit gas evolution and disengagement. Sodium chloride brine, substantially saturated, is fed to the anode compartment at a rate of about 300 cm^3/hr.

The effluent from the anode compartment is separated into a gas stream and a liquid stream. From about 1 to 10% of the effluent liquid stream is sent to waste; the remainder with additional water is resaturated with salt and used as feed to the anode compartment.

About 5% sodium hydroxide is fed to the cathode compartment. The feed rate is adjusted to produce an effluent from the cathode compartment having a concentration of about 10%.

The effluent from the cathode compartment is also separated into a gas stream and a liquid stream. Part of the liquid stream is diluted with water and used as feed to the cathode compartment.

After the flows to the electrode compartments have been established, a direct current of about 25 A is imposed on the cell. After several hours, the voltage of the cell stabilizes at about 4.5 V. The temperature of the effluents from the cell are adjusted to about 80°C by controlling the temperatures of the feeds to the electrodes.

The gas stream separated from the effluent from the anode compartment is analyzed by adsorption in cold sodium hydroxide and titration of the latter for available chlorine.

The current efficiency for chlorine evolution is found to be about 85%. The pH of the liquid stream separated from the effluent from the anode compartment is found to be substantially greater than 4.

Example 2: This example illustrates the improvements which can be obtained from a preferred embodiment using anolyte pH control. The cell of Example 1 was used.

The cell is operated as described in Example 1 except part of the gas separated from the effluent from the cathode compartment is admitted to the brine feed to the anode compartment. The rate of admission of the gas (substantially pure, but humid hydrogen) is adjusted to maintain the pH of the liquid separated from the effluent from the anode compartment in the range of from about 2 to 4. After several hours the voltage of the cell stabilizes at about 4.5 V.

The gas stream separated from the effluent from the anode compartment is analyzed as described in Example 1. The efficiency for chlorine evolution is found to be in the range of about 90 to 95%, higher values being associated with low pHs in the range.

Removable and Replaceable Dimensionally Stable Anodes

O. De Nora, G. Bianchi, and G. Meneghini; U.S. Patent 4,162,953; July 31, 1979; assigned to Oronzio De Nora Impianti Elettrochimici SpA, Italy describe a monopolar diaphragm electrolytic cell with dimensionally stable anodes in which the anodes rest freely in the cell and are spring-pressed toward the diaphragms by spring-loaded transverse arms on the positive current carriers, which in use contact, but are not mechanically connected to, the anodes.

The spring-pressed electrical contacts between the transverse arms and the anodes are sufficient to carry current to the anodes without substantial ohmic drop through these contacts and permit the anodes to be removed from the cells for recoating and other purposes without destroying any welds or other permanent mechanical connections between the anodes and other portions of the cell.

The cell illustrated in Figure 5.11a comprises a bottom **1** of steel, copper or other structural metal. The inner surface of the bottom is lined with a titanium or other valve metal sheet **20** (Figure 5.11b). The bottom constitutes the positive plate for distribution of current to the anodes of the cell and is connected by terminals **2** to the positive bars of the power distribution network. The cathodic box or cell can, preferably of steel, generically indicated by **3**, rests on the bottom. An insulating gasket electrically insulates the bottom of the cathodic box from the cell bottom and provides a hydraulic seal between the cathodic box and the cell bottom.

The cathodic box contains a mesh lining of low-carbon steel and includes a series of parallel tubular cathodes or fingers **4** welded at both ends to the side network **5**, which is welded all around to the inner surface of the cathodic box or can, at a distance from the inner surface of the box so as to provide a space **6**, which, together with the space inside the various tubular cathodes **4**, constitutes the cathodic compartment of the cell. The cathode box or cell can is connected by means of the terminals **8** to the negative bars of the power distribution network.

On the cathodes and on the network sides **5** of the cathodic box, the diaphragm, typically of asbestos fiber, is deposited by the known techniques, which essentially consist of sucking a suspension of asbestos fibers through the meshes of the network constituting the tubular cathodes and side wall, to deposit the asbestos on this network.

The bottom of the cell, of conductive metal (preferably copper or iron), is covered with a sheet **20** of titanium (Figure 5.11b). The current carriers or risers **9** of titanium or titanium-coated metal are permanently secured on the bottom of the cell and the base flanges **21** of titanium risers or current carriers **9** are welded to the titanium sheet **20**.

Two transverse arms **22** of titanium are welded on each side of each current carrier and at least two vertical contacts **23** of titanium (preferably platinum-plated on the contact surface) are welded on the two transverse arms.

The transverse arms are further provided with guides **24** for the insertion of suitable spreaders or retention elements, not shown.

Figure 5.11: Monopolar Diaphragm Cell

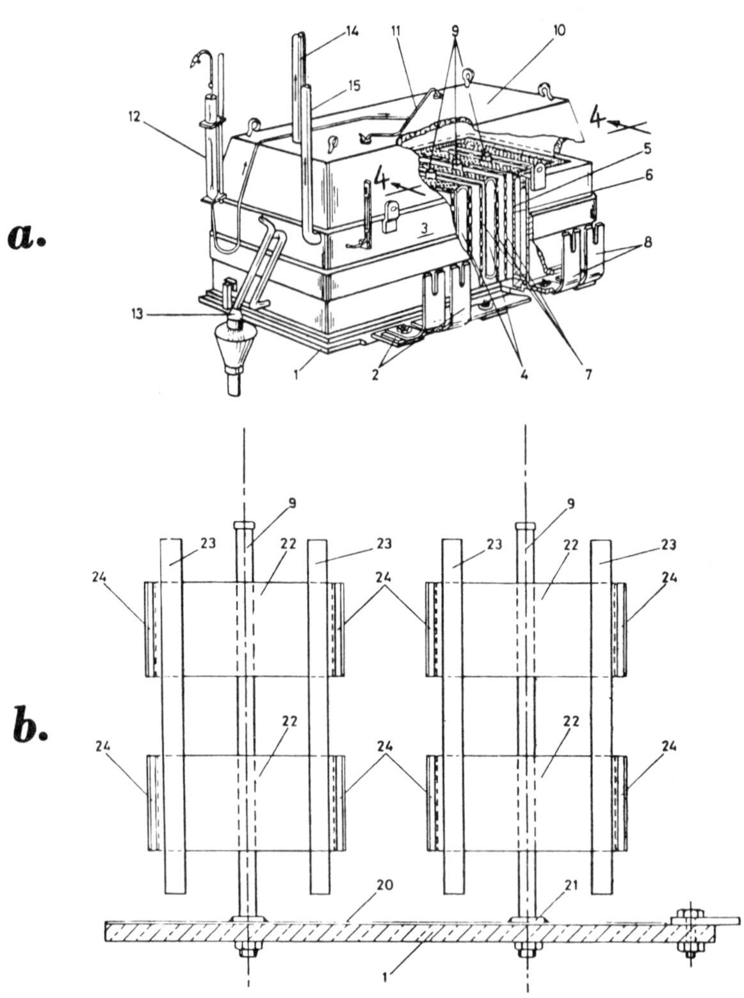

(a) Perspective view, partially broken away

(b) Symbolic view adjacent one end of the bottom of the cell

Source: U.S. Patent 4,162,953

The anodes **7** of the cell consist of expanded sheets or mesh of titanium, coated with an electrocatalytic coating preferably containing oxides of metals of the platinum group together with oxides of the valve metals. The anodes rest freely on the titanium sheet **20** at the bottom of the cell and are pressed toward the surfaces of the tubular cathodes, which are covered with diaphragms, by the action of the transverse arms welded on the current carriers **9**, which arms are preferably spread apart by the use of expanders. The transverse arms **22** are provided with two or more contacts **23** preferably extending over the entire height of the anodes. These contacts press against the inner surface of the anodes and distribute the electric current to the anodes. The cell is completed by a cover **10**, preferably of nonconductive plastic material, which rests on the upper flange of the cathodic box **3**.

The cell is filled with electrolyte (for example, concentrated sodium chloride brine) so that the electrode structures are completely immersed, and the level is maintained by adding electrolyte through the distributor **11** connected to a level indicator **12**.

The electrolyte percolates through the diaphragm and collects in the cathodic compartment, from which it is discharged together with the sodium hydroxide formed, through the outlet **13**. The position of the electrolyte outlet is adjustable so as to create and regulate the difference of hydraulic pressure through the diaphragm necessary to maintain the desired flow of electrolyte through the diaphragm.

The chlorine evolved on the anodes collects in the upper space of the cover above the electrolyte gas and flows through the outlet **14** to the chlorine recovery system, and the hydrogen evolved on the cathodes collects in the upper part of the cathodic compartment and flows through the outlet **15**.

Compressible Self-Guiding Electrode Assembly

An electrode assembly for an electrolysis cell is provided by *J.E. Ridgway; U.S. Patent 4,120,773; October 17, 1978; assigned to Hooker Chemicals & Plastics Corp.,* comprising at least two opposed electrode working faces, and placing between the working faces a resilient, compressible member against which compression of the electrode takes place during its insertion into the cell.

Upon inserting the electrode assembly into a smaller opening within the cell, which opening is defined by the adjacent electrodes, compression of the resilient, compressible elastic member through insertion of the electrode assembly into the opening takes place and automatically guides the assembly such that its working faces are correctly positioned relative to the adjacent electrodes. The degree of compression of each electrode assembly is determined by spacer assembly guides or appendages mounted on the electrode face. These guides determine the gap between the electrode face and adjacent electrodes.

Reference will be made hereinafter almost exclusively to the provision of a compressible anode assembly for use in electrolysis in a diaphragm cell. It should be understood, however, that while described as an anode structure, where conditions warrant the electrode assembly may also be used as a cathode, or in some instances, as a cathode and as an anode.

Figure 5.12a, which is a top view of an electrode assembly, contains two plastic frames **12** and one or more elastic devices **20**.

Figure 5.12b is a cross-sectional side view of an anode assembly and illustrates the spacer alignment guides of the leading **31** and trailing **30** edges of the frames. The leading edge of the frame, that is, the portion of the assembly which enters the cell first, and the trailing edge would be the opposite end of the assembly which enters the cell last. The spacer alignment guides, which are the appendages to the frame **12**, will have a slanted portion or chamfer to assist in insertion, and the slope of the chamfer would define, upon complete insertion of the frame into the cell, the gap between anode **10** and the adjacent electrode of the cell. Spacer alignment guides **30** and **31** will have a shape that will allow for the easy insertion of the electrode assembly into the cell. The shape is dependent upon whether or not the electrode assembly is placed into the cell prior to the installation of the opposite electrode, or after the installation of the opposite electrode unit and could be chamfered at the top or bottom edge, be spherical or elliptical in cross section.

In Figure 5.12b, the alignment guides are shown for the situation in which the cathode assembly is already in the cell and the anode assembly is to be placed into the cell after the cathode is positioned. The spacer alignment guides serve two functions; one to aid in insertion and the compression of the assembly into the cell, and the other, to provide the gap between anode and cathode surfaces.

Figure 5.12c illustrates the insertion of the electrode assembly into the cell. The slanted portion on spacer assembly guide **31** will assist in the initial insertion into the cell, and guide **31** will bear upon the cathode structure **42** as it goes into the cell.

Figure 5.12: Compressible Self-Guiding Electrode Assembly

a.

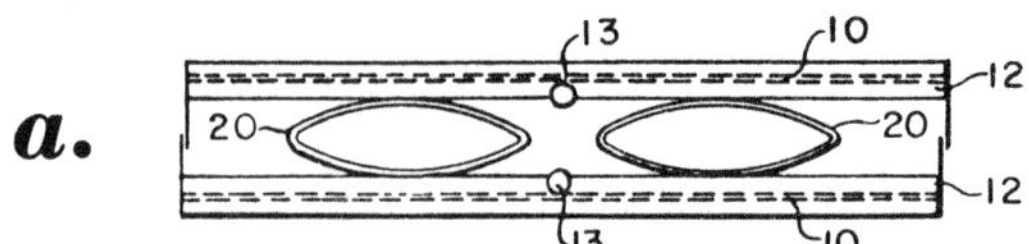

b.

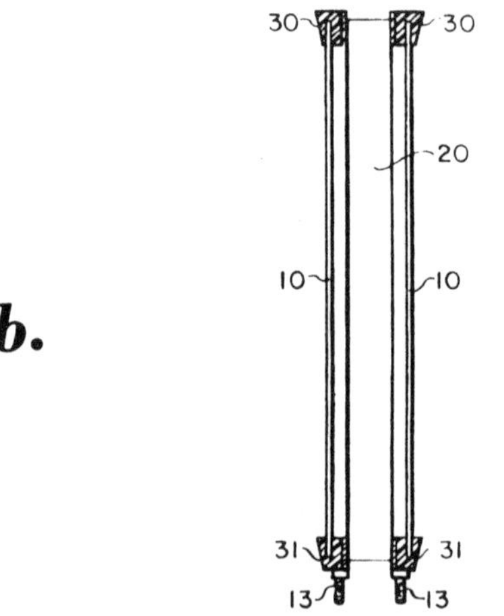

(continued)

Figure 5.12: (continued)

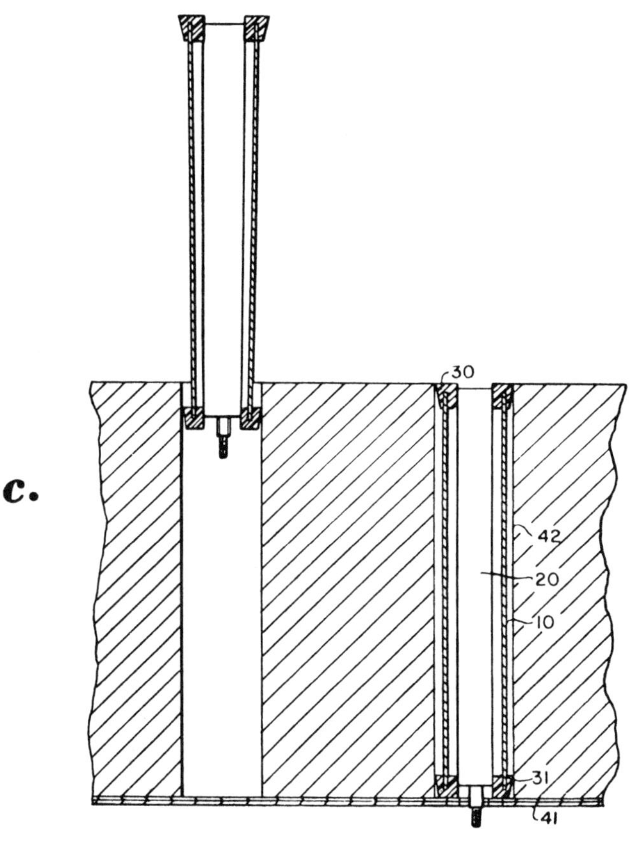

(a) Top view
(b) Cross-sectional side view
(c) Cross-sectional view showing insertion of electrode assembly into cell

Source: U.S. Patent 4,120,773

As the trailing edge of the frame with its slanted guide assembly portion **30** enters the cell, the leading edge of guide **31** is near the base of the cell **41** and the insertion of the assembly is completed with the frame resting on the cell base. The electrode **10** will now be aligned in its proper position relative to the cathode surface **42**. The cathode surface may be a simple cathode surface or as is more usually the case will be a diaphragm which has been deposited on the cathode and surface **42** will then be the diaphragm which has been deposited.

The degree of compression necessary for insertion will be determined by the elastic devices **20** within the electrode assembly. The degree of compression necessary should be such that the force being placed upon the leading guide assembly **31** will not damage the surface of the diaphragm **42** but be sufficient to provide, upon complete insertion of the electrode assembly, sufficient force against the frame **12** that electrodes **10** are held firmly in place to provide the proper gap distance.

After the anode assembly has been inserted into the cell, electrical contact between the anode surfaces **10** and the electrical circuit of the cell can be made and completed for later operation of the cell.

Connection Means for Anode Posts

An improved means for connecting anode posts and conductors to diaphragm cells is described by *P.A. Danna and R.D. Burt; U.S. Patent 4,070,266; Jan. 24, 1978; assigned to Olin Corporation*, which utilizes a conductor comprised of a pair of split and spaced-apart copper bars having a series of pairs of adjacent round grooves, each pair of grooves being adapted to receive an anode post. The pair of bars is firmly secured to the anode posts by bolting each pair of bars together.

A laminated cell base is employed comprised of a nonmetallic corrosion resistant upper layer, such as rubber, secured to a lower metallic supporting layer such as steel.

Sealing means are provided between a dished washer on the anode post and the upper layer, and between the anode post and wall of the cell base hole in the lower layer.

Example: A diaphragm cell of the type disclosed in U.S. Patent 3,485,730 issued to C.W. Virgil, Jr. on December 23, 1969, was modified to include conducting means and anode attachment means.

A cell base having an overall dimension of 63 inches by 56¼ inches was constructed of a 1 inch steel plate coated with a ¼ inch thick hard rubber interior liner. Two series of anode post holes were drilled in the cell base. Each series of holes was positioned in a straight line equidistant and parallel to the center line of the base; about 14 inches from the center line and 28 inches from each other, the center line being perpendicular to the 63 inch side of the cell. Each series of holes contains 16 holes, the centers of which were approximately 3 inches apart, the last hole in each series being approximately 4¾ inches from the edge of the cell base. The diameter of each hole was about 1²¹⁄₃₂ inches.

A flanged sleeve constructed of hard rubber of about ¼ inch thick having a sleeve height of about 1¼ inches, a flange outside diameter of about 2½ inches, and an inside diameter of about 1⁵⁄₃₂ inches was placed into each cell base hole. The flanged portion was sealed to the interior surface of the upper layer and the sleeve portion was sealed to the wall of the cell base hole.

Thirty-two anodes were placed in these holes, each anode being comprised of a mesh portion secured to a central anode post having a dished washer welded to the lower portion of the anode post. The mesh portion of each anode was approximately 24½ inches in width and 18¼ inches in height, being secured at the center of its short dimension to an anode post having a diameter of approximately 1⅛ inches. The length of these anode posts was approximately 27¾ inches and the upper edge of the mesh portion was located about ¾ inch from the top of the anode post.

Each anode post was constructed of a 1 inch diameter copper core clad with a 1/16 inch thick sheet of titanium on about the upper 22 15/16" portion of the anode post, leaving an O-ring recess of about 1/16 inch at the bottom of the cell base hole wall.

Each anode post had a titanium dished washer of about 2 inches outside diameter welded perpendicular thereto at a point about 21 inches from the top of the anode post. A neoprene washer having an outside diameter of 2⅜ inches and a thickness of about 3/16 inch was placed on each anode post with the top of the washer below the dished washer on the anode post. The anode post was placed in a hole in one of the rows of holes in the cell base. The bottom of each neoprene washer rested on the upper surface of the flanged portion of the flanged sleeve.

Each anode was placed parallel to each other and perpendicular to the abovementioned center line. A neoprene O-ring having a 1 inch inside diameter and a 1⅛ inch outside diameter was placed in the O-ring recess formed between the bottom of the interior of the sleeve portion and the 1 inch diameter copper core of the anode post extending below the end of the clad titanium sheet.

A pair of split spaced-apart copper bars was secured to each row of anode posts below the cell base with the upper surface of the copper bars pressed against the bottom of the cell base and O-ring. Each copper bar had a height of about 2 inches, a width of about 1 inch, a length of about 54 inches, and had 16 rounded or semicircular grooves with a diameter of about 1 inch positioned vertically along the bar spaced about 3 inches between centers and corresponding to the location of the anode posts.

On one exterior side of the split bars in each pair, extending from the edge of the cell base, there was secured a copper leaf having a length of about 2 feet 11½ inches, and to the exterior of that leaf was secured a leaf having a length of about 1 foot 2¼ inches. On the exterior side of the other split bar were similarly secured a leaf having a length of about 1 foot 11½ inches and another leaf of about 7¾ inches in length.

Each leaf was a copper bar having a width of about 1 inch and a height of about 2 inches. Each split pair of bars and the four leaves accompanying each pair were bolted together with a series of threaded bolts placed horizontally between each anode post. The centerlines of these horizontal bolts fall in a plane perpendicular to a plane through the centerline of the anode posts.

The bolts were ⅝ inch in diameter, the length varying according to the number of bars being secured together. Spring washers and threaded nuts were secured to the end of the threaded bolts and tightened until the round grooves firmly gripped the anode posts. A washer and threaded nut were placed on the end of each threaded anode post and tightened against the bottom of the split bars.

A bus bar attachment 10 inches by 10 inches was welded horizontally to the ends of the split bars and the four leaves of copper at the edge of the cell base. A bus bar was attached to the bus bar attachment for supplying current to the cell.

A cell can and top were secured to the cell base. Suspended from two sides of the cell can were 32 asbestos-coated steel mesh cathodes alternately spaced between and parallel to the anodes. Also secured to the exterior portion of the cell base were 3 I-beams in a horizontal position used as supports. Each I-beam was 44¼ inches in length positioned parallel to the abovedefined base center line about 6 inches from the edge of the cell base. The supports were positioned parallel to each other, one along the center line of the cell base and each of the others approximately 19½ inches between centers, and on opposite sides of the center line. The cell is operated for extended periods with a minimum of maintenance and corrosion problems, and with a high chlorine yield.

CATHODES

Silver-Titanium Intermetallic Coating Formed In Situ

R.C. Langley; U.S. Patent 4,186,066; January 29, 1980; assigned to Titanium Industries describes a cathode for use in an electrolytic process which comprises a substrate metal with a surface coating of a silver-titanium intermetallic formed in situ on the cathode.

Example 1: A silver solution was made, having the following composition: 2.0 g silver solution #9567 (25% Ag); 12.0 g rosin dissolved in essential oils (40% rosin); 12.0 g oil of lavender; and 24.0 g oil of rosemary.

The silver content of the solution was chosen to give a very thin metallic film per application. The solvent formulation was selected to give desirable flow properties when applied by brushing.

Titanium sheets in the form of flat mesh 7" x 12" x 0.080" in thickness were cleaned by degreasing. Surface oxide was removed and the surface was roughened by sandblasting. The mesh was cleaned in a detergent solution, rinsed with water and dried in air.

The titanium mesh was coated by brush application of the silver solution. The coated sample was fired by placing it on titanium supports on a continuous stainless steel belt which was moving through a large, production furnace open to the air. The sample was cooled in air upon leaving the furnace. Belt speed and temperature controls were selected to subject the sample to 550°C for 15 minutes. A single application gave a coating thickness of 2 to 3 microinches. A second application and the firing were made to give a total film thickness of 5 microinches.

Strips 1" x 7" were cut from the coated mesh and were tested in a small laboratory sodium chlorate cell. Conditions were: temperature, 65°C; electrolyte, NaCl, 300 g/ℓ; pH, 6.5 to 7.2; and voltage, 3.3 to 3.5.

The anode in the cell was titanium coated with ruthenium dioxide. Tests were conducted by measuring cell voltage using a silver coated cathode and replacing this with an uncoated titanium cathode of identical size and comparing the measurements. On average, the silver coated cathodes gave cell voltages of 0.10 to 0.12 volt lower than uncoated cathodes, for identical amperage. The maximum reduction of cell voltage observed was 0.15 volt.

After measuring coated samples, power was turned off and the samples were left in the electrolyte (300 g/ℓ NaCl) at 65°C for 48 hours. Cell voltages were then measured again and the initial values, in all cases lower than those obtained with uncoated titanium cathodes, were reproduced.

Example 2: A titanium alloy, having a composition by weight of Ti-Al-V 90-6-4, was cut into strips 8" x 2" x 0.078" in thickness. The strips were punched with a pattern of 0.0625" diameter holes evenly spaced, to give an open area of approximately 35%. The perforated strips were cleaned as described in Example 1. Half of the strips were coated with a single application of the silver solution of Example 1. These were fired in air to a peak temperature of 500°C and maintained at this temperature for 30 minutes.

Coated and uncoated strips were masked with plastic tape to leave an exposed area of one square inch. Asbestos paper was taped over exposed areas to simulate the asbestos diaphragm used in commercial chlorine cells. Anodes in the test cell were made of titanium mesh coated with an electrocatalytic alloy of Pd-Ir 70-30 by weight. Saturated brine was preheated with caustic and soda ash to remove calcium and magnesium. After filtering, brine was heated to 55°C and flowed through the test cell.

Coated samples were compared to uncoated samples by applying voltage until a current of 2.0 amps was indicated on an amp-meter. All coated samples achieved this current at a lower voltage than uncoated samples. On average, this was 0.1 volt lower.

Annealed Surface of Tungsten, Phosphorus, and Acicular Filaments of Cobalt Oxides

According to the method disclosed by *W.W. Carlin; U.S. Patent 4,181,586; January 1, 1980; assigned to PPG Industries, Inc.*, it has been found that the hydrogen overvoltage may be reduced, for example, by from about 0.2 volt to about 0.3 volt by using a cathode having an annealed surface of tungsten, cobalt, and phosphorus, where the cobalt is present as outwardly and upwardly extending acicular filaments rich in oxy compounds of cobalt.

The substrate has a cobalt-rich film thereon. The electrode has a more dense, less porous portion bearing upon the substrate and enriched in tungsten and depleted in cobalt relative to the original composition of the unannealed electroless plate. Atop the more dense, less porous inner portion is a less dense, more porous outer portion enriched in cobalt relative to the original electroless plate. This less dense, more porous, cobalt enriched outer portion of the film is characterized by outwardly and upwardly extending acicular filaments of elements of oxy compounds of cobalt.

The less dense, more porous outer portion is characterized by an electron spectroscopy analysis scan showing the substantial absence of tungsten within 3700 Å of the exposed surface thereof. The acicular elements themselves have been determined, by electron spectroscopy, to consist essentially of oxy compounds of cobalt, being Co_2O_3 and Co_3O_4 and predominantly Co_2O_3. It is believed that these acicular filaments are formed by the migration of cobalt out of the more dense portion during annealing in the presence of oxygen.

Example: An electrode having an annealed, electrolessly deposited coating of cobalt, tungsten, and phosphorus was prepared and utilized as a cathode in a chlor-alkali cell.

In preparing the cathode, a 5" x 7" by ⅛" (12.7 cm x 17.2 cm x 0.32 cm) mild steel perforated plate was sandblasted and then cleaned by dipping into a 1.1 normal solution of aqueous hydrochloric acid.

Thereafter, the plate was inserted in a bath containing:

	Grams
Cobaltous chloride ($CoCl_2 \cdot 6H_2O$)	4
Sodium hypophosphite ($NaH_2PO_2 \cdot H_2O$)	15
Sodium tungstate ($Na_2WO_4 \cdot 2H_2O$)	12
Sodium citrate ($Na_3C_6H_5O_7 \cdot 2H_2O$)	30
Sodium borate ($Na_2B_4O_7 \cdot 10H_2O$)	4
Water qs 24 ℓ	

The bath was at a temperature of about 78° to 87°C and a pH of 8.3 to 8.7. The perforated plate was placed in the bath for 6 hours and 20 minutes, rinsed in water, and dried in air for 16 hours. The perforated plate was then rinsed in 1.1 normal aqueous hydrochloric acid for 2 minutes, rinsed in water, and placed in a fresh electroless plating bath. The bath contained:

	Grams
Cobaltous chloride ($CoCl_2 \cdot 6H_2O$)	4
Sodium hypophosphite ($NaH_2PO_2 \cdot H_2O$)	15
Sodium tungstate ($Na_2WO_4 \cdot 2H_2O$)	12
Sodium citrate ($Na_3C_6H_5O_7 \cdot 2H_2O$)	30
Sodium borate ($Na_2B_4O_7 \cdot 10H_2O$)	4
Water qs 24 ℓ	

The bath was at a temperature of 83° to 90°C and a pH of 7.8 to 8.6. The perforated plate was placed in the bath for 4 hours and 40 minutes, rinsed in water, dried in air for 16 hours, inserted in 1.1 normal aqueous hydrochloric acid for 30 seconds, and rinsed in water.

Thereafter, 20 g of cobaltous chloride ($CoCl_2 \cdot 6H_2O$) and 60 g of sodium hypophosphite were added to the electroless plating bath. The perforated plate was inserted into the bath for 2 hours and 40 minutes with the bath maintained at 87° to 89°C and a pH of 8.3 to 8.5.

The electrolessly plated perforated plate was then removed from the bath, rinsed in water, and heated to 450°C for 48 hours in a resistance furnace that was open to the atmosphere.

The resulting cathode was then placed in a laboratory diaphragm cell having a 5" x 7" (12.7 cm x 17.2 cm) expanded metal mesh anode with an RuO_2-TiO_2

coating. The cathode was spaced ⅛ inch (3.2 mm) from the anode with a Du Pont Nafion 715 perfluorinated sulfonic acid polymer diaphragm therebetween.

Electrolysis was carried out at a current density of 190 A per square foot with a cell voltage of 2.8 to 3.2 volts for 353 days.

Nickel-Molybdenum Alloy Plated on Copper

H.C. Kuo; U.S. Patent 4,177,129; December 4, 1979; assigned to Olin Corporation has developed a highly conductive and corrosion-resistant low hydrogen overvoltage cathode. The cathode comprises a copper substrate plated with an alloy of nickel and molybdenum.

Example: A cleaned ¼" diameter copper rod was plated with Ni-Mo alloy in a bath having the following composition:

Ingredient	Concentration (g/ℓ)
Nickel sulfate	79.0
Nickel chloride	23.8
Sodium molybdate	4.0
Ammonium tartrate	73.6

pH = 9.8

at ambient temperature with current density of 4 A/dm^2 for 10 minutes. The hydrogen overvoltage of this plated alloy coating tested in 200 g/ℓ NaOH at 80°C and 20 A/dm^2 current density was about 0.18 V, which is about 0.15 to 0.2 V lower than that of a steel cathode.

Catalytically Active Raney Nickel

C.R.S. Needes; U.S. Patent 4,169,025; September 25, 1979; assigned to E.I. Du Pont de Nemours & Company has developed a process for making an electrode having a Raney nickel surface layer of thickness greater than 75 μ and an average porosity of at least 11%, comprising interdiffusing aluminum and nickel on a conductive metal core at a temperature of at least 660°C to form a nickel-aluminum alloy layer on the core from which layer aluminum is selectively dissolved. When used as a cathode in an electrolytic cell for producing hydrogen, chlorine and caustic from brine, the electrode exhibits low hydrogen overvoltage.

Example 1: Four groups of electrodes having porous Raney nickel surfaces are prepared. The preparation method for each group differs from that of the other groups only in the conditions of the interdiffusion step.

A 1.6 mm-thick nickel sheet, assaying at least 99% nickel, is cut into coupons measuring about 5.1 by 7.6 cm. The coupons, which are to become the cores of the electrodes are thoroughly cleaned by degreasing with acetone, lightly etching with 10% HCl, rinsing with water and, after drying, grit blasting with No. 24 grit Al_2O_3 at a pressure of 3.4 kg/cm^2 (50 psi).

The cleaned nickel coupons are aluminized by flame-spraying a 305 μ-thick coating of aluminum on the surface of each nickel coupon. A conventional plasma-arc spray gun operating at 13 to 16 kW at a distance about 10 cm from the coupon is used with aluminum powder of -200 to +325 mesh.

The aluminized nickel coupons are divided into four groups. Each group is heat-treated in a nitrogen atmosphere to interdiffuse the nickel and aluminum and form an Ni_2Al_3 layer. The times and temperatures of the interdiffusion heating are as follows: Group 1, 710°C for 12 hours; Group 2, 760°C for 8 hours; Group 3, 860°C for 4 hours; and Group 4, 910°C for 2 hours.

After heat treating, the coupons are allowed to cool in a current of nitrogen for about two hours. Samples are taken from each group for metallographic examination of the interdiffused layer.

The remaining coupons from each group are then subjected to a leaching treatment wherein the aluminum is selectively removed from the interdiffused layer to leave an active porous nickel surface on the coupon. The leaching treatment consists of immersing the interdiffused coupons in 10% NaOH for 20 hours, without temperature control, followed by 4 hours in 30% NaOH at 100°C. The coupons are then rinsed with water for 30 minutes.

Samples from each group of the electrodes are then subjected to metallographic and scanning-electron-microscope examination. The measured characteristics of the interdiffused Ni_2Al_3 layer, the final porous nickel surface layer and the cathode potential of each group of electrodes are summarized in the table below.

Comparison of Electrodes

	Ni_2Al_3 Precursor		. . . . . .Final Porous Nickel			
Electrodes	Layer Thickness (μ)	Grain Size (μ)	Layer Thickness (μ)	% Porosity	Agglomerate Size (μ)	Cathode Potential* (V)
Ex 1						
1	259	5	224	18.2	49	1.123
2	305	16	297	21.4	115	1.119
3	292	39	290	19.3	244	1.116
4	310	53	305	**	**	1.133
Ex 2***	58	2	53	7.4	28	1.165

*Cathode potential is measured versus a standard calomel reference electrode at 190 mA/cm^2 in catholyte containing 12% NaOH, 16% NaCl and 72% H_2O at 96°C.

**No measurement made.

***The precursor layer is a 29 micron thick layer of a nickel-zinc alloy in which the average grain size is less than 1 micron.

Example 2 (Comparative): An aluminized nickel coupon of the same dimensions and aluminized by the same technique as in Example 1 is heated for about 6 hours at 590°C in a stream of hydrogen to interdiffuse the aluminum and nickel. The interdiffused coupon is leached in 25% NaOH for 8 hours at 90°C to form a porous nickel surface and then is rinsed with water for 30 minutes.

Comparison of the measured characteristics of this electrode and the electrodes of Example 1, as shown in the table above, shows that the interdiffused Ni_2Al_3 precursor layer thickness, porosity and agglomerate size of this electrode are considerably smaller than the corresponding characteristics of the electrodes of Example 1, and, that the cathode potential of the comparison electrode is undesirably higher.

Raney Nickel Powder Imbedded in Nickel Plating

A cathode for use in electrolysis of an aqueous solution of an alkali metal halide or water is provided by *K. Kawasaki and T. Takeshita; U.S. Patent 4,170,536; October 9, 1979; assigned to Showa Denko KK, Japan,* which comprises a metallic cathode substrate and a powder of Raney nickel held on its surface partly embedded in a nickel layer deposited thereon from a nickel plating bath. The cathode has a considerably lower hydrogen overvoltage than ordinary cathodes. It can be produced by electrolytically depositing nickel on the surface of a metallic cathode substrate from an aqueous nickel plating bath containing a powder of Raney nickel suspended therein to form a codeposited layer of the nickel and the Raney nickel powder.

Example: A cathode was produced and tested as described below. The cathode substrate was a mild steel mesh (100 mm x 100 mm, wire diameter 2.4 mm, pitch 4.5 mm):

Plating bath:	
Nickel sulfate	240 g/ℓ
Nickel chloride	45 g/ℓ
Boric acid	30 g/ℓ
Undeveloped Raney nickel alloy powder (size 10-50 microns)	
No. 1-1	5 g/ℓ
No. 1-2	10 g/ℓ
No. 1-3	20 g/ℓ
No. 1-4	50 g/ℓ
pH	5
Temperature	70°C

Codeposit Plating Treatment – The above cathode substrate and a nickel plate anode were placed in the above plating bath, and while suspending the undeveloped alloy powder by stirring the bath with the aid of nitrogen gas bubbles, a direct current at 5 A was passed across the electrodes for 15 minutes.

Development – The resulting cathode having a codeposited layer composed of the undeveloped Raney nickel alloy powder and nickel was dipped in a 25% by weight aqueous solution of sodium hydroxide at 70°C until no hydrogen gas was seen to evolve. Thus, the aluminum ingredient was removed from the powder.

Application Test – Asbestos was caused to adhere to the mesh cathodes Nos. 1-1 to 1-4 produced by the above codeposition plating treatment and the development. Each of the asbestos diaphragm cathodes obtained was placed in an electrolytic cell in opposition to an anode composed of a titanium plate and a coating of ruthenium oxide. A saturated aqueous solution of sodium chloride was electrolyzed in the cell at a current density of 17 A/dm^2.

When the untreated cathode (control) was used, the cell voltage was 3.5 V. When the treated cathodes were used, the cell voltages decreased by the amounts shown below (a decrease in cell voltage from that of the control): No. 1-1, 0.04 V; No. 1-2, 0.14 V; No. 1-3, 0.15 V; and No. 1-4, 0.15 V.

Polymer Film Laminated on One Surface of the Cathode

K. Motani, T. Sata, and M. Nishimura; U.S. Patent 4,101,395; July 18, 1978; assigned to Tokuyama Soda KK, Japan provide a cathode-structure for liquid-phase electrolysis comprising a cathode and a polymer containing a cation exchange group, with the polymer being laminated in the form of a film on one surface of the cathode.

Example: A plain weave fabric of polytetrafluoroethylene was interposed between two 5-mil-thick films of a copolymer of tetrafluorethylene and perfluoro-(3,6-dioxa-4-methyl-7-octenesulfonyl fluoride) which has an ion exchange capacity corresponding to 0.91 meq/g of dry membrane (H^+ type, 1,100 equivalent weight) in the hydrolyzed state, and by melt-adhesion under heat, made into a single film structure. Furthermore, a 1.5 mil-thick film of the same copolymer having an ion exchange capacity corresponding to 0.67 meq/g of dry membrane (H^+ type, 1,500 equivalent weight) in the hydrolyzed state was superimposed on the resulting structure and melt-adhered to form a single polymeric membranous product.

One surface of a mild steel lath material was mechanically roughened, and a dispersion of polytetrafluoroethylene was coated on the roughened surface, followed by air drying and heating at 350°C. The coated mild steel material was dipped in an aqueous solution containing nickel rhodanide [$Ni(SCN)_2$] to plate the uncoated surface with nickel by electrolysis.

That side of the resulting polymeric membranous product which had a lower exchange capacity was pressed into the cathode by heating, whereby the mild steel lath electrode entered the polymeric membranous product. The assembly was immersed in an 8% methanol solution of potassium hydroxide at room temperature for 48 hours to convert the sulfonyl fluoride group to a potassium sulfonate group to form a cation exchange membrane and thus make the cathode-structure.

A two-compartment electrolytic cell was built by combining this structure with an anode composed of a titanium lath material coated with ruthenium dioxide and titanium dioxide. In this case, that side of the cathode-structure which was covered with the polymeric film was placed facing the anode, and the distance between the cathode and the anode was adjusted to 3 mm. A saturated aqueous solution of sodium chloride was fed into an anode compartment, and electrolyzed at a decomposition rate of 35%. Pure water was fed into the cathode so that 6.0 N NaOH could be steadily obtained from the cathode compartment. The electrolysis temperature was maintained at 85°C, and the current density was 30 A/dm^2.

This electrolysis was continued for three months. Results are shown below:

	Current Efficiency (%)	NaCl in NaOH* (ppm)	Cell Voltage (V)
Initially	85	38	3.65
3 months later	84	40	3.70

*Calculated on 48% NaOH.

Combined Cathode and Diaphragm Unit

An especially efficient and durable electrolytic cell is provided by *S.D. Argade and T.G. Coker; U.S. Patent 4,175,023; November 20, 1979; assigned to BASF*

Wyandotte Corporation by utilizing a combined cathode and diaphragm unit wherein the cathode is made by spray coating a ferrous metal substrate with a powder metal having lower hydrogen overvoltage than the substrate to form a protected cathode surface having a larger surface area than the substrate, and vacuum depositing fluorinated hydrocarbon polymer fibers onto the spray coated surface of the cathode to form a fibrous diaphragm securely adhered thereto.

Preferably, the cathode surface is made with nickel and the fluorinated hydrocarbon polymer is selected from the group consisting of homopolymers of chlorotrifluoroethylene and copolymers of chlorotrifluoroethylene and at least one compatible monomer with units of chlorotrifluoroethylene accounting for at least 80% of the monomeric units of the copolymer.

Example 1: A combined diaphragm and cathode unit was made as follows: The cathode used was a mild steel screen plasma sprayed on both sides with nickel in accordance with U.S. Patent 4,126,535. The composition of the diaphragm was Aclon 2000 polymer. The average cross-sectional dimensions of the fibers used to form the diaphragm were 1 μ by 4 μ, further branched into finer fibers, with a length of 0.25 to 0.5 mm. Such fibers were suspended in water, to the extent of 12.7 g/ℓ (dry weight of fiber employed), along with 4 g/ℓ of dioctyl sodium sulfosuccinate and 2 g/ℓ of a fluorine-containing surfactant, namely, Fluorad FC-170.

Fiber dispersion and slurry agitation were performed with the use of a propellor-type mechanical agitator driven by a "Lightnin" mixer.

A two-layered web was formed by drawing two successive volumes of slurry through the cathode screen at a ratio of 8.3 ml of slurry per cm^2 of screen area per layer according to the following schedule: 2 minutes at 25 mm of mercury difference from atmospheric pressure, 3 minutes further at 50 mm of mercury difference in pressure, and 2 minutes further at 100 mm of mercury difference in pressure.

The second layer was then applied: 3 minutes at 50 mm of mercury difference from atmospheric pressure, 8 minutes further at 100 mm of mercury difference in pressure, and 2 minutes further at 150 mm of mercury difference in pressure. The full vacuum of 615 mm of mercury was then applied for 20 minutes. There was obtained a diaphragm having a gross thickness of 2.7 mm and having a permeability coefficient of 1.7×10^{-9} cm^2. After being dried at 100°C for 16 hours, the unit is ready for use in a chlor-alkali cell.

Example 2: A steel cathode screen was plasma spray coated on one side with nickel powder to a thickness of 0.002 inch. A fiber diaphragm was then vacuum deposited on the coated side of the cathode in accordance with the procedure of Example 1.

After 40 days of operation in a chlor-alkali cell, the electrode potential was 1.23 to 1.27 volts versus standard calomel electrode (SCE) at 80°C and a current density of 160 mA per cm^2. In previous measurements, an uncoated steel cathode with an asbestos diaphragm had an electrode potential of 1.35 to 1.40 volts versus SCE under the same conditions. A cathode plasma coated with nickel on one side with an asbestos diaphragm had an electrode potential of 1.27 to 1.29 V versus SCE.

In a related disclosure by *S.D. Argade and T.G. Coker; U.S. Patent 4,182,670; January 8, 1980; assigned to BASF Wyandotte Corporation,* a polymer-impregnated asbestos fiber is vacuum deposited onto the cathode.

Preferably, the cathode surface is made with nickel and the polymer-impregnated asbestos diaphragm contains from about 10 to about 30% based on the weight of asbestos of an organic cementing agent such as a fluorohydrocarbon polymer.

Example 1 (Comparative): To a conventional 2%, by weight, asbestos slurry containing 0.5% by weight, based on the weight of the asbestos, of Fluorad FC-126 fluorohydrocarbon surfactant, was added 20%, by weight, based on the weight of the asbestos, of polyethylenechlorotrifluoroethylene powder. The powder was mixed with the slurry to render the slurry uniform.

The slurry was then deposited onto a mild steel cathode screen by drawing the slurry through the cathode screen under vacuum in accordance with conventional techniques.

Thereafter, the diaphragm was washed with distilled water by vacuum suction of the water through the diaphragm. The diaphragm was then dried in an oven at about 100°C, to remove excess water, for eight hours. Thereafter, the diaphragm was heat treated to bond the asbestos fibers, at 260°C for one hour.

The diaphragm was then mounted in a test chlor-alkali cell and brine electrolysis was carried out therewithin. The performance of the unit is given in the table below.

	. . Example 1. . .		. . .Example 2 . . .		. . .Example 3 . . .	
Days of Operation	Cell (V)	Cathode (V)	Cell (V)	Cathode (V)	Cell (V)	Cathode (V)
24	3.16	1.37	3.08	1.23	2.99	1.19
28	3.16	1.36	3.22	1.27	3.01	1.20
32	3.06	1.33	3.01	1.27	3.00	1.21
36	3.31	1.39	3.05	1.29	2.98	1.21
38	3.35	1.39	3.07	1.28	3.00	1.21

Example 2: A mild steel cathode was plasma spray coated on one side with a powder nickel to a thickness of about 0.002 inch. The powder nickel employed was Metco Nickel Powder 56 N-FS. A polymer impregnated asbestos diaphragm was deposited on the nickel coating of the cathode in accordance with the procedure of Example 1. The diaphragm was then mounted in a test chlor-alkali cell and brine electrolysis was carried out therein. The performance of the unit is also given in the table.

Example 3: A procedure of Example 2 was repeated except that the mild steel cathode was plasma spray coated with nickel on both sides thereof. The diaphragm was then mounted in a test chlor-alkali cell and brine electrolysis was carried out therewithin. The performance of this unit is also given in the table.

In the table above, the cathode volts were measured against a saturated calomel electrode to obtain the potential of the cathode. As is seen from the data, the cells of Examples 2 and 3 operated with good voltage efficiency both through the diaphragm and at the cathode, and this is particularly true of the preferred

Example 3. However, the cells of Example 1, a comparison example, were inferior. In the examples tested, the asbestos diaphragms were made with the usual thickness used in such diaphragms.

The results of both the cell and cathode voltage are also better than with conventional asbestos diaphragms at the same thickness. However, because of increased strength, the diaphragms made according to the disclosed process may be made thinner, and it is expected that with thinner diaphragms further increased cell energy efficiency will be provided.

Cellular Metal Structure

The improved cell developed by *F.E. Towsley; U.S. Patent 4,121,992; Oct. 24, 1978; assigned to The Dow Chemical Company* comprises a cathode having a cellular metal structure comprising a continuous interconnected network of electrolytically deposited metal defining therebetween a plurality of substantially convex and substantially electrically nonconductive cellular compartments. The arrangement of the compartments is adapted to permit passage of the oxidizing gas to the catholyte. The cellular metal structure is further characterized in that the deposited metal interfaces the cellular compartments within the cellular metal structure.

Example: A circular type electrolytic metal deposition cell containing a centrally located cathode rod surrounded by packed beads and the anode, in a circular or cylindrical symmetric arrangement, was employed to produce a cellular copper structure.

The electrolytic metal deposition cell assembly contained a cathode rod ¼ inch in diameter and 6 inches in length. The cathode rod was 99.49% by weight copper, 0.50% by weight tellurium, and contained a trace amount of phosphorus. The cathode rod was cleaned to remove oxide coating with abrasive paper to a uniform bright color level and then stirred in CH_3CCl_3 solvent. The rod was subsequently immersed and stirred in a solution of 250 ml 0.1 N NaOH mixed with 1.25 g Na_2CO_3 for 20 to 30 minutes.

The cathode was inserted in the center of a cylindrical Alundum round bottom thimble with an outside diameter of 26 mm and an outside height of 60 mm. The thimble material contained sintered aluminum oxide particles and formed a porous, electrically insulated and mechanically strong container. The pores of the thimble were of a size no greater than that sufficient to contain about -45 mesh (U.S. Standard) polystyrene beads, but were large enough to permit flow of electrolyte between an electrolyte reservoir and the interior of the thimble.

A helical coil, hand-wound from ⅛ inch outside diameter copper tubing, was placed around the outside wall of the thimble to form the anode. The central hole in the copper tubing was about ¹⁄₁₀ of the tube diameter, and the winding mandrel was a 1⁵⁄₁₆ inch diameter steel pipe. The copper tubing was cleaned with abrasive paper before winding, and treated with CH_3CCl_3 solvent and $NaOH/NaCO_3$ in substantially the same manner as the cathode.

Silicone rubber gaskets ⅛ inch in thickness, were adapted to fit around the cathode rod near the top and bottom ends of the rod. The washers were of sufficient diameter to fit in the barrel of the thimble and form a tight fit, especially at the bottom end of the cathode. The clearance between the upper washer and the lower washer was about 1⅜ inches.

The interior of the thimble was packed with substantially spherical beads. The beads were polystyrene with 4.0% by weight divinylbenzene and traces of isopentane. The beads passed through a U.S. Standard #45 sieve, but were caught on a U.S. Standard #50 sieve. The average size of the bead was about 330 μ. The beads were stirred with deionized water in a small beaker and then poured into the thimble with the cathode rod and the lower washer inserted in place. The beads were manually pressed down from above to pack the beads in the thimble space.

When sufficient beads were added to fill the thimble to about ⅓ inch from the top, the upper washer was added to the thimble.

The electrolyte contained 900 ml deionized water, 135.5 g of $CuSO_4 \cdot 5H_2O$, 60 ml concentrated H_2SO_4 (density 1.84 g/cc and 110 mg gelatin powder). The electrolyte was placed in the interior of the thimble and in the electrolyte reservoir.

The electrolytic metal deposition cell circuitry contained a direct current power source, a 50 ohm resistor and a 0 to 20 ohm variable resistor connected in series between the power source and a 0 to 300 mA meter. A high impedance multirange voltmeter was connected between the cathode rod and the anode.

The cell was allowed to equilibrate for 1.5 hours, and then a direct current potential of 0.100 V was applied across the cell and the current was adjusted to about 50 mA. This level corresponded to about 4.8 A/ft^2 current density at the cathode rod surface. Copper metal was deposited at the cathode rod surface and the plating interface advanced through the packed beads toward the walls of the thimble.

When copper metal had substantially filled the available interstitial spaces between the beads within a ⅜ inch radius of the cathode rod, the cell was disconnected and the cellular copper structure was removed.

The product was a cellular copper structure comprising a continuous network of electrolytically deposited copper defining a plurality of substantially spherical compartments containing polystyrene therebetween. The rod may be removed from the cellular metal by suitable means well-known in the art. The lightweight cellular metal product formed is of sufficient strength to withstand aluminum machining speeds without cracking. The polystyrene may be removed from the compartments, and the cellular copper structure utilized as an oxidizing gas depolarized cathode.

OXYGEN AND AIR ELECTRODES

A number of various types of electrochemical devices have been developed over the past few years for the production of electrical energy by electrochemical reaction and obversely for the consumption of electrical energy to effectuate electrochemical reactions. Many of these devices rely upon a reaction involving oxygen (or air) as part of the mechanism to accomplish the desired result. For example, such devices may contain oxygen electrodes which are oxygen-reducing cathodes in which oxygen is catalytically electroreduced. Alternatively, such devices may contain oxygen electrodes which catalyze the evolution of oxygen from water. In general, these electrodes are known in the art as oxygen electrodes. Thus, metal-oxygen batteries, metal-air batteries, fuel cells, electrolyzers, metal electrowinning devices, etc., are among the well-known electrochemical devices which may contain oxygen electrodes.

Pyrochlore Electrocatalyst

An electrochemical device is described by *H.S. Horowitz, J.M. Longo, and J.I. Haberman; U.S. Patent 4,146,458; March 27, 1979; assigned to Exxon Research & Engineering Co.* which has an oxygen electrode which contains a pyrochlore type electrocatalyst material. This electrocatalyst material is one or more electrically conductive pyrochlore compounds selected from the compounds of the following formula:

$$(1) \qquad A_2B_2O_{7-y}$$

wherein A is any of the known pyrochlore structure metal cations, B is a pyrochlore structure metal cation at least a major portion of which is selected from the group consisting of one or more of Ru, Rh, Ir, Os, Pt, Ru-Pb mixtures and Ir-Pb mixtures, and wherein $0 \leqslant y \leqslant 1.0$. Desirably, A is a pyrochlore structure metal cation at least a major portion of which is selected from the group consisting of one or more of Pb, Bi and Tl. A preferred group of pyrochlore compounds which are included in the above formula are those having the following formula:

$$(2) \qquad Pb_2(M_{2-x}Pb)O_{7-y}$$

wherein M is selected from the group consisting of Ru and Ir, wherein $0 \leqslant x \leqslant 1.2$ and $0 \leqslant y \leqslant 1.0$.

Example: The electrocatalyst compound $Pb_2Ru_2O_{7-y}$ is prepared as follows: A mixture of powdered, mechanically blended $Pb(NO_3)_2$ and RuO_2 in amounts so as to achieve a lead to ruthenium molar ratio of approximately 1.5:1.0 is reacted at 850°C for 20 hours in air with one interruption for regrinding. The resulting reacted powder is washed with an alkaline solution to leach out any excess PbO. X-ray diffraction indicates that the reacted and washed powder is single phase $Pb_2Ru_2O_{7-y}$. The surface area is determined, by the BET N_2 absorption method, to be about 3 m^2/g.

To illustrate the utility of this $Pb_2Ru_2O_{7-y}$ compound, performance curves for the electrocatalytic reduction and evolution of oxygen at 75°C in 3 N KOH are obtained using the $Pb_2Ru_2O_{7-y}$ compound and using platinum supported (10% by weight) on active carbon. The platinum on carbon electrodes is typical of conventionally used supported noble metal electrocatalysts. In these tests, the material is fabricated into test electrodes consisting of the catalyst, a binder, a wetproofing agent and a support. Teflon serves as both a binder and wetproofing agent for all the electrodes tested. Gold expanded metal screen is used as the support.

Electrodes are fabricated by mixing a weighed amount of catalyst with a few drops of water, adding a measured volume of Teflon 42 suspension, and mixing vigorously to precipitate the Teflon. The gummy product is then spread on a weighed gold Exmet screen and is pressed dry between filter paper. The electrode is then cold pressed for 0.5 minute at 200 psi, is allowed to air dry for 30 minutes, and is then hot pressed at 325°C, 500 psi for 0.5. minute. After cooling, the electrode is weighed to determine its loading and then placed in the electrochemical cell for testing.

The electrochemical half-cell used for testing is of the interface maintaining type and consists of a jacketed liquid phase cell compartment and a gas phase cell compartment. The liquid side contains the platinum wire counterelectrode, a saturated calomel reference electrode (in contact by Lugin capillary), and magnetic stirrer. The gas side contains the gas (oxygen) inlet and outlet and a stopcock to drain off any condensate. The working electrode is held in place (between the two compartments) between two Teflon disks with a gold current collector pressing against it.

The cell is connected to a Princeton Applied Research Model 173 potentiostat with a programmer and logarithmic current converter. Constant rate potential sweep measurements are conducted. Outputs of potential and log of current are recorded on an x-y plotter, and the resulting potential vs log current density plot, referred to as performance curve, is used to evaluate the electrode activity. These results are shown in the Voltage-Current Density Graph of Figure 5.13a which shows an electrocatalytic performance curve for the reduction of oxygen at 75°C in 3N KOH using $Pb_2Ru_2O_{7-y}$, prepared as above, as the catalyst.

Curve 1 represents the activity of the electrode when oxygen is continually supplied to it. When the cell is purged with nitrogen and the performance curve is run with nitrogen continually supplied to the electrode, the catalytic activity is eliminated as shown by curve 2. This demonstrates that the activity exhibited by the electrode in oxygen (curve 1) is truly catalytic in nature (i.e., it is not just the electrochemical reduction of the active material taking place).

Figure 5.13: Voltage-Current Density Graphs

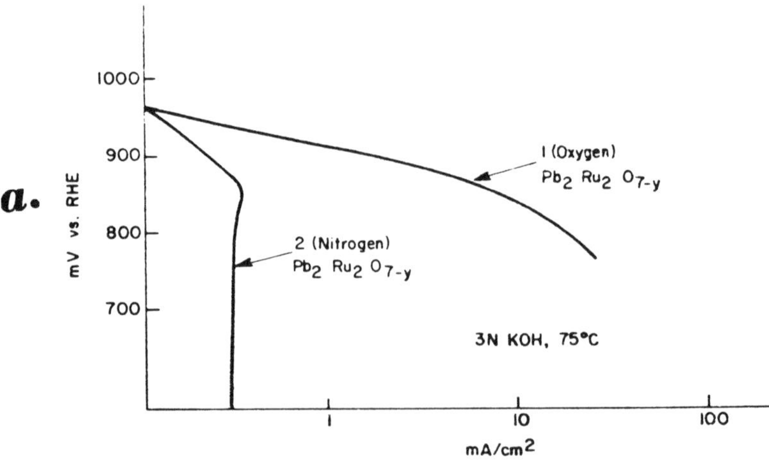

(Continued)

Figure 5.13: (continued

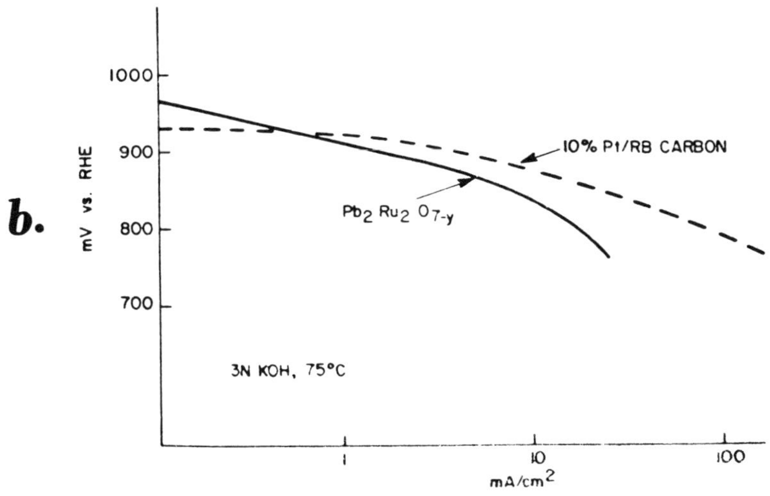

(a) $Pb_2Ru_2O_{7-y}$ in oxygen and in nitrogen
(b) $Pb_2Ru_2O_{7-y}$ compared to platinum on carbon

Source: U.S. Patent 4,146,458

Figure 5.13b (Voltage-Current Density) compares the oxygen electroreduction performance curve of the $Pb_2Ru_2O_{7-y}$ with that of 10% (by weight) platinum supported on active carbon (Pittsburgh RB Carbon). The platinum on carbon electrode is typical of conventionally used supported noble metal electrocatalysts. As the graph illustrates, the lead ruthenate electrode is competitive with the platinum on carbon electrode.

Intercalation Compound of Carbon and Fluorine

C.N. Welch; U.S. Patent 4,135,995; January 23, 1979; assigned to PPG Industries, Inc. has developed a cathode having a surface of an intercalation compound of carbon and fluorine in contact with an aqueous alkaline liquor, for example, an aqueous alkaline liquor containing alkali metal ions such as potassium ions or sodium ions, and the hydrophobic portion of the cathode in contact with gas, which gas contains an oxidizing material such as oxygen.

By an intercalation compound of carbon and fluorine is meant a carbonaceous material crystallized in a graphitic layer lattice with the layer atoms being approximately 1.41 Å apart, the layers being a greater distance apart, e.g., at least about 3.35 Å, and with fluorine atoms present between the layers.

Intercalation compounds of carbon and fluorine may be prepared by reacting graphite with a Lewis acid fluoride and chlorine trifluoride in the presence of hydrogen fluoride.

Example: An air cathode was prepared and utilized as a cathode in a laboratory cell. A slurry was prepared containing 1.0 g of Ozark-Mahoning Fluorographite $(CF_{0.6})_n$ and 0.15 g of Triton surfactant in 1.5 g of water. To this slurry was slowly added 0.2 g of Du Pont Teflon 30B solution of 60 weight percent solid polyperfluoroethylene, thereby forming a gummy paste.

The paste was spread over a 2.5 inch by 2.5 inch nickel wire cloth screen having 0.007 inch diameter nickel wire woven at 20 mesh by 20 mesh. The screen, with the slurry spread over it, was then pressed between 2 sheets of 0.005 inch thick aluminum sheets which were sandwiched between a pair of polyperfluoroethylene-coated aluminum plates at a temperature of 275°C and a pressure of 3,100 pounds per square inch gauge for 10 minutes. The aluminum was then dissolved off in 2 normal caustic soda and the cathode was dried at a temperature of 80°C for 60 minutes. Thereafter, a porous polyperfluoroethylene sheet 6.5 mils thick and having 2 to 5 μ pores was applied to one side of the cathode at a temperature of 190°C and a pressure of 50 pounds per square inch.

The resulting cathode having hydrophilic $(CF_{0.6})_n$ in and on one side of a nickel screen current collector and hydrophobic polyperfluorethylene on the opposite side of the nickel screen current collector was utilized as an air cathode in a laboratory electrolytic cell containing a 1 normal NaOH electrolyte. An electrical current was passed from an anode of the cell to a cathode of the cell, oxygen was bubbled along the polyperfluoroethylene surface of the cathode, and gas was seen to be evolved at the anode of the cell. The results shown below were obtained.

Cathode Current Density (A/ft^2)	Cathode Voltage (vs Saturated Calomel Electrode)	Cumulative Time (minutes)
25	-1.36	0.16
25	-1.15	1
25	-1.04	5
10	-0.72	6
50	-1.38	6
25	-1.01	15
25	-0.92	25
25	-0.93	35
25	-0.92	60
25	-0.84	90
25	-0.91	120
10	-0.76	180
25	-1.05	180
50	-1.29	180

Oxygen Electrode Rejuvenation Methods

F. Solomon, D.F. Lieb, and R.L. LaBarre; U.S. Patent 4,185,142; January 22, 1980; assigned to Diamond Shamrock Corporation disclose methods for rejuvenation of oxygen electrodes which maximize the power efficiency available from such oxygen electrodes while minimizing the voltage necessary to operate such oxygen electrodes over extended periods of time. These methods include in situ and out of cell techniques using hot water washing followed by air drying and dilute acid washing followed by air drying.

Example 1: An oxygen electrode according to U.S. Patent 3,423,245, was installed into an electrolytic cell as the cathode and run at 2 A/in^2 and 60°C until the voltage reached -0.982 V as measured against a Hg/HgO reference electrode, when it was considered to have decayed beyond commercial usefulness. The oxygen electrode was then taken out of the cell and was soaked in deionized water for several days.

An uncracked partially delaminated section of the oxygen electrode was then washed for 15 minutes in dilute acetic acid at 50°C. It was then rinsed with deionized water, dried and then pressed between two plates for 90 seconds at approximately 2,000 pounds per square inch pressure. Upon restart, the following potentials were evident, showing a voltage savings initially of 0.742 V and, finally, after 60 days, a savings of 0.589 V over the cathode at the time of initial failure.

Day	Potential	Day	Potential
1	-240	25	-301
2	-208	26	-309
3	-214	27	-319
4	-226	28	-331
5	-238	29	-332
6	-262	30	-331
8	-270	31	-341
9	-273	34	-350
10	-290	35	-354
11	-291	36	-357
12	-295	37	-358
14	-305	38	-364
15	-304	41	-371
16	-304	42	-376
17	-306	43	-383
18	-306	44	-375
19	-221	45	-383
22	-381	49	-393
23	-289	60	failure
24	-295		

Example 2: An oxygen electrode according to U.S. Patent 3,423,245 was run in an electrolytic cell as the cathode at 1 A/in^2 and 60°C until the voltage reached -0.577 V as measured against a Hg/HgO reference electrode. The oxygen electrode was then washed in situ with warm distilled water while the electrolytic cell was shut down. The cell was then slowly started up to attain the same current density and temperature after 24 hours. The potential then was -0.497 for a savings of 0.080 V.

Example 3: An oxygen electrode according to U.S. Patent 3,423,245 was run in an electrolytic cell as the cathode at 1 A/in^2 and 60°C until the voltage reached -0.830 V. The oxygen electrode was removed from the cell and cleaned ultrasonically in 0.1 N HCl solution. Some delamination was apparent so the cathode was then pressed between two nickel plates at about 200 lb/in^2, heated to 115°C and left overnight. The oxygen electrode was replaced into the electrolyte cell which was started up slowly. The potential then was -0.760 at 1 A/in^2 for a savings of 0.070 V.

OTHER

Monopolar Filter Press Electrode

T.W. Boulton; U.S. Patent 4,126,534; November 21, 1978; assigned to Imperial Chemical Industries, Limited, England has devised an electrode for use in a monopolar electrolytic cell of the filter press type which allows very small or even zero anode/cathode gaps to be used in such cells without damage to the diaphragms or membranes.

The electrode may be suitable for use as one of the electrodes, that is, either an anode or cathode or both, positioned alternately in the electrolytic cell between the terminal anode and cathode. In this case both surfaces of the metal sheet suitably carry a group of elongated metal members electrically conductively mounted on and projecting from the surface of the sheet and lying in planes substantially parallel to and laterally spaced from the surfaces of the sheet, at least one of the groups of elongated members being flexible.

The electrode shown in Figure 5.14a, which is an electrode suitable for use as a terminal cathode in a monopolar electrolytic cell, comprises a sheet **1** of iron or steel and a plurality of 3 mm thick iron or steel wires **2**, capacitor discharge stud welded at **3** to the sheet.

Figure 5.14: Monopolar Filter Press Electrode

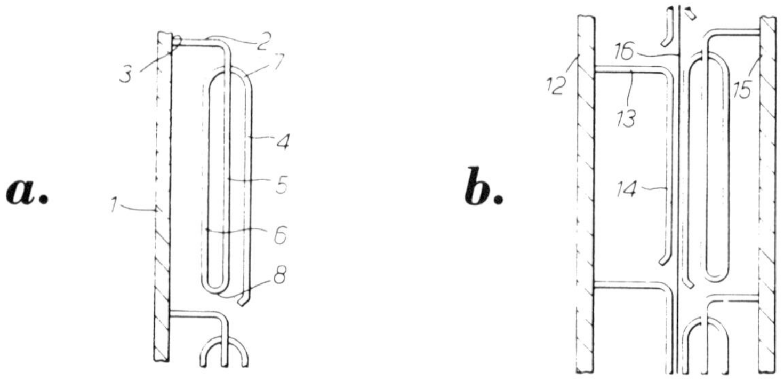

(a) Elevation in cross section of a part of an electrode

(b) Cross section in elevation of a part of of a monopolar electrolytic cell containing the electrode

Source: U.S. Patent 4,126,534

The wires have straight portions **4, 5, 6** and have bends at **7** and **8** to form a loop which provides the wire with flexibility. The straight portions each lie in a plane and are substantially parallel to the face of the sheet and are laterally displaced therefrom.

The electrode shown in Figure 5.14a may be an anode suitable for use in a monopolar electrolytic cell in which case the sheet **1** and the wires **2** will be of a film-forming metal, e.g., titanium, and at least the straight parts **4** will carry an electrocatalytically active coating.

Figure 5.14b shows a part of a monopolar electrolytic cell having an anode in the form of a titanium sheet **12** carrying a plurality of substantially rigid 3 mm thick titanium wires **13** having straight portions **14** lying in a plane laterally displaced from and substantially parallel to the sheet. The wires do not have a loop and are relatively rigid. The straight portions carry an electrocatalytically active coating. The cell also comprises a cathode **15** of iron or steel of the type described with reference to Figure 5.14a.

A separator **16** is positioned between and may be in contact with the wires of the anode **12** and the wires of the cathode **15** and provides the cell with distinct anode and cathode compartments. The separator may be a porous diaphragm or a cation exchange membrane.

Example: A laboratory membrane cell as shown in part of 5.14b was assembled, the anode comprising a titanium sheet of dimensions 300 mm x 970.5 mm and having on one face thereof six rows of titanium wires, each row containing 32 wires and each wire having a straight portion 154 mm long by 3 mm diameter. The wires were provided with an electrocatalytically active coating.

The cathode comprised a mild steel sheet having five rows of flexible looped mild steel wires of 3 mm diameter, each row containing 32 wires. The loops provided flexibility to the wires. The distances between the titanium sheet **12** and the membrane **16**, i.e., the width of the anode compartment, and between the mild steel sheet **15** and the membrane **16**, in the width of the cathode compartment, were each 28 mm.

The membrane was a perfluorosulfonic acid membrane based on copolymers of tetrafluoroethylene and fluorinated vinyl ethers, Nafion. The membrane was adjacent to both the cathode and anode, i.e., the anode/cathode gap was zero. Sodium chloride brine (concentration 300 g/ℓ of NaCl) was fed to the anode compartment at a rate of 6 ℓ/hr. Deionized water was added to the cathode compartment. The temperature of the cell was maintained at 85°C.

A current of 300 A (equivalent to a current density of 1.8 kA/m^2) was passed through the cell. The cell operating voltage was 2.9 V. The chlorine produced contained 94% by weight of Cl_2 and less than 0.1% by weight of H_2. The sodium hydroxide produced contained 10% by weight of caustic soda. The cell operated at a sodium hydroxide current efficiency of 86%. The membrane was undamaged by the wires of the anode and cathode.

Electrolyzer with Released Gas

P. Jonville; U.S. Patent 4,086,155; April 25, 1978; assigned to Battelle Memorial Institute, Switzerland describes an electrolyzer comprising a cathode and an anode, disposed in an aqueous electrolyte bath, the electrodes each having at least one active surface directed substantially in facing relation to at least one active surface of the electrode of opposite polarity. At least one of the electrodes is permeable to gas and is the source of a gas release at the time of operation of the electrolyzer. At least one portion of the active surface of the electrode permeable to the gas is covered by a porous layer constituted by at least one refractory oxide which is electrically insulative and chemically inert with regard to the electrolyte and to the products formed at the time of electrolysis. The layer has a homogeneous distribution of pore sizes of a value sufficient for the electrolyte to traverse this layer and impregnate the electrode, the mean value of the radii of the pores of the insulating refractory oxide layer being at least as small as one-tenth of that of the pores of the electrode that it covers. The electrode and the oxide layer form an element of self-supporting structure, the oxide layer constituting a surface portion in contact with the electrolyte.

The electrolyzer shown in Figure 5.15 comprises a vessel **1** divided into a cathode-compartment **2** and an anode-compartment **3** by a porous diaphragm **4**. The interior of the vessel is filled with a liquid electrolyte bath in which are completely immersed a porous cathode **5** and a porous anode **6**.

Figure 5.15: Electrolyzer with Released Gas

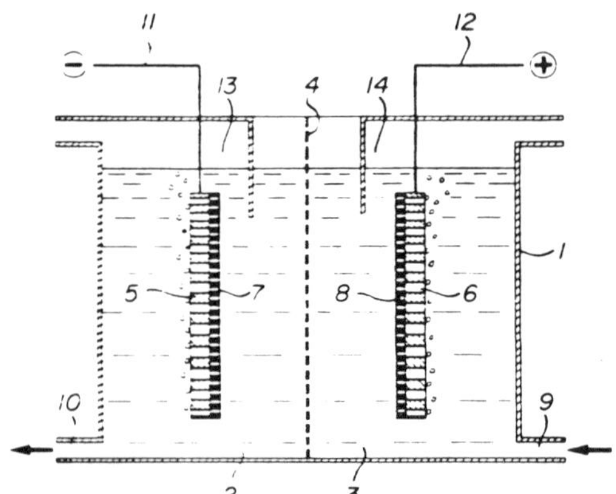

Source: U.S. Patent 4,086,155

The cathode and the anode are constituted by rectangular plates having substantially the same dimensions, placed with their faces parallel and opposite one another. The face of the cathode turned towards the anode is covered by a porous layer **7** of refractory oxide solid with the cathode. Similarly, the face of the anode turned towards the cathode is coverd by a porous layer **8** of a refractory oxide solid with the anode.

A conduit **9** which opens into the anode-compartment permits introduction of electrolyte into the electrolyzer. The electrolyte can be constituted by an aqueous solution of sodium chloride.

A conduit **10** extending from the cathode compartment permits removal, in continuous or intermittent manner, of a portion of the liquid contained in this compartment. This liquid can be constituted by an aqueous solution of sodium chloride and sodium having a sodium chloride content less than that of the fresh electrolyte which is introduced into the electrolyzer through the conduit **9**.

Suitable regulation means (not shown) permit adjustment of the flow and/or the introduction of the electrolyte through the conduit **9** and the removal of liquid from the cathode compartment through the conduit **10** in order to obtain a suitable concentration of sodium chloride in the liquid removed from the latter conduit.

At the time of operation a suitable difference of potential is applied between cathode **5** and the anode **6** by means of electrical conductors **11** and **12** respectively, these conductors being connected to an electrical energy source, not shown in the figure.

In the course of electrolysis, there is produced a gaseous release at each of the electrodes. The gas which is released at the cathode can be hydrogen and that which is released at the anode can be chlorine. These gaseous releases are in the form of bubbles exclusively at the external faces of the electrodes, that is to say, at the faces which are not turned towards the electrode of opposite polarity.

The gas released at the cathode is collected in a collector compartment **13** and that which is released at the anode is collected in a collector compartment **14.**

Cathode Busbar Structure

W.W. Ruthel and L.G. Evans; U.S. Patent 4,178,225; December 11, 1979; assigned to Hooker Chemicals & Plastics Corp. provide a cathode busbar structure for an electrolytic cell which can enable the electrolytic cell to be designed to operate as a chlor-alkali diaphragm cell at high current capacities of about 150,000 A and upward to about 200,000 A while maintaining high operating efficiencies while in normal or jumpered operation. These high current capacities provide for high production capacities which result in high production rates for given cell room floor areas and reduce capital investment and operating costs.

Referring now to Figure 5.16, electrolytic cell **11** comprises corrosion resistance plastic top **12,** a cathode walled enclosure **13** and cell base **14.** The top is positioned on cathode walled enclosure **13** and is secured to the cathode walled enclosure by fastening means (not shown). A seal is maintained between the top and the cathode walled enclosure by means of a sealing gasket. The cathode walled enclosure is positioned on the cell base and is secured to the cell base by fastening means (not shown). A seal is maintained between the cathode walled enclosure and the cell base by means of an elastomer sealing pad. The electrolytic cell is positioned on legs **15** which are used as support means for the cell.

The cathode busbar structure comprises copper lead-in busbar **18** and a plurality of copper busbar strips **19, 21** and **22** which have different relative dimensions and are positioned in such a configuration wherein lead-in busbar **18** and the busbar strips are adapted to carry an electric current and to maintain a substantially uniform current density through cathode busbar structure to electrical contact points on sidewall **17** of the cathode walled enclosure whether the cell is connected in normal operation with the other cells in a circuit or whether the cell is connected in a jumpered position with another cell.

The cathode busbar structure may also be provided with cooling means **23** which comprises steel plates **24, 25** and **26** and entrance and exit ports **27** and **28** fabricated in any suitable manner, as by welding, to form the cooling means.

Figure 5.16: Cathode Busbar Structure

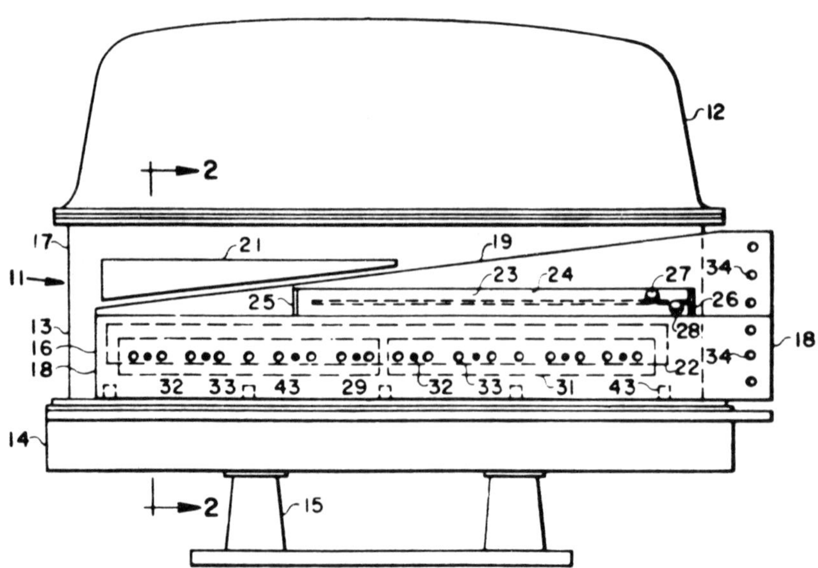

Source: U.S. Patent 4,178,225

Cooling means **23** may be attached in any suitable manner, as by welding, suitably to lead-in busbar **18** and busbar strip **19**. Coolant, preferably water, is circulated through the cooling means by passage through the entrance and exit ports. Cooling means **23** is provided primarily for use when an electrolytic cell adjacent to electrolytic cell **11** is jumpered and is removed from the electrical circuit. The use of the cooling means permits considerably less copper to be used in the cathode busbar structure which results in a substantial reduction in capital investment costs for cathode copper.

While the cooling means is provided primarily for use when an electrolytic cell adjacent to electrolytic cell **11** is jumpered, the cooling means can be used during routine cell operation either to cool the cathode busbar structure during any periodic electric current overloads or to continuously cool the cathode busbar structure, thereby permitting further reductions in the use of copper in the cathode structure with an accompanying reduction in capital investment costs for cathode copper.

The lead-in busbar is provided with steel contact plates **29** and **31** which serve as contact means. Steel contact plates **29** and **31** are attached to the lead-in busbar in any suitable manner, as by means of screws **32**. The lead-in busbar and steel contact plates **29** and **31** are provided with holes **33** which serve as means for attaching intercell connectors carrying electricity from an adjacent cell or leads carrying electricity from another source to lead-in busbar **18**. Lead-in busbar **18** and busbar strip **19** are suited to use as a cathode jumper busbar, the busbar **18** and busbar strip being provided with holes **34**, which serve as means for attaching cathode jumper connectors when an adjacent electrolytic cell is jumpered and is removed from the electrical circuit. It is during this jumpering operation that cooling means **23** can provide its greatest utility by preventing the temperatures in the cathode busbar structure from rising to levels whereby damage to cathode busbar structure or other components of electrolytic cell **11** may occur.

Example: The following data are typical of the performance of an electrolytic cell provided with the disclosed cathode busbar structure operating at a current capacity of 150,000 A. The performance is compared with the performance of a smaller electrolytic cell of the prior art. Both cells were equipped with metal anode blades. The smaller cell was operated at a current capacity of 84,000 A. Both electrolytic cells are chlor-alkali diaphragm cells.

	84,000 A Cell of the Prior Art	150,000 A Cell Provided with the Cathode Busbar
Current efficiency	96.4	96.4
Average cell voltage (including busbars)	3.84	3.83
Power, kWhdc/ton Cl_2	2,735	2,725
Cell liquor temperature, °C	100.5	100.7
Anolyte temperature, °C	94.5	94.7
Percent NaOH in cell liquor	11.5*	11.5*
Chlorine production, tons/day	2.83	5.06
NaOH production, tons/day	3.20	5.71
Brine feed, g/ℓ	324	325
Current density, A/in²	1.5	1.5

*The cells can be operated at lower caustic content in the cell liquor. This will result in greater current efficiencies.

The above data show that the electrolytic cell provided with the cathode busbar structure disclosed operates at essentially the same current efficiency, voltage and operating conditons as the smaller electrolytic cell of the prior art at the same anode current density. The electrolytic cell provided with the cathode busbar structure has a higher production rate for a given cell room floor area, uses less operating labor and also has a lower capital investment per ton of chlorine produced.

Protection of Catalytic Coatings

Recently, various catalytic low hydrogen overvoltage alloy cathode coatings have been developed, such as Ni-Mo, Ni-Mo-V, Ni-Ti, precious alloys and others for chlor-alkali cell application. These coatings are generally plated or coated on steel or copper substrates. When the cells are shut down for any reason, the catalytically coated cathodes are immersed in the strongly caustic catholyte and corrosion of the coatings and bare copper or steel occur. If the cell is then restarted after such shutdown, the dissolved copper or iron ions are replated on top of the catalytic coatings and the catalytic activity of the catalytic coatings is degraded.

H.C. Kuo; U.S. Patent 4,169,775; October 2, 1979; assigned to Olin Corporation has designed an electrolytic cell having auxiliary circuit means for applying a cathodic protection current to the cathode and exposed metal parts to help prevent corrosion thereof during cell shutdown.

Figure 5.17 shows schematically a cell **10** which comprises a cell housing **12**, an electrolyte **14**, an anode **18**, a cathode **16**, a primary circuit **20**, an auxiliary circuit **22** and a membrane or diaphragm **24**.

Figure 5.17: Schematic Circuit Diagram Showing Auxiliary Circuit Means

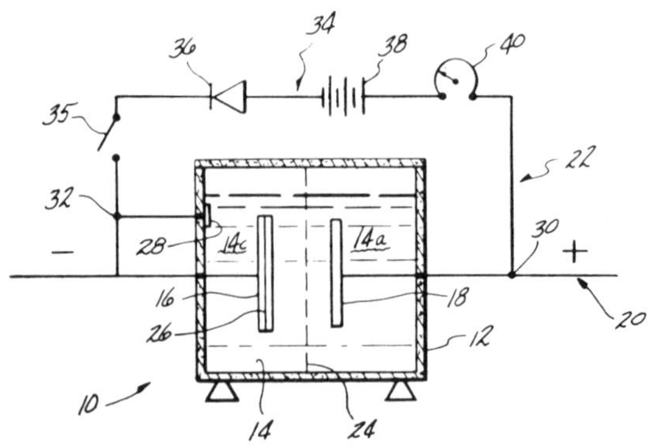

Source: U.S. Patent 4,169,775

The cathode is of the type having a conductive substrate such as steel, titanium, or copper and a catalytic coating **26**, such as, for example, a Ni-Mo alloy having more than 50% by weight molybdenum or a Ni-Mo-V catalytic coating such as, for example, that disclosed in U.S. Patent 4,033,837, issued July 5, 1977, to Kuo et al.

Auxiliary circuit means **22** is any means for applying and maintaining a negative potential on cathode **16** and the coating as well as on any exposed metal parts **28** of housing **12** relative to the potential of the anode, by supplying a dc cathodic protection current of a density of from about 0.01 to 100 A/m^2 of cathode surface area, and preferably of from about 0.01 to 0.1 A/m^2. The circuit means includes a positive dc terminal **30**, a negative dc terminal **32** and a dc power source **34** connected between the terminals. The preferred dc power source comprises in series a switch **35**, a diode **36**, a battery **38**, and a rheostat **40**. Terminal **30** connects to the anode while terminal **32** connects to the cathode and any exposed metal parts of the housing. The battery must have a higher voltage than the open circuit cell voltage. Any voltage that is sufficiently high works since rheostat **40** provides a means for controlling the current through circuit means **22**. Battery **38** can preferably be a normal 12 volt lead-acid type car battery for economy or can be a more elaborate battery of an ac current rectified into dc. The type of power source **34** is of secondary importance to the primary feature of an auxiliary circuit means functionable during cell shutdown. The battery has a great advantage that if normal domestic current to the cell is halted by a power failure, the battery will, by design, be able to cathodically protect the catalytic coating **26** of the cathode **16**.

Switch **35** and diode **36** serve to protect auxiliary circuit means **22** against damage by the very large currents normally passing through primary circuit means **20** during normal cell operation. The switch could be opened during normal cell operation to minimize battery discharge and is closed immediately after any power loss in primary circuit **20**. This closure of the switch could even be automated to assure immediate cathode protection.

Example: A 2½" x 3¼" sheet of copper mesh was cleaned and then plated at 400 A/m^2 for about 1 hour with a nickel molybdenum alloy in a citrate bath having the following composition: 24 g nickel chloride per ℓ of bath solution, 6 g/ℓ sodium molybdate, 80 g/ℓ nickel sulfate, 88 g/ℓ sodium citrate at a pH of 9.5. The Ni-Mo plated Cu sheet was then operated as the cathode in a membrane cell continuously for about 10 weeks at 2,000 A/m^2 current density and 85°C temperature with a cation exchange membrane of the Du Pont Nafion sulfonic acid resin type. The cell showed a low overvoltage of about 210 mV (about 200 mV less than a steel cathode of comparable shape).

Another copper mesh of the same size as above was plated in a bath of composition: 45 g/ℓ nickel sulfate, 180 g/ℓ sodium pyrophosphate, 10 g/ℓ sodium molybdate, 30 g/ℓ ammonium chloride and ammonium hydroxide to adjust pH to 9.0. This mesh was then operated as the cathode in a membrane cell at a current density of 3,000 A/m^2 and 85°C temperature for about 12 weeks. Initially, the cathode showed an overvoltage of about 230 mV and increased slowly to about 300 mV within the first three weeks of operation. The cathode then showed an overvoltage of about 300 to 350 mV during the next nine weeks of operation, which is about 100 to 150 mV below the overvoltage of a comparable steel cathode.

Both of the above two cells with the Ni-Mo alloy coated cathodes were occasionally subjected to cell shutdowns of various periods of from a few hours to a few days. Every time the cell was shut down, the current density was kept within a range of 10 to 100 A/m^2 by use of an auxiliary circuit means similar to that of Figure 5.17.

Their consistently low overvoltage indicates corrosion protection of the cathodes was achieved as a result of the low current density applied to the cathode by the auxiliary circuit means.

Intermittent Reversal of Cathode Polarity to Remove Deposits

Electrolysis of impure solutions containing hardness impurities such as calcium and magnesium causes deposits to form on the cathodes of such an electrolysis cell which rapidly reduces the efficiency of the cell. The electrolysis cell developed by *I. Malkin; U.S. Patent 4,088,550; May 9, 1978; assigned to Diamond*

Shamrock Corporation contains a plurality of cathodes. The efficiency of such a cell is rejuvenated periodically by changing the polarity of less than the total number of cathodes in the cell so as to clean such cathodes while continuing normal operation of the cell by means of the remaining cathodes.

In Figure 5.18, the electrolytic cell consists of a housing **1** in which a single anode **2** and a plurality of cathodes **3** are placed with an appropriate electrodic gap between them and an aqueous electrolyte **4** is placed therein. Depending upon the specific electrolysis reaction, any gases formed during the reaction may be vented to the atmosphere or collected by appropriate means if the gases are toxic or otherwise dangerous.

Figure 5.18: Intermittent Reversal of Cathode Polarity to Remove Deposits

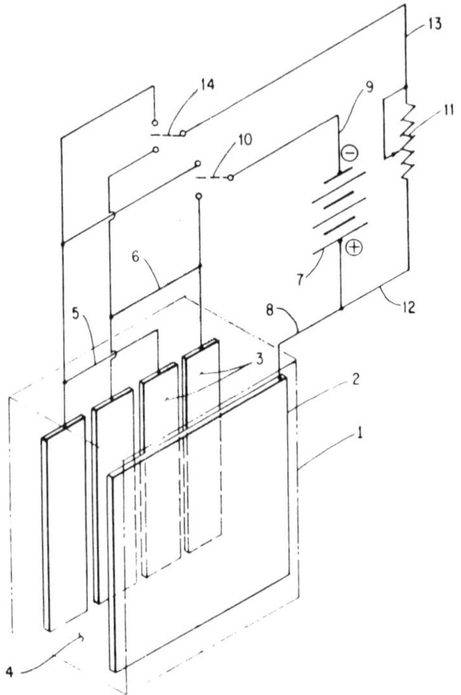

Source: U.S. Patent 4,088,550

Cathodes **3** in the illustrated cell all are alternately interconnected by means of leads **5** and **6**. Both leads can be connected through switch **10** and bus **9** to the negative pole of the power source. The leads may also be connected back to the positive pole of the power source via switch **14**, bus **13**, variable resistor **11** and bus **12**. The anode of the cell is permanently connected to the positive pole of the power source **7** via bus **8**. In actual practice, two of the four illustrated cathodes **3** would be operating in the cleaning mode while the remaining two cathodes **3** would be operational as cathodes in the electrolytic cell.

As shown, the first and third cathodes would be interconnected through leads **5** back through switch **10** and bus **9** to the negative pole of the power source **7**. In this mode the first and third cathodes are operating as cathodes in conjunction with anode **2** to effect the electrolysis. While this electrolysis is taking place, the second and fourth cathodes would be interconnected via lead **6** through switch **14**, bus **13**, variable resistor **11** and bus **12** to the positive pole of the power source **7**.

Thus, a portion of the current from power source **7** would flow back through bus **12**, variable resistor **11**, bus **13**, through switch **14** to make the second and fourth cathodes become slightly anodic. After the lapse of sufficient time to allow the cleaning of the hardness deposits from the second and fourth cathodes, switches **10** and **14** would be changed to have the second and fourth cathodes act as cathodes in the electrolytic cell and have the first and third cathodes then become slightly anodic to effect the cleaning of these cathodes.

Example: A cell of the type shown in Figure 5.18 was assembled. It consisted of a foraminous anode (11 cm x 5 cm x 1½ mm thick) of titanium metal having a coating thereon of ruthenium oxide-titanium oxide in a mol ratio of 2:1 and 6 cathodes of Hastelloy C, each 1 cm x 5 cm x 1½ mm thick. The cathodes were lined in a plane parallel to the anode and spaced approximately 1 cm from each other with the gap between the anode and cathode plane being 2 cm. Means for continuously introducing and removing electrolyte to the cell containing the electrodes was provided.

The electrolyte consisted of 10 g/ℓ sodium chloride which had been dissolved in tap water containing 100 ppm of calcium and 40 ppm of magnesium ion. The anode was operated at a current density of 0.075 A/cm^2.

Those cathodes operating anodically are adjusted to a current density of 0.005 A/cm^2. The mode of cathode operation is reversed for each cathode every 8 hours (3 cathodes operating normally and 3 cathodes operating anodically at any given time). After 1 week the cathodes remained clean of hardness deposits whereas when the same cathodes are operated in the standard manner heavy buildup of hardness deposits occur on the cathode.

PURIFICATION AND TREATMENT

BRINE PURIFICATION

Nearly all commercial salts and brines are contaminated to some degree by various impurities, and while the level at which such impurities may be tolerated is dependent upon the specific process and method of operation, it is desirable in electrochemical uses to obtain a brine of comparatively high purity.

The presence of metallic impurities is of considerable concern in the electrolysis of brine to produce chlorine and alkali metal hydroxide, since trace metals encourage the evolution of hydrogen in the electrochemical cell. Such hydrogen generation is undesirable, as the combination of hydrogen, chlorine and oxygen forms an explosive mixture over a wide range of proportions.

In practice it is preferred to limit the amount of hydrogen present in the chlorine to less than 1%. This is especially critical where the chlorine product is to be liquefied or absorbed, since the proportion of hydrogen in the remaining gas may quickly rise into the explosive range.

The effect of metallic impurities on hydrogen gas evolution in chlor-alkali cells has been extensively studied, and it has been determined that heavy metals such as vanadium, antimony, molybdenum and arsenic have a considerable catalytic effect on hydrogen formation in the cell.

Many other metals such as aluminum, calcium, and magnesium also promote hydrogen evolution, and it has been found that combinations of two or more metals often have more effect than do the same metals taken separately, e.g., magnesium and iron form a synergistic pair. It has thus been the practice to reduce metallic impurities in electrolytic brine to the lowest possible level.

Addition of Magnesium Followed by Magnesium Hydroxide Precipitation

Conventional purification techniques often fail to reduce the level of metals such as aluminum and heavy metals such as antimony, arsenic, molybdenum, and vanadium to the degree necessary for satisfactory use of the brine in electrochemical cells.

A process for the removal of trace metals from alkali halide brines is provided by *Z. Nagy; U.S. Patent 4,073,706; February 14, 1978; assigned to Diamond Shamrock Corporation.* The addition of controlled amounts of magnesium ions to brine and subsequent precipitation of magnesium hydroxide removes metal contaminants, and provides a brine suitable for use in the electrolytic production of chlorine and alkali metal hydroxide.

Example: The purification process was carried out on the brine feed of a full-scale electrolytic chlor-alkali cell room of the mercury type. Using the conventional $NaOH/Na_2CO_3$ purification treatment, the normal NaCl brine feed to this cell room contained about 0.2 ppm Mg and 0.1 ppm Al. This feed brine allowed the cell room to operate normally and produce chlorine gas with an acceptable hydrogen content of about 0.6%.

It was the normal practice to mix a small stream of waste brine into the main stream of brine before purification. During a period of several weeks, the aluminum level in this waste stream increased, while other parameters remained constant. The normal brine treatment was not satisfactory to remove the aluminum, and its concentration built up to about 1.2 ppm in the treated feed brine. Concurrently, the hydrogen in the cell gas increased to the undesirable level of 2.4% even after a current reduction to 75% of full capacity.

At this point an aqueous solution of $MgCl_2$ was continuously metered into the brine at a rate calculated to provide from 5 to 25 ppm Mg^{+2} (or an average of about 10 ppm Mg^{+2}) in the brine. The $MgCl_2$ solution was injected after addition of NaOH and prior to the addition of Na_2CO_3 to the brine. After mixing, the brine was allowed to settle for 8 to 10 hours and was then filtered through a down flow sand and gravel system.

Within 24 hours after the Mg^{+2} treatment was begun, the Al level in the feed brine had been reduced by 90%, and the cell room could again be operated at full capacity with normal hydrogen evolution. Within 72 hours the Al concentration had decreased to less than 0.02 ppm.

Recycle of Precipitated Strontium Carbonate, Calcium Carbonate, and Magnesium Hydroxide

A.B. Gancy and C.J. Kaminski; U.S. Patent 4,115,219; September 19, 1978; assigned to Allied Chemical Corporation provide an improved process for the purification of raw sodium chloride brines containing dissolved impurities including strontium, calcium and magnesium which comprises (a) contacting the raw brine with sodium carbonate and solids recycled from step (d) for formation of strontium carbonate and calcium carbonate solids, (b) contacting the treated brine containing the carbonate solids with sodium hydroxide for formation of magnesium hydroxide solids, (c) removing a major portion of the strontium carbonate, calcium carbonate and magnesium hydroxide solids to provide a purified brine, and (d) recycling a portion of removed solids for admixture with the raw brine and the sodium carbonate.

The quantity of dissolved strontium impurity in the final filtered brine is significantly less when muds recycle is employed, and the quantity of calcium is generally also reduced.

As shown in Figure 6.1, raw brine containing strontium, calcium and magnesium impurities is fed via line **10** to first treatment zone **2** wherein the brine is contacted, preferably with continuous stirring, with sodium carbonate introduced therein via line **12** and with muds which are introduced to treatment zone **2** via line **14** as recycle from settling zone **6**, thereby forming insoluble strontium carbonate and calcium carbonate solids.

The liquor containing these solids is withdrawn from first treatment zone **2** via line **16** and passed to second treatment zone **4** wherein the liquor is contacted with sodium hydroxide, which is introduced to zone **4** via line **20**, resulting in the formation of magnesium hydroxide solids. Liquor containing strontium carbonate, calcium carbonate and magnesium hydroxide solids is withdrawn from zone **4** via line **18** and passed to settling zone **6** wherein the solids are allowed to separate from the liquor by settling.

The settled solids are removed from zone **6** as muds via line **24**. A portion of these muds is recycled to zone **2** via line **14** and the remainder is passed to waste via line **26**.

The clarified liquor, which is substantially free of solids, is removed from zone **6** via line **22** and may be optionally passed to further solids recovery apparatus **8** for removal of any residual solids remaining in the liquor, thereby allowing recovery of purified brine via line **28** from apparatus **8**.

Figure 6.1: Recycle of Precipitated Strontium Carbonate, Calcium Carbonate, and Magnesium Hydroxide

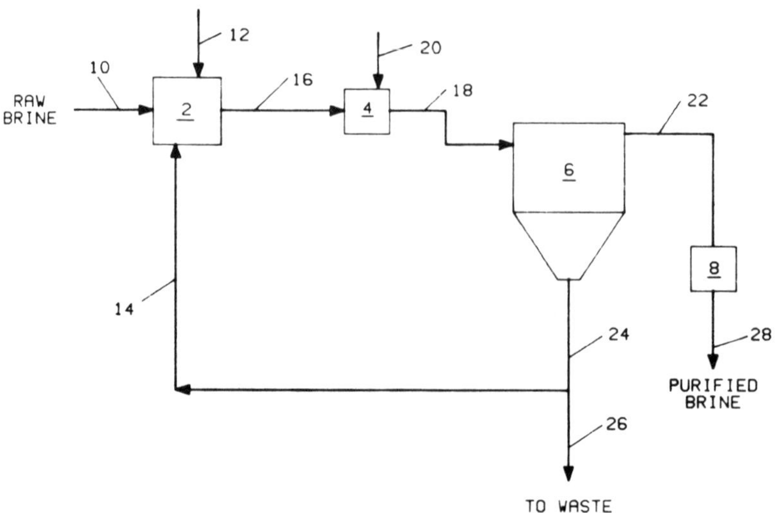

Source: U.S. Patent 4,115,219

Example: To a stirred (about 30 rpm) 1,500 gallon mixing tank are continuously added (1) 150 gpm of a raw brine containing about 25% by weight sodium chloride and 173 ppm calcium, 72 ppm magnesium and 16 ppm strontium dissolved impurities (calculated as elemental calcium, magnesium and strontium, respectively), (2) sufficient brine containing about 25% by weight sodium chloride, saturated with sodium carbonate and containing 25% by weight sodium carbonate solids, to provide a sodium carbonate excess relative to the raw brine of 0.5 gpl, and (3) muds recycled from a clarifier in an amount of about 600% solids.

The solids thus recycled have the following analysis (dry basis): Sr^{++}, 3% by weight; Mg^{++}, 11% by weight; Ca^{++}, 27% by weight; $CO_3^{=}$, 43% by weight; and OH^-, 16% by weight.

Brine is withdrawn from the first tank at a rate sufficient to provide an average residence time therein of 9.5 minutes. The withdrawn brine, to which the sodium carbonate and muds have been added, is passed to a series of two 3,000 gallon reaction tanks, each of which is stirred at about 12 rpm. Brine is allowed to remain in each of these two tanks for an average residence time of 19 minutes, to complete the formation of strontium carbonate and calcium carbonate.

Brine containing these strontium and calcium solids is withdrawn from the last 3,000 gallon reaction tank and passed to a trough wherein an aqueous sodium hydroxide solution containing 10% by weight NaOH is added in an amount sufficient to provide a sodium hydroxide excess of about 0.2 gpl.

The brine to which the sodium hydroxide is added is passed to a 37,500 gallon clarifier which is stirred at a rate of about 4 rpm, producing an essentially clear brine containing a total of 1.8 ppm calcium, 1.2 ppm magnesium and 0.4 ppm strontium (dissolved plus suspended solids; calculated as the elements).

The clear brine is passed to a pressure tube filter for removal of substantially all residual solids, thereby producing a final, filtered brine containing a total of 1.0 ppm calcium, 0.3 ppm magnesium and 0.3 ppm strontium, calculated as the elements.

Removal of Heavy Metals by Adding Iron and Adjusting pH

In the method disclosed by *W.W. Carlin; U.S. Patent 4,155,819; May 22, 1979; assigned to PPG Industries, Inc.* heavy metal impurities are removed from brine by adding a flocculent such as iron to the depleted brine, fortifying the brine while increasing its pH into the basic range, and thereafter adjusting the pH of the brine to approximately pH 7 in order to flocculate vanadium.

Figure 6.2 shows a flow chart for the process. A brine line **3** containing depleted brine goes from the cell **1** to a dechlorinator **5** where hydrochloric acid is added to the depleted brine in order to remove the chlorine therefrom. The dechlorinated, depleted brine then passes to a dissolver or resaturator **7** where solid salt refortifies the brine, e.g., to a level of 315 to about 325 gpl sodium chloride. This refortified brine then passes from the dissolver or resaturator **7** to a neutralizer **9** where sodium hydroxide addition occurs.

The neutralized brine, having a pH of from about 7 to 10 next goes to a precipitator where barium ions, typically in the form of barium carbonate, carbonate ion, typically in the form of sodium carbonate, and sodium bicarbonate are added to precipitate the calcium ion and sulfate ion. In the case of a flowing mercury amalgam electrolytic cell, sulfide ion, normally in the form of sodium sulfide, is added to precipitate mercury.

Figure 6.2: Flow Chart for Removal of Heavy Metals

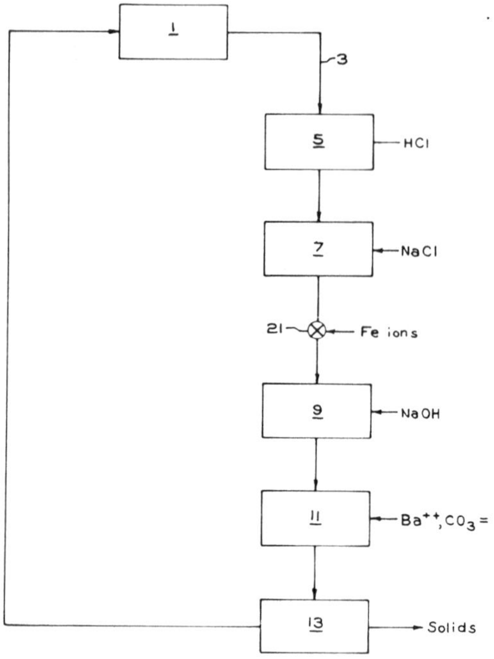

Source: U.S. Patent 4,155,819

The resulting slurry is then passed from the precipitator **11** to filter **13** where it is separated into a solid portion and a liquid portion. The liquid portion is a resaturated, dechlorinated brine reduced in calcium content, sulfate content, and heavy metal content. The brine so treated is typically recirculated to the electrolytic cell **1**.

Iron ion, typically iron chloride, is added at **21** and hydrochloric acid is subsequently added, whereby to maintain the pH of the brine at approximately pH 7, after the brine is refortified in the dissolver or resaturator **7**.

Example: A series of tests were conducted to determine the effect of pH on the removal of vanadium using iron. A 3-liter sample of brine was prepared. 250 ml of water were added to the brine in order to simulate the sodium chloride concentration of dilute effluent brine from a mercury cell. The resulting dilute brine had a sodium chloride content of 275 gpl and a vanadium content of 0.240 mg/ℓ. Three samples of 800 ml each were then taken.

Each sample was heated to 75° to 78°C and its pH adjusted to pH 3 by the addition of HCl. Twenty ml of a 5 gpl solution of $FeSO_4 \cdot 7H_2O$ were added to each sample and the pH was then adjusted to pH 7.5 by the addition of aqueous sodium hydroxide.

One hundred grams of solid sodium chloride were added to each sample. Each sample was stirred and maintained at 50°C for 15 minutes. At this point, the pH of each sample was pH 8.0.

The pH of the first sample was adjusted to pH 7 by the addition of hydrochloric acid. The sample was then allowed to settle for 20 minutes and was thereafter filtered. The resulting filtrate had the following analysis: V, $<$0.01 mg/ℓ; Cr, $<$0.005 mg/ℓ; Ni, 0.02 mg/ℓ; Mn, 0.005 mg/ℓ; and Fe, 0.03 mg/ℓ.

No adjustment was necessary for the second sample, which had a pH of 8. The sample was filtered and a filtrate containing 0.02 mg/ℓ of vanadium was obtained. The pH of the third sample was adjusted to pH 9.0 by the addition of NaOH. The sample was allowed to settle for 20 minutes and was then filtered. The filtrate contained 0.06 mg/ℓ of vanadium.

Alkaline Phosphate Precipitation

In a method developed by *W.B. Darlington; U.S. Patent 4,176,022; Nov. 27, 1979; assigned to PPG Industries, Inc.* brine is first filtered to remove particulates and insolubles, following which it is treated with soda ash and alkali metal hydroxide to lower both the concentration of the calcium ion which frequently is on the order of a fraction of a percent, by calcium carbonate precipitation and that of magnesium ion, by magnesium hydroxide precipitation, down to a total alkaline earth metal ion content of about 2 or 3 ppm, i.e., about 2,000 or 3,000 ppb, of total alkaline earth metal ion impurity.

After the precipitate is removed, as by filtration, the brine is passed through a chelating ion exchange resin to further reduce the alkaline earth metal ion impurity concentration to about 200 to 400 ppb (0.2 to 0.4 ppm) total alkali metal ion impurity concentration.

Thereafter, the brine, at a pH of about pH 10 or higher, is contacted with phosphate ion and, optimally with fluoride ion, to form a calcium phosphate precipitate believed to be an apatite. The precipitate is physically separated from the brine, for example, by filtration, decantation, or centrifugation, to produce an alkali metal chloride brine containing less than 20 ppb of alkaline earth metal ion impurities which alkaline earth metal ion impurities are primarily calcium ion impurities.

Thereafter, the brine, containing less than 20 ppb of alkaline earth metal ion impurities, may, additionally, be passed through a chelating ion exchange resin to reduce the alkaline earth metal ion impurities to below about 2 ppb.

Example: Brine was obtained that had previously been treated by sodium carbonate and sodium hydroxide, followed by passage through a chelating ion exchange resin column. The resulting brine had the following composition: NaCl,

300 to 315 gpl, and Ca^{++}, 400 ppb. To 1.2 liters of the brine were added 0.5352 g of 85% H_3PO_4 followed by 0.3377 g of NaF, and, as a seed, 0.1543 g of $Ca_3(PO_4)_2$ at 74°C.

The brine was mixed for 16 hours with a magnetic stirrer and then filtered through a Millipore filter having 0.45 μ pores. The resulting precipitated filtered brine had a calcium ion content of 24 ppb.

The precipitated filtered brine was then passed through a 30.5 cm x 2.5 cm diameter Amberlite XE 318 chelating ion exchange column at a flow rate of 16 cm^3/min. The resulting brine had a calcium ion content of 14 parts per billion.

Precipitation of Calcium and Sulfate Ions as Glauberite

According to *R. Schäfer; U.S. Patent 4,132,759; January 2, 1979; assigned to Bayer Aktiengesellschaft, Germany* in the process for the production of chlorine and alkali metal hydroxide by electrolysis according to the amalgam process using calcium- and/or sulfate-containing crude salt, the calcium and/or sulfate contents introduced into the brine circuit by the crude salt are removed from the brine by precipitation of the double salt $Na_2SO_4 \cdot CaSO_4$. Small particles of glauberite may be introduced to the brine to initiate and accelerate precipitation.

$Na_2SO_4 \cdot CaSO_4$, which occurs in nature as glauberite, is stable under conditions which can be established in brines for electrolysis purposes. The process is particularly advantageous in cases where a small calcium content is acceptable in the brine and where the salt being processed contains, in molar terms, more sulfate than calcium. This is always the case when, in addition to anhydrite, potassium sulfate, magnesium sulfate and/or polyhalite, for example, are present as impurities in the salt.

Example 1: A depleted brine containing 280 g/ℓ of NaCl, 30 g/ℓ of $SO_4^=$ and 1,120 mg/ℓ of Ca^{++} is strengthened with 30 g/ℓ of salt. The salt contains 1% of sulfate of which half is present as anhydrite and the other half as polyhalite. After the salt has been dissolved, approximately 200 mg/ℓ of anhydrite can be separated off. In other words, the anhydrite had remained largely undissolved under these conditions. The brine contains 30.15 g/ℓ of $SO_4^=$ and 150 mg/ℓ of Ca^{++}.

After the addition of glauberite seeds and a residence time of 1 hour at 75°C the brine is restored to its original composition in terms of Ca^{++} and $SO_4^=$. The deposit consists of well-crystallized rhombi. It can be easily identified under a microscope. X-ray analysis confirms that it is glauberite.

Although the solubility product in the presence of glauberite was not reached during this precipitation, it is nevertheless lower than in the presence of anhydrite.

If it is desired to produce a brine substantially free from calcium (precipitation with excess soda), the sulfate is removed by glauberite precipitation. Example 2 describes a procedure typical of this embodiment.

Example 2: The depleted brine only contains 4 mg/ℓ of Ca^{++}. The other conditions remain the same. Approximately 50% of the anhydrite dissolves, the rest being separated off. The brine now contains 30.2 g/ℓ of $SO_4^=$ and 65 mg/ℓ of Ca^{++}. 5% of the brine circuit is branched off as a component stream to which 2.6 g/ℓ of $CaCl_2$ are added. The sulfate level drops to 25.8 g/ℓ with precipitation of glauberite.

Sufficiently long residence times may readily be adjusted for the small component stream and/or precipitation may be accelerated by the addition of seeds. The glauberite is filtered off (~6 g/ℓ), the solution is combined with the main stream, precipitated with soda and NaOH, filtered and used for electrolysis.

Removal of Silica by Coprecipitation

S. Ogawa, T. Nishimori and T. Kanke; U.S. Patent 4,155,820; May 22, 1979; assigned to Asahi Kasei Kogyo KK, Japan have found that, even in an aqueous sodium chloride solution with a concentration of 10% or more, silica can be adsorbed on precipitates of magnesium hydroxide, calcium carbonate, iron hydroxide, barium sulfate, etc., at the time of precipitation thereof so as to be coprecipitated. The amount of silica adsorbed and coprecipitated therewith can be increased by circulation of these precipitates.

Example: In the flow sheet shown in Figure 6.3, **1** is a cation exchange membrane, **2** an anode chamber, **3** a cathode chamber, **4** an anolyte tank, **5** a catholyte tank, **6** a chlorine gas line, **7** a hydrogen gas line, **8** a purified aqueous sodium chloride solution line containing sodium chloride with a concentration of 310 g/ℓ, and **9** a pure water line for controlling the caustic soda concentration in the cathode chamber. Elements **4** and **2** are under circulation with a part of the dilute aqueous sodium chloride solution being discharged through line **10**.

Figure 6.3: Flow Sheet for Electrolysis Process Which Includes Removal of Silica by Coprecipitation

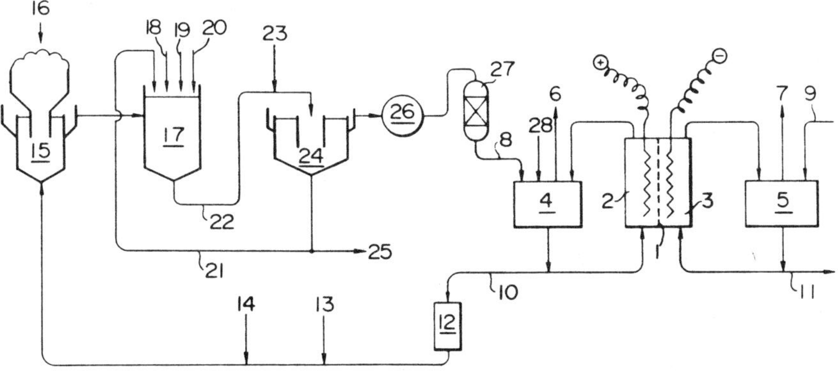

Source: U.S. Patent 4,155,820

Elements **5** and **3** are also under circulation, and the caustic soda formed is discharged through line **11**. Element **12** is a dechlorination tower and **13** is a caustic soda line from which caustic soda is added so that the pH in the sodium chloride dissolving tower **15** may be from 4 to 9.

Element **14** is a line for water from which there is supplemented water to be consumed in the system such as water migrating from the anode chamber to the cathode chamber through a cation exchange membrane or water accompanied with chlorine gas. Element **15** is a sodium chloride dissolving tower.

Element **16** is solid sodium chloride, **17** a reaction vessel, **18** caustic soda, **19** sodium carbonate, **20** barium chloride or barium carbonate, **21** a line for circulating precipitates of the impurities, **22** a feed line to thickener, **23** a line for addition of agglomerating agent, **24** thickener, **25** the precipitates of impurities to be discharged out of the system, **26** a filter for filtration of overflow from thickener, **27** a cation exchange tower filled with chelate resin and **28** a feed line of hydrochloric acid for maintaining the pH at a constant value in the anode chamber.

In this flow, an aqueous sodium chloride solution purified so as to contain 310 g/ℓ of sodium chloride, 20 ppb of calcium ion, 10 ppb of magnesium ion and 1.5 g/ℓ of sulfate ion is added from line **8** and 35% of hydrochloric acid from line **28** into anolyte tank **4** to maintain the concentration of the aqueous sodium chloride solution in the anolyte tank at 180 g/ℓ and at pH 2.

The aqueous sodium chloride solution with the same composition as mentioned above is discharged through line **10**, adjusted at pH 4 to 9 by line **13** and conveyed to the sodium chloride dissolving tower.

The sodium chloride added from **16** has an average composition as follows: calcium, about 0.05%; magnesium, about 0.04%; SO_4, about 0.15%; silica, etc., about 0.02%; and NaCl, about 96.5%.

The above sodium chloride is dissolved in water and allowed to react in the reaction vessel with addition of caustic soda, sodium carbonate and barium carbonate so that the components dissolved in the filtrate from filter **26** may be as follows: calcium, about 10 ppm; magnesium, about 0.3 ppm; and SO_4, about 1.5 g/ℓ.

Accordingly, the following precipitates are formed per liter of saturated aqueous sodium chloride solution from the outlet of the reaction vessel: $CaCO_3$, about 0.200 g/ℓ; $Mg(OH)_2$, about 0.155 g/ℓ; $BaSO_4$, about 0.583 g/ℓ; others, about 0.032 g/ℓ.

When a slurry containing about 80 g/ℓ of precipitates is circulated from the underflow of the thickener to the reaction vessel to vary the concentrations of the precipitates at the outlet of the reaction vessel, the concentrations of soluble silica or other heavy metals in the filtrate at the outlet of the filter **26** are measured to give the results as shown below.

In the above experiment, the reaction vessel is maintained at 60°C with a residence time of about 10 minutes and pH of about 10.2. From the line **23**, 0.7

ppm of an acrylamide-type high molecular agglomerating agent is added. As the result of operation of the thickener at an elevation speed of about 1 m/hr, the amount of precipitates in the overflow of the thickener is about 10 ppm.

	Amount of Slurry Circulated*			
	0	10	20	30
pH	10.2	10.2	10.2	10.2
SiO_2, mg/ℓ	19	11	6	4
V, mg/ℓ	0.067	0.066	0.051	0.044
Cr, mg/ℓ	0.108	0.013	0.0075	0.008
Fe, mg/ℓ	0.08	0.03	0.08	0.03

*Times as much as the amount of impurities to be precipitated.

As apparent from the table, soluble silica and heavy metals are coprecipitated with the increase in amount of the slurry circulated. Thus, maintaining the soluble silica concentration at about 4 ppm and further subjecting the resultant aqueous sodium chloride solution to purification in a cation exchange tower **27** filled with chelate resins to a calcium ion content of 20 ppb and magnesium ion content of 10 ppb, there is obtained a purified aqueous sodium chloride solution.

By adding this solution into the anolyte tank, electrolysis is carried out. When electrolysis is conducted using a cation exchange membrane of the perfluorosulfonic acid type at a current density of 50 A/dm^2 at 90°C, the electrolysis voltage is found to be 4.2 V. On the other hand, when no slurry is circulated from line **21**, the soluble silica concentration is increased to about 19 ppm and the electrolysis voltage is found to be about 4.5 V.

Chelation

N. Yokota, S. Tokuda, Y. Ito and K. Itaya; U.S. Patent 4,119,508; October 10, 1978; assigned to Osaka Soda Co. Ltd., Japan disclose a method of purifying the raw brine used in the electrolysis of the aqueous NaCl or KCl solution by removing the Mg^{++} and Ca^{++} ions from the brine, which comprises contacting the brine with a member selected from the group consisting of:

(A) The chelate-forming water-insoluble resins capable of forming an intramolecular complex with the Mg^{++} and Ca^{++} ions, the resins being selected from the class consisting of the styrene-butadiene copolymer resins containing the group $>N-CH_2COOH$, the epichlorohydrin polymer resins containing the group $>N-CH_2COOH$, the N-phenylglycine-glycidyl methacrylate copolymer resins containing the group $>N-CH_2COOH$ and the styrene-divinylbenzene copolymer resins containing the group $>N-CH_2COOH$; and

(B) The water-insoluble adsorbent solids adsorptively supporting a chelate-forming compound selected from the group consisting of the aminoacetic acids containing at least one $>N-CH_2COOH$ in their molecular structure and the oligomers and the alkali metal salts thereof.

Example 1: A salt-saturated raw brine (NaCl 310 g/ℓ, Mg^{++} 50 mg/ℓ, Ca^{++} 250 mg/ℓ), after adjusting its pH to 5.5, was flowed down at a space velocity of 15 hr^{-1} through a column of 600 mm diameter packed with 200 liters of a chelate resin with 20 to 50 mesh, containing the iminodiacetic acid groups and epichlorohydrin polymer as the main chain. The Mg^{++} and Ca^{++} contents of the purified brine, after passage through the column, were 0.7 mg/ℓ and 1.2 mg/ℓ, respectively.

When this brine was then fed to a mercury type alkali salt electrolysis cell and electrolyzed with a current density of 70 A/dm^2 and a temperature of 75°±2°C, the current efficiency for forming caustic soda was 98.2%.

Example 2: *(Control)* – The starting salt-saturated raw brine used in Example 1 was adjusted to pH = 10.4 by adding caustic soda and sodium carbonate, following which the solution was allowed to stand to settle the resulting precipitate. The supernatant brine was then filtered. The content of Mg^{++} was 3.4 mg/ℓ and that of Ca^{++} was 5.8 mg/ℓ. After adjusting this brine to a pH = 7.2, it was electrolyzed as in Example 1; the current efficiency for forming caustic soda was 94.2%. When the caustic soda used for purifying the brine is considered, the efficiency in obtaining the caustic soda corresponds to 92.4%.

Cation Exchange

According to *E. Zirngiebl; U.S. Patent 4,078,978; March 14, 1978; assigned to Bayer Aktiengesellschaft, Germany* in the electrolysis of an alkaline earth metal-containing alkali metal brine wherein the alkaline earth metals contained in the brine are removed, the brine is subjected to electrolysis, and the residual brine is concentrated and recycled for further electrolysis along with make-up fresh brine. The improved process comprises contacting the alkaline earth metal-containing brine with a weakly acidic cation exchanger in the Na+ form, the exchanger comprising units of at least one of acrylic acid and methacrylic acid, whereby the alkaline earth metals are adsorbed on the cation exchanger, and periodically regenerating the cation exchanger.

The cation exchanger is regenerated by contact with a hydrochloric acid solution of about 5 to 10% concentration, the regenerating solution effluent from the column being combined with the redissolved sodium sulfate which is precipitated during concentration of the cell liquor thereby to precipitate any sulfate contained in the brine as the alkaline earth metal sulfate. Advantageously the precipitation is effected at a pH above about 8 whereby any magnesium present in the brine or regenerating effluent is also precipitated as magnesium hydroxide.

Figure 6.4 is a flow chart of the process. The meanings of the reference numerals used are as follows:

1. Crude brine;
2. Brine recovered by the concentration of cell liquor by evaporation;
3. Brine for use in the electrolysis process;
4. Ion exchanger;
5. Diaphragm electrolysis;
6. Concentration of cell liquor by evaporation;

7. 5 to 10% hydrochloric acid for regeneration;
8. Regeneration solution containing calcium chloride and magnesium chloride;
9. 1 to 10% sodium hydroxide for conditioning;
10. Conditioning solution containing sodium chloride;
11. Electrolysis product hydrogen;
12. Electrolysis product chlorine;
13. 50% sodium hydroxide obtained by concentration by evaporation;
14. Further purification of the sodium hydroxide;
15. Sodium chloride contaminated by sodium sulfate and precipitated during concentration by evaporation;
16. Calcium sulfate and magnesium hydroxide for dumping.

The sodium chloride brine **3**, consisting of substantially equal parts of mined crude brine **1** and brine **2** recovered during concentration of the cell liquor by evaporation, is freed from calcium and magnesium ions in the ion exchanger **4**.

Figure 6.4: Flow Chart for Purification Process

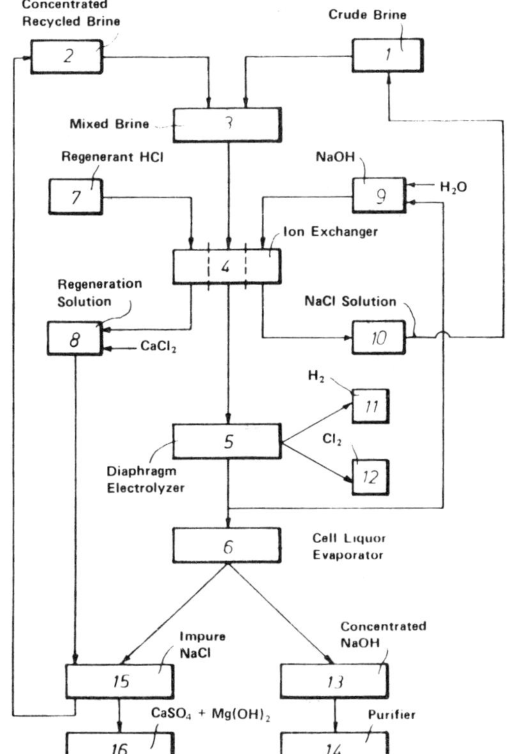

Source: U.S. Patent 4,078,978

The 11% cell liquor issuing from the electrolysis cell, which still contains up to 16% of sodium chloride, is concentrated by evaporation **6** into 50% sodium hydroxide **13**, most of the sodium chloride and, finally, the sodium sulfate being precipitated **15**.

Following separation of the sodium hydroxide **13**, the precipitation product **15** is redissolved and the sulfate ions are precipitated in the form of calcium sulfate by addition of the regeneration solution **8**. Sometimes further addition of $CaCl_2$ is necessary, if there exists a Ca^{++} deficit with respect to $SO_4^{=}$ content in the crude brine. The precipitation reaction is preferably carried out at a pH value above 8 so that the magnesium ions introduced with the regeneration solution **8** are precipitated together with the calcium sulfate in the form of magnesium hydroxide **16**. The calcium sulfate acts as a filtration aid and addition nucleus.

The approximately 25% sodium chloride solution **2** obtained is fed back into the process. Since the throughflow of brine through the ion exchanger **4** during regeneration and conditioning, i.e., conversion into the Na^+ form, is continuous, it is best to use several ion exchangers arranged in parallel.

Dual Chelating Ion-Exchange Towers

An electrolysis of an aqueous solution of sodium chloride is carried out by *Y. Oda, M. Suhara, S. Goto, T. Hukushima, K. Miura and T. Hamano; U.S. Patent 4,202,743; May 13, 1980; assigned to Asahi Glass Company, Limited, Japan* by employing a fluorinated cation exchange membrane having an ion-exchange capacity of 0.8 to 2.0 meq/g dry polymer and having carboxylic acid groups as functional groups and maintaining a concentration of an aqueous solution of sodium hydroxide in a cathode compartment in a range of 20 to 45% by weight and maintaining a calcium concentration in the aqueous solution of sodium chloride at 0.08 mg/ℓ or lower.

The purification of the aqueous solution of sodium chloride to reduce the calcium concentration is carried out in a system comprising the first and second chelate ion-exchange towers which are connected through a detector for detecting impurities in the aqueous solution.

Example: The purification of the aqueous solution of sodium chloride was carried out in the system comprising two ion-exchange towers shown in Figure 6.5.

The first and second ion-exchange towers **1** and **2** were connected in series by a passage in which the detector **3** (Water Analyzer II type, Nippon Technicon KK) and the automatic switching type valves **5, 6, 7, 8, 10, 11, 12** and **13** were connected in the passages to form the purification system.

As the chelate resin, 150 liters of granular ion-exchanger of styrene-divinylbenzene copolymer having iminodiacetic acid groups (diameter of 297 to 1190 μ) (Diaion CR-10, Mitsubishi Chemical Ind. Ltd.) was filled in the towers.

The saturated aqueous solution of sodium chloride having a calcium concentration of 7.8 ppm and a magnesium concentration of 1.5 ppm which was purified by the sedimentation-separation process (concentration of NaCl, about 300 g/ℓ at about 60°C) was fed to the system at a feed rate of about 4.1 m^3/hr and it was continuously treated as follows:

(1) The NaCl aqueous solution was fed from the pipe **4** through the valve **5**, the ion-exchange tower **1**, the valve **6**, the detector **3**, the valve **7**, the ion-exchange tower **2** and the valve **8** and the purified NaCl aqueous solution was discharged through the pipe **9**.

It took about 60 hours until deflection of Ca^{++} and Mg^{++} in the NaCl aqueous solution by the detector **3**. The time was about 75% of the theoretical ion-exchange capacity life of the chelate resin in the ion-exchange tower **1**. (The concentrations of Ca^{++}, Mg^{++} and Fe^{++} in the purified NaCl aqueous solution could not be detected.)

Figure 6.5: Dual Chelating Ion-Exchange Towers

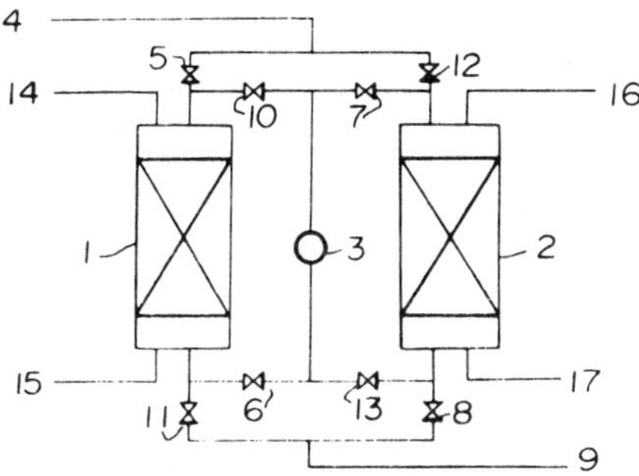

Source: U.S. Patent 4,202,743

(2) At the time of detection of the impurity ions by the detector **3**, the valves were switched to feed the NaCl aqueous solution through the valve **12**, the ion-exchange tower **2** and the valve **8** and the purified NaCl aqueous solution was discharged.

The NaCl aqueous solution in the ion-exchange tower **1** was purged with water and the chelate resin in the ion-exchange tower **1** was regenerated by feeding the 1 N HCl and then 1 N-NaOH for about 5 hours.

(3) After regenerating the ion-exchange tower **1**, the valves were switched and the NaCl aqueous solution discharged from the ion-exchange tower **2** was discharged through the valve **13**, the detector **3**, the valve **10**, the ion-exchange tower **1** and the valve **11** as the purified solution.

It took about 60 hours from the switching of the step (1) to the step (2) to the detection of the impurity ions by the detector **3**. This time was about 75% of the theoretical ion-exchange capacity life of the chelate resin in the ion-exchange tower **2**.

(4) At the time of detection of the impurity ions by the detector **3**, the valves were switched to feed the NaCl aqueous solution through the valve **5**, the ion-exchange tower **1** and the valve **11** to discharge the purified NaCl aqueous solution.

During the operation, the chelate resin in the ion-exchange tower **2** was regenerated in the same manner as that of the ion-exchange tower **1** for 5 hours. After the regeneration of the chelate resin in the ion-exchange tower **2**, the operation was returned to the step (1).

Precipitation by Multistage Evaporation and Cooling

G. Zabotto, J.-M. Guichard and D. Fournier; U.S. Patent 4,087,253; May 2, 1978; assigned to Rhone-Poulenc Industries, France have developed a method of obtaining caustic soda and sodium chloride from an electrolytic cell liquor containing sulfate ions by concentrating the liquor through multiple-effect evaporation, cooling the concentrate obtained, separating the sodium chloride and the salt containing sulfate and recovering the caustic soda, characterized in that during a first stage the solution is treated under conditions of temperature and concentration such that only sodium chloride precipitates from it and is collected.

In a second stage, a solid phase is formed of sodium chloride, sodium sulfate and triple salt of caustic soda, sodium chloride and sodium sulfate, the solid phase being treated with a caustic soda solution of a concentration below 35% by weight, so as to decompose the triple salt, and the salt containing sulfate is separated.

In the third stage, the solution from the second stage is finally cooled so as to precipitate the triple salt of caustic soda, sodium chloride and sodium sulfate, and the triple salt is separated from the soda obtained, which is extracted, and that the solid phase is collected at a stage during which it is dissolved. It is advantageous for the solid phase to be recycled to an upstream stage in the process, to a point such that it is dissolved.

As can be seen from Figure 6.6, the method is carried out in a triple action arrangement comprising three evaporators **1**, **2** and **3**. The heating vapor circuit comprises a boiler (not shown) which supplies vapor under pressure to the first evaporator through a pipe **4**. A set of pipes **5**, **6** and **7** permits vapor circulation countercurrent to the flow of caustic soda liquor to be concentrated, which progresses through pipes **8**, **9**, **10**, **11**, **12**, **13** and **14**.

The initial electrolytic cell liquor is fed through pipe **8** into a dissolving tank **15**, then through pipe **9** into an evaporator **3**. Sodium chloride is withdrawn from evaporator **3** through drain **16**. The remaining liquor passes through pipe **10** into evaporator **2** and pure sodium chloride is again withdrawn through drain **17**. Then the liquor is withdrawn through pipe **11** and fed to a decomposer **18** where the triple salt is decomposed.

Figure 6.6: Flow Diagram for Multistage Purification Process

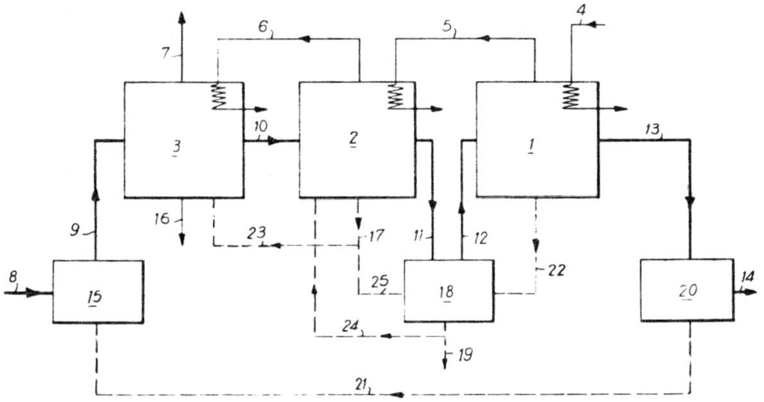

Source: U.S. Patent 4,087,253

The salt containing sulfate is separated from decomposer **18** through an outlet **19** while the liquor is passed through pipe **12** to evaporator **1**. Lastly, the final liquor is removed from evaporator **1** through pipe **13** into cooler **20**. The caustic soda is then cooled, separated from the mixture of salts and particularly triple salt which crystallizes during the cooling process, and the caustic soda collected through pipe **14**.

Apart from this circuit for the main circulation of the liquor, the arrangement further comprises a circuit **21** for recycling the triple salt and sodium chloride from cooler **20** to tank **15** and a circuit **22** for recycling all the solid phase from evaporator **1** to decomposer **18**. It is also possible to provide a circuit **23** for recycling the solid phase from evaporator **2** to evaporator **3**, a circuit **24** for recycling the sulfate salt collected in decomposer **18** to evaporator **2** and a circuit **25** for removing the salt precipitated in **2** to decomposer **18**.

These various circuits, in accordance with the general description, make it possible to maximize the quantity of pure sodium chloride collected, by recycling salts such as the triple salt or a mixture of the salts through circuits **24** and **21** or **25**.

The initial liquor operating conditions are as follows: NaOH, 11.2% by weight; NaCl, 14.5% by weight; Na_2SO_4, 0.6% by weight; and H_2O, 73.6% by weight. The concentration of caustic soda and the operating temperatures of evaporators **1**, **2** and **3** are, respectively: evaporator **1**, 44%, 143°C; evaporator **2**, 22%, 85°C; and evaporator **3**, 16%, 50°C.

Finally, the concentration of caustic soda withdrawn at **14** is 50% by weight and the pure sodium chloride collected contains less than 0.05% of Na_2SO_4. It is thus possible to obtain pure sodium chloride, optimize yields and avoid the hazards represented by the presence of the triple salt.

Utilization of High-Sulfate Salt Bodies

It is becoming increasingly necessary to consider utilizing extant salt bodies which do not meet electrolytic cell requirements but which can be upgraded. Such deposits are generally more remote from electrochemical plants and their utilization necessarily entails greater freighting costs. It is then essential to devise a minimally expensive method of upgrading the raw salt obtained from these lower purity deposits.

Exemplary of existing impure salt bodies of considerable volume are those which have been deposited in primary evaporation ponds, adjacent the western edge of the Great Salt Lake Basin, Utah, during processing of local brines for potassium chloride recovery.

These deposits have sulfate ion contents of from about 0.6 to 1.2% by weight, as compared to typical sulfate contents of 0.15 to 0.3% for salt produced by evaporation of seawater. In view of the distance of these deposits from the majority of potential consumers, a cheap method of sulfate removal is imperative to their utilization.

According to *B.H. Bieler; U.S. Patent 4,094,956; June 13, 1978; assigned to The Dow Chemical Company* the content of sodium sulfate (and less soluble sulfates) in rock salt (halite) can economically be reduced from levels as high as 5% by weight to levels as low as about 0.1% by weight by a process in which the halite is crushed to a certain particle size range and particles less than 0.5 mm in effective diameter are removed while (or after) the crushed material is subjected to attrition washing with a low sulfate, high NaCl brine. The washed, coarse particles are rinsed with a low sulfate brine, drained and dried to an extent appropriate to their contemplated use.

Example: A laboratory attrition washer was made by cutting the top off of a 500 ml separatory funnel (resultant opening ~5 cm in diameter) and cementing the periphery of a circular disc of 35 mesh U.S. Standard screen to the walls of the funnel at a level about half-way between the mouth and the stopcock. A small propeller-type stirrer blade was mounted at the end of a shaft driven by a variable speed motor and extending into the funnel about ⅔ of the distance to the screen.

The washing brine was prepared by saturated deionized water (at about 25°C) with some of the same (representative) primary pond salt (halite) which was to be processed. In two tests, this brine was diluted to 90% saturation.

Each of five essentially identical splits of halite (from primary evaporation pond deposits, Great Salt Lake Basin) was dried under a heat lamp and subjected to just sufficient crushing, with a mortar and pestle, so that all particles would pass through a 4, 8 or 12 mesh U.S. Standard Sieve. Intermediate cuts of the crushed material were retained on either a 16 or 35 mesh sieve.

A 50 g sample of one of the intermediate cuts and 250 ml of the washing brine were introduced to the funnel (stopcock closed) and the propeller speed adjusted until the salt particles were moving across the screen actively enough to make frequent and relatively energetic contact with each other. Stirring was continued for 2 minutes, then the stopcock was opened to permit brine and the fines (the particles which had passed through the screen) to run out. Stirring was discontinued when the brine level dropped to the level of the stirrer blade. The wet, coarse salt crystals above the screen were dumped out onto an 8" diameter, 42 mesh U.S. Standard sieve screen, sampled for analysis and washed with a spray of brackish water (25 ml) containing 0.02% by weight of sulfate and a total of 0.6% of dissolved solids. The product crystals were dried in an evaporating dish, weighed and sampled for sulfate analysis.

Two otherwise essentially identical runs were made (with -8+35 mesh crushed halite cuts) in which the periods of attrition-washing were 0.5 and 5 minutes.

The effects of particle size range in the crushed halite on product crystal yield (recovery) and sulfate content are apparent from the data in Table 1. The effects of varying the duration of the attrition-washing are evident from Table 2. The effects of spray washing are apparent from both tables.

Table 1: Effect of Particle Size Range on Yield and Sulfate Content of Product*

Particle size	-4 + 16	-8 + 16**	-8 + 16**	-8 + 35	-12 + 35
Wt % sulfate in product***					
Before spray wash	0.40	0.40	0.35	0.50	0.29
After spray wash	0.40	0.20	0.22	0.33	0.29
% NaCl recovery from					
Uncrushed halite	63	56	51	79	66
Crushed cut	76	ND	90	91	92

*Recovered from halite initially containing 0.8% Na_2SO_4.
**Washing brine only 90% saturated in NaCl.
***$SO_4^=$, determined by gravimetric analysis.

Table 2: Effect of Duration of Attrition-Washing on Sulfate Content and Yield of NaCl Product

Attrition wash duration, minutes	0.5	2.0	5.0
Weight % sulfate in product*			
Before attrition wash	0.84	0.75	0.86
Before spray wash	0.55	0.50	0.30
After spray wash	0.45	0.33	0.28
%NaCl recovery**			
After spray wash	91	90	84

*$SO_4^=$, determined by gravimetric analysis.
**From crushed cut.

On the basis of the data in the foregoing tables, a reasonable compromise between high NaCl recovery and low sulfate contents in the recovered product would appear to be achievable by maximizing the proportion of the halite pieces converted to -8+35 mesh material and attrition-washing for about 2 to 4 minutes. It is also evident that rinsing is essential if it is desired to attain minimal sulfate contents in the product crystals.

Although any suitable drying method may be employed to reduce the moisture content of the NaCl product, the most economical, and therefore the most preferred, method is by simple evaporation from windrows exposed to the atmosphere and warmed by sunlight. However, where the requisite heat can be provided at a tolerable cost, as by waste process steam, solar heating, wind generators or combustion of organic wastes, resort may be had to forced circulation of heated air over windrows (in appropriate sheds) or in rotary driers.

Removal of Chlorate from Recirculating Brine Stream

In the production of chlorine and caustic by the electrolytic decomposition of brine in a membrane cell, depleted anolyte is often recirculated with salt resaturation. Chlorate build-up in this recirculating brine results from membrane inefficiencies, and has in the past required purging.

P. Lai, S. Szymanski and N.L. Christensen; U.S. Patent 4,169,773; October 2, 1979; assigned to Hooker Chemicals & Plastics Corp. have found that a portion of the recirculating brine stream may be reacted with strong acid, such as HCl, to reduce the chlorate, resulting in production of additional chlorine, water, and salt. Such chlorine may be joined with the cell product, while the salt may be utilized in the resaturation of the remainder of the recirculating brine stream.

Example: A Hooker MX membrane cell is utilized for the manufacture of chlorine and caustic, as illustrated in Figure 6.7.

Figure 6.7: Removal of Chlorate from Recirculating Brine Stream

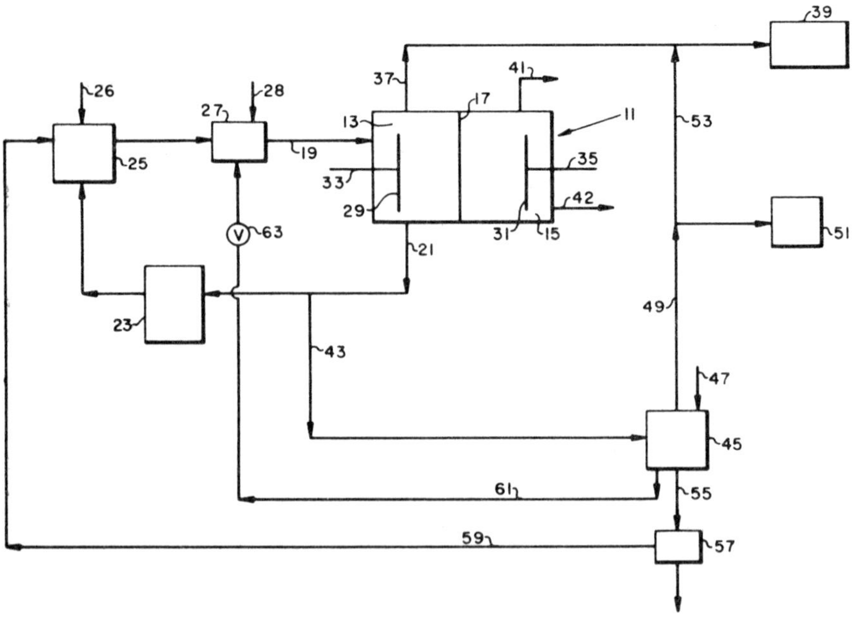

Source: U.S. Patent 4,169,773

After equilibrium is reached, the following product streams and approximate material balances result, in pounds per hour per ton of Cl_2 produced. The anolyte recirculation stream **21** consists of about 15,732 pounds NaCl, 1,096 pounds $NaClO_3$, 115 pounds NaOCl, and 49,806 pounds H_2O. A treatment stream **43**, comprising 1.8% of the volume of the brine recirculation stream, and containing about 280 pounds NaCl, 19 pounds $NaClO_3$, 2 pounds NaOCl, and 886 pounds H_2O, is fed to reaction vessel **45**, where it is reacted with about 21 pounds HCl, and 48 pounds H_2O. The reaction vessel yields 818 pounds H_2O, 22 pounds Cl_2, 2 pounds O_2.

In addition, 293 pounds NaCl and 15 pounds water are taken from the crystallizer **57** to the resaturator **25**, where they are joined with 2,985 pounds of crystalline salt to resaturate the brine. In the pH adjustment, 6 pounds HCl and 13 pounds H_2O are added to the brine prior to reentry into cell **11**.

Also exiting the cell, from the anolyte compartment, are 16 pounds of O_2, 454 pounds of H_2O, and 1,898 pounds of Cl_2, which when joined by 102 pounds of Cl_2 recovered by the dechlorination unit **23**, yields one ton of chlorine per hour. The dechlorination unit is fed 131 pounds HCl and 306 pounds H_2O per hour, yielding 37 pounds H_2O in addition to the aforementioned Cl_2. From the catholyte compartment of cell **11**, 60 pounds H_2, 2,162 pounds NaOH, and 9,002 pounds H_2O are withdrawn.

The chlorate reduction reaction takes place under ultraviolet radiation, insuring that no ClO_2 is produced in reaction vessel **45**. Operating at a temperature of from about 70° to 90°C, essentially complete removal of chlorate from the treated stream is achieved.

From the above, it is seen that a substantial reduction of chlorate content in the recirculating brine system is possible, with the resulting products being either used directly in the chlor-alkali system itself, or being joined with the cell products to increase yield.

Pretreatment with Sodium Hypochlorite Formed from Reaction Products

Electrolysis of impure saline solutions containing dissolved iron and manganese causes deposits to form on the anodes of such electrolysis cell which rapidly reduces the efficiency of the cell. According to *J.E. Bennett and J.E. Elliott; U.S. Patent 4,085,014; April 18, 1978; assigned to Diamond Shamrock Corporation* pretreating the impure saline solutions with sodium hypochlorite solution formed from the reaction products of the electrolysis cell precipitates the iron and manganese dissolved therein and effectively prevents the formation of the deleterious deposits on the anode of the electrolytic cell.

Example: A Sanilec (Diamond Shamrock Chemical Co.) seawater cell was set up for operation on New York harbor seawater. This seawater cell had a capacity for producing 150 lb/day of available chlorine in the form of sodium hypochlorite solution. The Sanilec cell actually consists of two separate cells each capable of producing 75 lb/day of available chlorine.

Each cell consists of three anodes and two cathodes. The electrodes are arranged parallel to one another and are made of expanded mesh titanium with a dimensionally stable coating. The active electrolytic area was 1,250 in^2. The cell was operated at 1 A/in^2 and seawater was fed to the cell at a rate such that the effluent from the cell contained 0.05% sodium hypochlorite.

At start-up, the current efficiency of the cell was 76%. The efficiency rapidly declined until it had decreased to 44% current efficiency after a lapse of only 10 days. Examination of the cell showed essentially no hardness build-up on the cathode but considerable iron and manganese build-up on the anode due to the presence of high amounts of manganese in the New York seawater.

Analysis at various times showed the manganese content of the New York harbor seawater to range from 50 to 200 ppb manganese. The operating temperature of the cell was in the range of 17° to 18°C and the incoming seawater had a salinity of 69 to 74%, that of normal ocean water. Acid washing of the contaminated anode resulted in the recovery of current efficiency for the cell to 82%.

PRODUCT PURIFICATION AND CONCENTRATION

Purification of Sodium Hydroxide at Low Cost

A.D. Babinsky and L.L. Benezra; U.S. Patent 4,093,531; June 6, 1978; assigned to Diamond Shamrock Corporation describe an improved method and apparatus for purification and concentration of a spent diaphragm cell liquor to remove sodium chloride, chlorates, metal impurities of various ionic character, and asbestos fibers from the cell liquor in a one-step process at a savings of costs to obtain the marketable sodium hydroxide commodity.

This method employs the use of a three-compartment cell having a porous catalytic anode, a porous asbestos diaphragm separating the anode compartment from a central compartment where the cell liquor is added, a cation-exchange membrane separating the central compartment and the cathode compartment and feeding of hydrogen gas emanating from the anode and cathode compartments into the porous catalytic anode to decrease the potential across the cell below the evolution potential for chlorine gas and coincidently reduce the amount of electrical power necessary for the purification and concentration process.

In a typical operation of electrolytic cell **10**, shown in Figure 6.8, for the purification and concentration of a cell liquor such as that from a chlorine and caustic production system, comprising mainly sodium hydroxide or potassium hydroxide, a three-compartment cell is assembled, having an anode compartment with an anode therein, a central compartment, and a cathode compartment with a cathode therein.

The electrolytic cell **10** must be effectively sealed with compressive force or gasketing. If bolts are used, they must be insulated to prevent short circuiting of the electrolytic cell **10**. The porous catalytic anode **24** could be a platinum oxide gauze supported by a titanium screen such that hydrogen may be supplied to the hydrogen supply line **26** and present flow through the anode compartment **12** to the escape port **30**.

A porous diaphragm **18** is used to divide the anode compartment **12** from the central compartment **14** and can be made of an asbestos type material. A hydraulically impermeable cation-exchange membrane **20** is used to divide the central compartment **14** from the cathode compartment **16** and preferably would have a sulfonic acid pendant group equivalent weight in the range of 1,000 to 1,400.

Figure 6.8: Side Section View Taken on Vertical Plane of Cell Used to Purify Cell Liquor

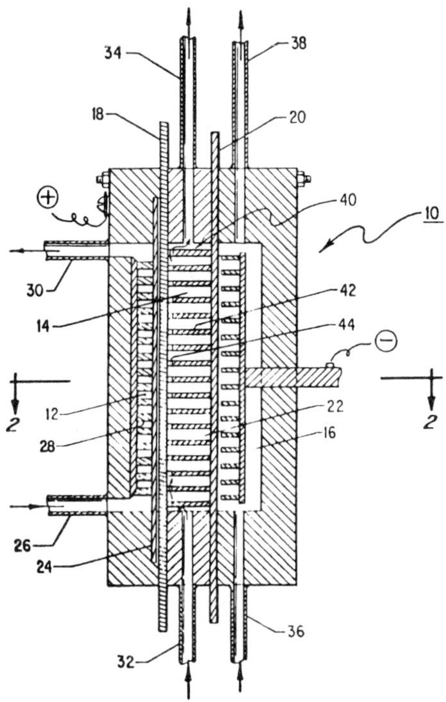

Source: U.S. Patent 4,093,531

The cathode compartment **16** is initially charged through the charge opening **36** with a solution of pure alkali metal hydroxide. The particular concentration of this initial charge does not seem to be of great significance to the overall operation with the exception that it will take a longer period of time to bring a lower concentration of this initial charge up to the resultant concentration. Generally, as a matter of convenience, this solution will be approximately 30 to 45% sodium hydroxide or potassium hydroxide as the case may be.

The central compartment **14** is charged with the cell liquor to be purified and concentrated such as a spent cell liquor from a chlorine and caustic electrolytic cell through cell liquor ingress opening **32**. As electrolyzing current is impressed upon the electrolytic cell **10** such that the anode potential will be maintained below the chlorine gas evolution potential. The cell operation temperature can range from 75° to 100°C with a preferred range of 80° to 95°C.

Hydrogen gas is inserted into the hydrogen supply at the gauge pressure required to maintain the hydrogen-electrolyte interface in the catalytic region of the porous catalytic anode **24**, and any necessary pumps and circulation equipment shall be hooked up to egress **38** and escape port **30** to recirculate the hydrogen escaping from either of those two ports back into the hydrogen supply **26** maintaining this pressure at all times. The cell liquor solution in the central compartment **14** is recirculated at a rate sufficient to allow essentially complete depletion of alkali metal cations by the time the solution reaches egress **34**.

Generally the spent cell liquor inserted into the central compartment **14** will have a sodium hydroxide concentration of approximately 11% and a sodium chloride concentration of approximately 14%. The solution exiting from egress **34** of the central compartment **14** will generally have a sodium hydroxide concentration of approximately 6% and a sodium chloride concentration of approximately 18% plus whatever impurities may be present in cell liquor which are being separated from the alkali metal hydroxide. With careful control sodium hydroxide concentrations as low as 0.5% can be achieved.

The purified and concentrated alkali metal hydroxide exiting from the egress **38** of the cathode compartment **16** will generally have a resultant sodium hydroxide concentration of 30 to 40% and a resultant sodium chloride concentration of approximately 0.1 through 0.5%. The current efficiencies found for electrolytic cells operated in this manner run from a low of approximately 60% to a high of approximately 80%.

Ion Exchange

Highly purified alkali metal hydroxide and carbonate solutions are obtained by *R.M. Wheaton; U.S. Patent 4,154,801; May 15, 1979; assigned to The Dow Chemical Company* by treating self-contaminated solutions thereof with composite ion exchange resin bodies having low residual amounts of quaternary ammonium cation not intimately associated with carboxylate anions to minimize inorganic chloride and chlorate anion leakage.

In the polymerization of the carboxylate carrier with a crosslinked quaternary ammonium anion exchange resin, a certain balance of excess + and - sites (on a molecular scale) which have been found to exist as a perfect balance of sites (on a gross or localized basis) cannot be achieved.

Such excesses can be measured in ion-exchange capacity units (milliequivalents per wet milliliter, meq/ml of resin) and reported as $\Delta C+$ and $\Delta C-$, both components ordinarily being present in any given preparation because of localized excesses. The magnitudes of these $\Delta C+$ and $\Delta C-$ aspects have been found to have significant impact on the performance of the composite resin in removal of the salt impurities from certain alkali metal solutions as well as on the regenerative properties of the resin. Accordingly, it is critical and essential that levels of quaternary ammonium ion capacity in excess of the carboxylate ion capacity of the composite resin be minimized at any locale within a given composite resin body.

Thus, the $\Delta C+$ must be about or less than 0.02 meq/ml, preferably about or less than 0.01 meq/ml of bulk resin volume. While the $\Delta C-$ level is not as critical as the $\Delta C+$ level, it is preferably maintained at less than about 1.0, most preferably less than about 0.5 meq/ml of bulk resin volume.

Example 1: This example illustrates the preparation of a composite ion-exchange resin having the desired $\Delta C+$ level. The absorption capability of the resin is defined as the separation factor (S.F.) which is expressed in terms of resin bed volumes and is determined by the difference in bed volumes between the midpoints of the leading edges of sodium hydroxide and salt effluent concentration waves derived by plotting the ratios of concentration out/concentration in versus bed volumes (effluent volume/resin volume). The greater the separation factor, the greater the capacity of the resin to absorb salt impurities.

175 g of water, 250 g (0.5 mol) of an amine-derived styrene divinylbenzene copolymer (styrene, 85% by weight; divinylbenzene, 7.5% by weight; ethylvinylbenzene, 7.5% by weight) and 60 g (0.83 mol) glacial acrylic acid (equivalent ratio acrylic acid: styrene copolymer = 1.66:1.0) were mixed with stirring for a period of about 85 minutes at ambient temperatures.

6.0 g of $Na_2S_2O_8$ catalyst were added and the mixing continued for about 40 minutes. The reaction mixture was then heated gently. Exotherm heating commenced at about 55°C, attaining a maximum at about 80°C. The reaction mixture was then maintained with stirring at about 70°C for 17 hours. Water and sodium carbonate were then slowly added until the reaction mixture was neutralized, after which the mixture was stirred for about 150 minutes.

The reaction mixture was then filtered and the resin solids washed, dried and analyzed for $\Delta C+$ and $\Delta C-$. Since the resin was found to have a $\Delta C+$ of 0.2 meq/ml, the resin was further heated in a 4% sodium hydroxide solution at 75° to 85°C for about 4¼ hours and then at about 100° to 105°C for about 2 hours to neutralize localized, unassociated excesses of quaternary ammonium ions. As a result of such operations, a resin having a $\Delta C+$ of 0.010 and a $\Delta C-$ of 0.46 meq/ml, based on bulk resin volume, was obtained.

About 100 ml of the resin thus obtained were placed in a fixed bed column 91 cm x 1.2 cm in diameter and a 25% sodium hydroxide solution containing 7 mg/ℓ chloride was fed through the resin column at a rate of 4 ml/min and at a temperature of about 50°C. No salt was detected in the NaOH product stream after passage of the first 1.2 bed volume of solution, with a salt level of about 60 ppm being noted after passage of about 1.5 bed volumes. The resin had a high S.F. of greater than about 3.0.

Example 2: A number of NaOH solutions containing appreciable levels of NaCl were separately fed to a fixed resin bed column as described in Example 1 and containing a composite ion-exchange resin having a $\Delta C+$ of about 0.02 meq/ml or less. Introduction of each feed solution to the resin column was continued until the effluent volume equaled at least one-half the volume fed. A separation factor (S.F.) expressed in resin bed volumes as in the above example was determined for each feed solution.

Based on such studies, it was discovered that good salt removal from the NaOH solutions was achieved at sodium hydroxide concentrations ranging from 5 to

35% by weight, while optimum separation was observed in the 25 to 30% range (S.F. >4.0). The separation factor declined rapidly above about 30% by weight NaOH (S.F. of 50% NaOH being about 0.3).

Elimination of Hypochlorite

The commercial use of membrane cell has been hindered by the relatively small choice of membranes available that will withstand the environment of an electrolytic cell. Many types of ion-exchange membrane materials that would otherwise be extremely useful are susceptible to deterioration and distortion under cell operating conditions.

A major cause of such deterioration is the presence in the electrolyte of minor amounts of hypochlorites generated by the cell which react with the membrane material to its detriment. The method developed by *E.H. Cook, Jr. and D. Pouli; U.S. Patent 4,146,445; March 27, 1979; assigned to Hooker Chemicals & Plastics Corp.* eliminates this problem and makes a wide variety of ion-exchange materials available for use in membrane cells, which previously were deemed unsuited for such use.

A method for electrolytically producing a purified alkali metal hydroxide solution by electrolyzing an alkali halide solution in a membrane cell is described. The membrane cell includes at least one buffer compartment separating an anolyte and a catholyte compartment. The barrier separating the anolyte and the buffer compartment is a porous or permeable membrane. The membrane separating the catholyte compartment and the buffer compartment is a cation-active, hydrocarbon, preferably nonporous, membrane.

A minor amount of a compound selected from the group consisting of alkali metal sulfites, bisulfites, sulfides, oxalates, or mixtures thereof, is added to the electrolyte, preferably with the brine feed, to protect the membranes from hypohalites formed in the cell.

Amounts of protective compounds selected from alkali metal sulfites, bisulfites, oxalates, sulfides, or mixtures thereof, between about 0.01 and 2.0% by weight of electrolyte are found to be aptly suited to use. The protective compounds are preferably added along with the sodium chloride feed solution.

Removing Nitrogen Trichloride from Chlorine Gas

In a process for making chlorine electrolytically in which a build-up of nitrogen trichloride occurs in the bottoms of cooling apparatus, a method and apparatus is provided by *E.N. Balko and S.D. Argade; U.S. Patent 4,138,296; February 6, 1979; assigned to BASF Wyandotte Corporation* wherein the nitrogen trichloride is dissolved in an organic solvent such as carbon tetrachloride, the chlorine removed, and the solution treated to destroy the otherwise hazardous nitrogen trichloride. The solvent is then separated and recycled to avoid environmentally undesirable waste products.

There is shown in Figure 6.9 a flow diagram illustrating a typical apparatus **11** for the processing of electrolytically produced chlorine from a brine, wherein nitrogen trichloride appears as a contaminant in the chlorine, and an auxiliary apparatus **12** for the removal and safe destruction of the nitrogen trichloride.

Figure 6.9: Flow Diagram for Process Which Includes Removal of Nitrogen Trichloride from Chlorine Gas

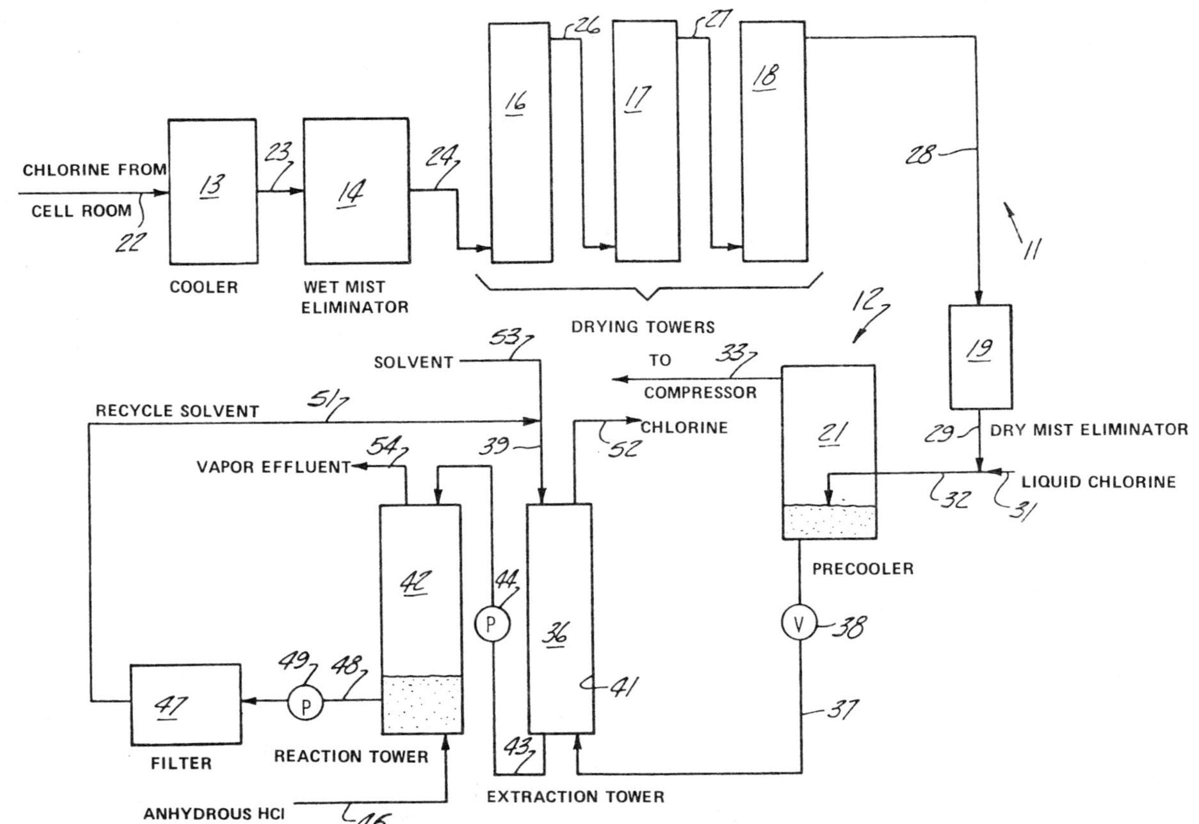

Source: U.S. Patent 4,138,296

Auxiliary apparatus **12** comprises an extraction device **36**, means for moving condensate from the precooler **21** to the extraction device **36** in the form of line **37** equipped with a valve **38** (for gravity feed) or a pump, means for adding an inert solvent to the extraction tower through line **39**, means for heating the extraction device to remove much of the chlorine therefrom (ambient temperatures are sufficient to supply heat through walls **41** and vaporize the chlorine), reactor means **42**, means for moving the contents of the extraction tower **36** to the reactor means **42** in the form of line **43** and a pump **44**, means for adding a reaction component capable of reacting with nitrogen trichloride and converting it to ammonium chloride through line **46**, means (filter **47** or possibly a centrifuge) for separating the solvent from the reaction mixture after the conversion to ammonium chloride, and means for recycling the separated solvent to the extraction device.

The separation and recycling is provided by line **48** communicating between the reaction tower **42** and filter **47**, with pump **49** located in line **48** to move the slurry from the reaction tower **42** to the filter **47**; and the solvent passing through the filter through recycle line **51** to line **39** and back into the extraction tower.

In operation, a typical chlorine stream in line **22** from the cell room will have a temperature of about 190°F. After cooling, it will enter the drying towers at about 55° to 60°F and leave the drying towers at about 80° to 90°F. This stream is then joined with liquid chlorine (the boiling point of which is -34°C or -29°F).

The partial pressure of nitrogen trichloride at -10°C is 60 torr and at -20°C is about 20 torr. Thus it is seen that the nitrogen trichloride condenses into the liquid chlorine and accumulates in the precooler bottoms. The amount of nitrogen trichloride accumulation will depend somewhat on the efficiency of removal thereof at the final separation from the chlorine streams.

For example, the concentration of nitrogen trichloride in the final chlorine product may range from about 1 to 10 ppm depending upon the efficiency of the final purification. This, in turn, will vary with the expected use of the chlorine, and certain direct in-plant use will allow for the higher concentrations while generally commercial sales should be at the lower end of the scale or less. Use of a good purification tower, of course, lowers the nitrogen trichloride in the final product.

Thus it is seen that the rate of accumulation of nitrogen trichloride will depend upon the amount generated in the system, and the efficiency of removal. In any event, nitrogen trichloride accumulates in the condensate in most systems and must be disposed of. Various methods of checking the nitrogen trichloride in the condensate have been used.

In some plants, a program was developed to dispose of the condensate before build-up gets too high. This procedure has the disadvantage that a change in operations could make the scheduled disposal dangerous. For example, a plant explosion was caused by the change in the concentration of ammonia in the cooling water.

It is therefore recommended that the condensate be monitored by analysis to determine the safe optimum time for disposing of each batch of condensates.

Such analytical data can then be used to minimize the excess of material used to destroy the nitrogen trichloride.

Referring again to Figure 6.9 and the procedure of operation, valve **38** is opened to drain a batch of condensate from precooler **21** or equivalent. It is recommended that the condensate have less than about 2% nitrogen trichloride therein. The condensate flows by gravity to the extraction tower **36** wherein contact by simple liquid mixing or by a spray of organic solvent moving countercurrent to the vapor scrubs out the nitrogen trichloride from the chlorine as it vaporizes.

The chlorine gas is then removed through line **52** to chlorine disposal. If desired, this chlorine could be recycled back to the purification system, but the amounts of chlorine evaporated here are so small it may be economically more feasible to dispose of it.

The organic solvent enters the extraction tower from recycle line **51**, or in the case of start-up from line **53**, or in the case of make-up from both lines **51** and **53**.

In general, any organic solvent which is inert or substantially inert to the chlorine may be used. Carbon tetrachloride is preferred since it is completely inert, and trichloromethane or chloroform is eminently satisfactory although some chlorination thereof to carbon tetrachloride may occur. Hexachloroethane, or any highly chlorinated lower aliphatic hydrocarbon may be used so long as it is a good solvent for the nitrogen trichloride. Thus, chlorinated methanes, chlorinated ethanes and chlorinated propanes in which molecular proportion of chlorine to hydrogen exceeds 3:2 are all suitable.

The amount of solvent required to keep the nitrogen trichloride safe is not clearly known. However, the more concentrated the nitrogen trichloride, the more hazardous the solution. In general, it is believed that the concentration of nitrogen trichloride should not exceed 5% by weight in solution and it is preferred to keep the concentration below about 2.0% by weight. Accordingly, the chlorine removed in the extraction tower **36** is replaced completely by solvent to retain a safe dilution. In this operation, substantially all of the chlorine is also preferably removed, because chlorine operates to reverse the equilibrium of the reaction used to destroy nitrogen trichloride.

After the extraction operation is complete, the solution is pumped to the reaction tower **42**, by means of pump **44** located in line **43**. Anhydrous hydrogen chloride is then bubbled through the solution to react with the nitrogen trichloride therein. The reaction proceeds quantitatively as shown below:

$$NCl_3 + 4HCl\ (gas) \rightarrow NH_4Cl + 3Cl_2 \uparrow$$

The anhydrous hydrogen chloride may be added as a substantially pure gas or mixed with other dry gases. Preferably, dry nitrogen or dry air is mixed with the hydrogen chloride and added therewith to provide greater agitation and to strip chlorine from the solvent in order to minimize the back reaction.

The chlorine thus formed in the reaction is removed through line **54** and disposed of. As was the case of the chlorine in line **52**, the amounts of chlorine are too small to be worth recovering. In addition, the chlorine removed in line

54 will contain excess hydrogen chloride together with any nitrogen or air that was added. However, in certain brine plants, such as mercury cell, this chlorine-hydrochloric acid mixture may be conveniently recycled to the plant.

The solvent from the reaction tower **42**, is removed through line **48** and pumped through filter **47** by pump **49**. The solvent passing through the filter is then returned to the extraction tower **36** through recycle line **51**. The amount of hydrogen chloride to be used in the reaction tower is preferably determined by the analyses of the nitrogen trichloride removed from the precooler. In this way, a small excess, about 20% excess, is sufficient to prevent a build-up of nitrogen trichloride in the solvent recycle system. If desired, this solvent may also be checked by analysis.

Energy-Efficient Concentration of Caustic

Heated substances, e.g., electrolytes or gases from electrodes, formed during electrolysis of an alkali halide in an electrolytic cell using a cation exchange membrane to separate anode and cathode chambers are used effectively by *A. Kazihara, S. Ogawa, T. Kobayashi and M. Seko; U.S. Patent 4,090,932; May 23, 1978; assigned to Asahi Kasei Kogyo KK, Japan* as a heat source for heating an aqueous caustic alkali solution to be concentrated in an evaporator.

Example 1: Electrolysis of sodium chloride is carried out according to the flow sheet as shown in Figure 6.10a. As cation exchange membrane, there is used a membrane having carboxylic acid groups only on the cathode side on a base polymer of fluorocarbon resin having pendant sulfonic acid groups.

In a bipolar system electrolytic cell which is divided by this cation exchange membrane into cathode and anode chambers, electrolysis is carried out at a current density of 55 A/dm^2. The starting saturated aqueous sodium chloride solution is fed through lines **23** and **24**.

From the anode, chlorine gas is generated and discharged through lines **26, 27** and **28**. From the cathode, hydrogen gas is generated and discharged through lines **19** and **20**.

As the catholyte, 25% by weight aqueous caustic soda solution is obtained at a current efficiency of 90%. The electrolysis voltage per one cell is 4.71 V. From line **33**, 50% by weight aqueous caustic soda solution is obtained at the flow rate of 15 tons/hr.

The temperatures, flow rates and exchanged heat at each heat exchanger at principal points are shown in Figure 6.10b. In this example, heat exchange is not substantially conducted in the heat exchangers **7, 9, 10** and **30** in Figure 6.10a.

Example 2: Electrolysis is carried out according to the flow sheet as shown in Figure 6.10a wherein the system of double-effect evaporator as shown in Figure 6.10c is arranged at the line **18**. The temperatures, flow rates and exchanged heat in the heat exchangers at principal points are shown in Figure 6.10d.

The same cation exchange membrane and electrolytic cell as used in Example 1 are used.

Figure 6.10: Flow Sheets and Heat and Mass Balance for Concentrating Caustic

a.

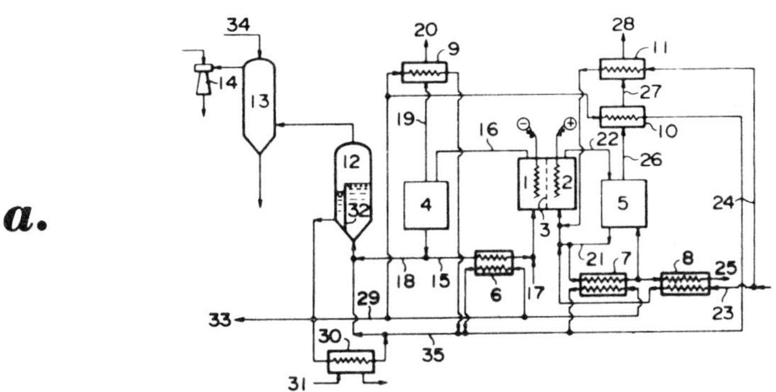

b.

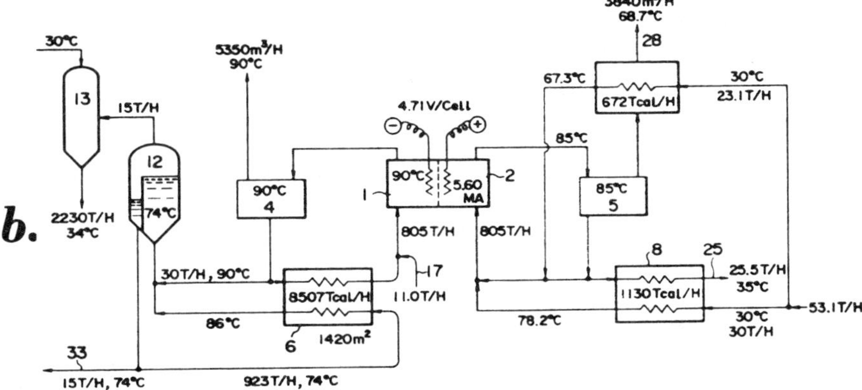

c.

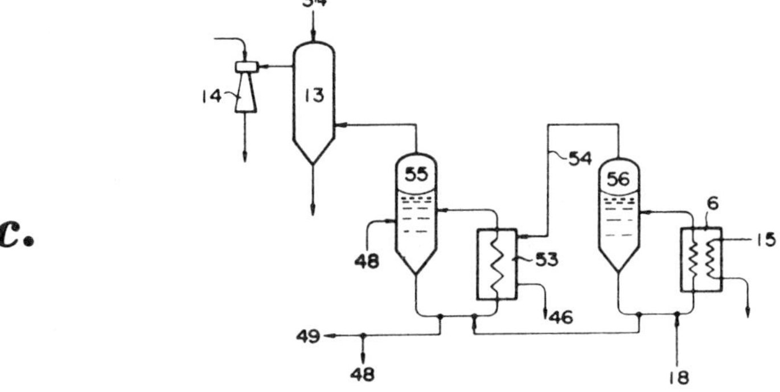

(continued)

Figure 6.10 (continued)

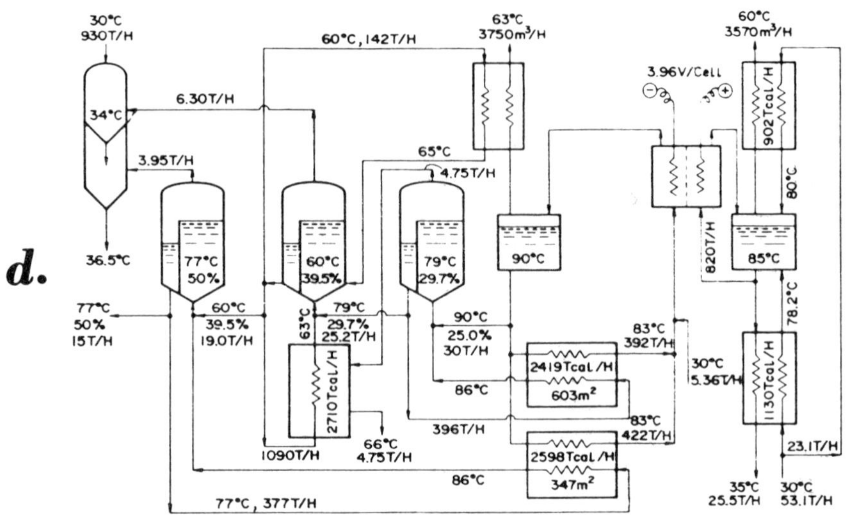

(a) Flow sheet of process
(b) Heat and mass balance for Figure 6.10a
(c) Flow sheet with multiple-effect evaporator
(d) Heat and mass balance for 6.10c

Source: U.S. Patent 4,090,932

As the result of reuse of 4.75 tons/hr of steam due to the effect of double-effect evaporator, the current density is decreased to 36 A/dm^2 and the electrolysis voltage to 3.96 V per one cell. When electrolysis is carried out under otherwise substantially the same conditions as in Example 1, the heat balance can be kept in the entire system. Accordingly, while maintaining low electric power unit without use of boiler steam, 50% aqueous caustic soda solution can be obtained.

Three-Unit Process for Concentrating Caustic

According to *J. Nakata, Y. Chiba, T. Seto and Y. Fukuhara; U.S. Patent 4,145,265; March 20, 1979; assigned to Asahi Glass Company, Ltd. and Ishikawajima-Harima Jukogyo KK, both of Japan* in the electrolysis of an alkali metal chloride in an ion exchange membrane type electrolysis cell the dilute caustic alkali solution produced in the cathode compartment is concentrated in a series of heating and evaporating operations wherein it is heated by heat exchange with the hot dilute metal chloride solution removed from the anode compartment. The caustic alkali solution and metal chloride solutions flow countercurrently through the heat exchange steps of the concentrating operations.

Example: In the concentration apparatus having three unit concentration processes in series as shown in Figure 6.11, the concentration of an aqueous solution of NaOH obtained by an ion exchange membrane type electrolysis of an aqueous solution of NaCl was carried out.

Figure 6.11: Flow Sheet for Three-Unit Concentration Process

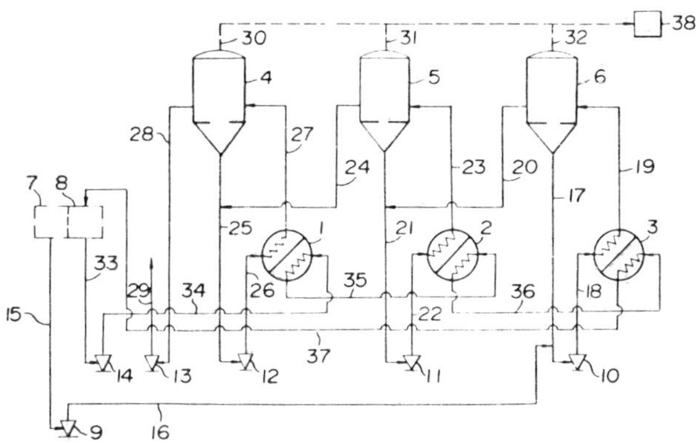

Source: U.S. Patent 4,145,265

The aqueous solution of NaOH obtained from the cathode compartment of the electrolytic cell (30% by weight of NaOH; 95°C) was passed consecutively through the No. 3, No. 2 and No. 1 unit concentration processes and a concentrated aqueous solution of NaOH (48% by weight) was obtained from the evaporator of the No. 1 unit concentration process.

The dilute aqueous solution of NaCl obtained from the anode compartment of the electrolytic cell (20% by weight of NaCl; 100°C) was passed consecutively through the No. 1, No. 2 and No. 3 concentration units. The dilute aqueous solution of NaCl discharged from the heat exchanger of the No. 3 unit concentration process is saturated with the raw material of NaCl and the saturated solution is recycled to the anode compartment. The temperatures of the aqueous solutions of NaOH and the dilute aqueous solutions of NaCl and the concentrations of the aqueous solution of NaOH and the pressures in the No. 1, No. 2 and No. 3 evaporators in the No. 1, No. 2 and No. 3 concentration processes are shown below.

The No. 1, No. 2 and No. 3 evaporators were made of nickel. The No. 1, No. 2 and No. 3 heat exchangers each had titanium-nickel composite partitions.

	. . Concentration Process Unit . .		
	1	2	3
Conc. of aq. soln of NaOH (wt %)	36.8	43.0	50.0
Temp. of aq. soln of NaOH (°C)	65.5	72.5	80.0
Temp. of aq. soln of NaCl (°C)	75.0	82	87
Pressure in evaporator (mm Hg abs)	61.5	61.5	61.5

MEMBRANE CELLS

The commercially used equipment for the large-scale electrolysis of brine has traditionally been either a mercury cell or a diaphragm cell. Both the mercury cell and the diaphragm cell have particular drawbacks. Although the mercury process yields a caustic of low salt content, it gives considerable difficulty in respect to compliance with pollution control laws and regulations. The diaphragm process inherently produces a caustic product which contains a proportion of salt, and while there are some purposes for which such a product may be used without difficulty, there are a number of other purposes which require a substantially salt-free product.

Various ion-exchange membrane materials have been available, but it has long been apparent to those skilled in the electrolysis of brine that it would be desirable to have a process in which a permselective membrane is used in a cell to separate the anolyte compartment from the catholyte compartment, in place of the asbestos diaphragms customarily used in the diaphragm process. In laboratory tests, cells operating with permselective membranes as separators have been made and tested, but there are many difficulties which must be overcome before the idea of using a permselective membrane as a separator in a chlor-alkali cell can be brought to commercial reality.

Some of the early membranes did not have adequate physical strength, and others did not have adequate inertness with respect to the brine and the caustic solutions with which the membrane was required to be in contact for a substantial service life. Still other materials, although satisfactory in regard to the requirements just mentioned, gave disappointing results in respect to the cell voltages required and the current efficiencies observed. Commercial cells for the electrolysis of brine deal with immense quantities of liquids and use immense amounts of electrical power, while producing products (chlorine and caustic) which are relatively inexpensive.

A process which must compete with the mercury and diaphragm processes must be relatively low in its consumption of electrical power. A reduction in cell voltage of 0.15 V, or an increase in current efficiency of 2%, is certainly significant in terms of the costs of practicing the process. The costs are also significantly affected by the frequency with which it is necessary to interrupt the operation of the process. In general, interruptions any more frequent than about once a month are not tolerable. The art has been advanced by the development of dimensionally stable anodes, made of ruthenium-, oxide- or platinum-coated titanium or the like, in place of the graphite anodes previously used.

One particular problem, in connection with the operation of membrane-type cells, has been that the cell voltage tends to rise with time, and it has been appreciated that this increase in cell voltage is associated with the precipitation of calcium hydroxide particles within the membrane. This implies that the control of the calcium content of the brine fed to the process is of considerable importance.

Those skilled in the art of operating diaphragm cells are familiar with the practice customarily employed to reduce the calcium content of raw brine to a satisfactorily low level, such as about 10 ppm of calcium. The raw brine is treated with sodium carbonate to precipitate calcium ions as calcium carbonate. The treated brine is freed of calcium carbonate by settling and/or filtration.

Some fine particles of calcium carbonate survive this treatment, so that the treated or polished brine ordinarily contains, as mentioned above, about 10 ppm of calcium. The volumes of liquid dealt with are so immense, however, and the times of operation are so long that even this small proportion of calcium is sufficient to give difficulty. Those skilled in the art have not known, prior to this process, how this difficulty may be overcome, and a reliable, economical membrane-cell process thus practiced.

In general, there is good reason for those skilled in the art to look away from the improving of the quality of the brine and reducing or interrupting the supply of current to the cell as means to maintain a low cell voltage. Although other things being equal, a purer brine is obviously desirable, it has not been apparent that the expense of practicing a better calcium-removal method could be justified, and that the control of calcium, in a membrane electrolysis process, is as important as it is. There has been instead a tendency to try other membrane materials, in the hope of finding one that will work better and not make it necessary to change the existing brine-treatment practices.

PROCESS DESIGN

Low-Calcium Brine and Periodically Reduced Current

M. Krumpelt; U.S. Patent 4,115,218; September 19, 1978; assigned to BASF Wyandotte Corporation makes available to the chemical industry the benefits of a commercial process for the electrolysis of sodium chloride brine, using permselective membranes as separators of the anolyte and catholyte compartments. A principal benefit is that a caustic is produced which does not contain sodium chloride. Another benefit is that, unlike with the diaphragm cells of the prior art, there is a more productive use of floor space, since with cells of a given height, substantially all of the cell height may be devoted to electrolysis, whereas with the diaphragm cells of the prior art, it was always necessary to have some considerable proportion of the cell height devoted to providing the necessary head to force the anolyte through the diaphragm into the catholyte compartment.

The disclosure particularly concerns the improvement which is obtained when the brine processed is limited to a calcium content of 6 ppm or less and, as a measure for periodically reducing the cell voltage, there is practiced (a) at least the reduction to a value 20% of normal or less the current through the cell, or the complete interruption of such current, for a period of time from approximately 5 to 30 minutes, alone or together with (b) a flushing of the catholyte compartment of the cell, by the use of a practice involving adding water to the catholyte compartment at a volumetric rate of approximately five to twenty times normal. These steps, taken in conjunction with what is already known, yield a practical process.

Example: A laboratory-scale chlor-alkali membrane cell having a membrane area of 18 in^2 (116 cm^2) had been operated for 170 days at 25 A current and with a brine containing between 1 and 5 ppm of calcium and between 1 and 3 ppm of magnesium. The cell had dimensionally stable anodes, and it had a membrane made of a copolymer of tetrafluoroethylene and perfluorovinylether sulfonyl fluoride and having an equivalent weight of 1,350, such as material commercially available as Nafion resin. The cell potential, initially 3.6 V, had risen to 4.09 V.

The catholyte compartment was normally fed with demineralized water at a rate of 75 ml/hour. The current to the cell was completely interrupted for a period of 2 minutes, and during that time, the water feed rate was increased to 500 ml/minute. Samples of the catholyte effluent were taken before, during, and after the flushing, and they were analyzed for calcium, magnesium, and total organic carbon. After the flushing, the cell potential stabilized at 3.73 V. The data are presented below.

	Ca	Mg	TOC
		(ppm)	
Feed water	1.3	0.4	0
Catholyte before flush	6.2	4.0	9
Catholyte during flush	11.3	2.2	18
Catholyte 1 hour after flush	76	16.3	43
Catholyte 4 hours after flush	6.8	1.3	11

Turbulence-Inducing Cathode

According to *E.D. Creamer, M. Krumpelt and J. Jorné; U.S. Patent 4,142,950; March 6, 1979; assigned to BASF Wyandotte Corporation*, high current efficiency can be obtained in an electrolytic cell by inducing turbulence in the catholyte preferably by utilizing a gas-directing cathode and cation-permselective membrane combination. There is disclosed a process for electrolysis, particularly, the electrolysis of an alkali metal chloride such as sodium chloride to produce chlorine and sodium hydroxide.

The cell has a cathode and an anode divided into catholyte and anolyte compartments by a cation-permselective membrane. Turbulence-inducing means such as a gas-directing cathode provides turbulence in the catholyte at the surface of the membrane by directing gas evolving on the cathode toward or away from the membrane. Multicell arrangements are also disclosed wherein the cells are connected in series.

Referring to Figure 7.1, there is illustrated in schematic form a cross-sectional view of an electrolytic cell comprising a cell wall **20**, a flattened, expanded metal anode **23**, a gas-directing expanded metal cathode **10** and a cation-permselective membrane **24**. Conductive means for connecting the anode and cathode to sources of positive and negative electrical potentials, respectively, are not shown.

Figure 7.1: Electrolytic Cell with Turbulence-Inducing Cathode

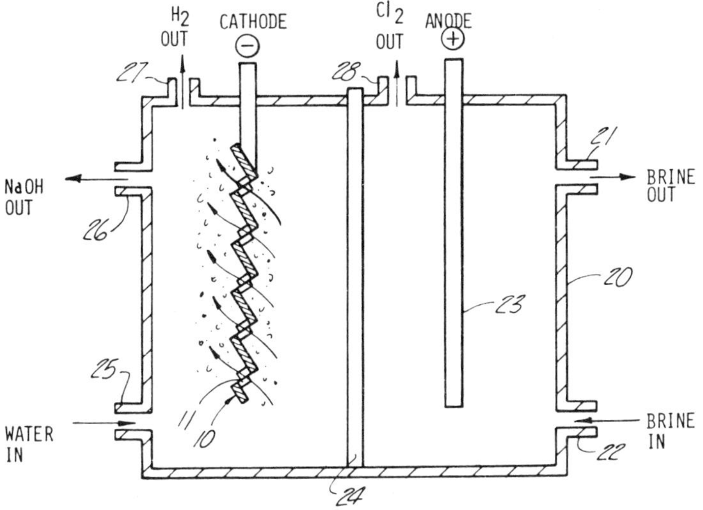

Source: U.S. Patent 4,142,950

An aqueous solution of alkali metal chloride, preferably acidic, is fed through line **22** and exits from line **21**. Water is fed in through line **25** and sodium hydroxide solution is obtained through line **26**. During electrolysis, chlorine gas is removed through line **28** and hydrogen gas is correspondingly removed through line **27**. The electrolysis is conducted at high caustic current efficiency by maintaining the gas-directing, expanded metal cathode **10** in relation to the cation-permselective membrane **24** such that the hydrogen gas evolved on the cathode is directed rearwardly (as shown) or forwardly toward the cation-permselective membrane so as to induce turbulent flow between the membrane and cathode. Thus a high concentration sodium hydroxide solution can be obtained through line **26** while at the same time maintaining high caustic current efficiency.

Example 1: An electrolytic cell body was constructed of chlorinated polyvinyl chlorine plastic containing 20% by weight of asbestos based upon the total weight of the filled plastic. The cell is schematically shown in Figure 7.1.

The cell contained a flattened expanded metal anode made of ruthenium oxide-coated titanium and a cathode made of nickel coated steel. The electrodes communicate with current sources by means of steel members. The cathode was shaped into a turbulence-inducing form by expanding a metal sheet by stamping out openings between the remaining webs or filaments of the mesh which measure ⅜ inch high by 1¼ inches wide, the remaining metal filaments being about 2 mm in thickness. The electrodes were mounted in the cell on either side of a cation-permselective membrane so as to provide an electrode spacing of 0.1 inch with the cathode installed so as to direct cathodic gases away from the membrane.

The membrane was Nafion, type 313. The membrane was joined to a backing or supporting layer network of polytetrafluoroethylene filaments woven into a cloth having an area percentage of openings therein of about 22% by volume. The membranes which were initially flat are fused onto the polytetrafluoroethylene cloth under conditions of high temperature and pressure with some of the membrane portions actually being caused to flow around the filaments of the cloth during the fusing so that the membrane and cloth become an integral unit. Before being sold, the membrane was hydrolyzed by boiling in water.

It has been found that heating the membrane at about 200°C for about 2 hours is required to allow the attainment of a desirable base level of current efficiency after installation of the membrane in the electrolytic cell. The cation-permselective membrane utilized was in two layers each bonded together and consisting of a hydrolyzed copolymer of a perfluorinated hydrocarbon and a fluorosulfonated perfluorovinyl ether, the outer layer being 2 mils in thickness and having an equivalent weight of about 1,350 and the inner layer being 4 mils in thickness and having an equivalent weight of about 1,100.

Ruthenium-oxide-coated titanium anodes were used which were prepared by coating a titanium mesh having about 2 mm thickness filaments with about 50% by volume open area with ruthenium oxide to a thickness of about 10^{-3} mm.

The cell was operated under the following conditions: current density, 200 A/ft^2; cell voltage, 3.25 to 3.85 V; temperature 82° to 88°C; pH in the anolyte, 3 to 3.5.

During the operation of the cell, saturated brine was fed to the anode compartment at a rate to consume 60% by weight of the brine with no recycling of the brine used. Water was fed to the cathode compartment at a rate to produce approximately 5 normal sodium hydroxide and caustic concentration was determined accurately within ±0.5% by weight by repeatedly accumulating known volumes of catholyte in the amount of 0.22 ℓ over a time interval of about 2 hours. Concurrent with the collection of these known volumes, an integrated sample was accumulated using a metering pump and subsequently titrated to determine the normality of the sodium hydroxide solution within ±0.5%. The caustic current efficiency was calculated utilizing the following equation:

$$\text{Current efficiency (NaOH)} = \frac{\text{(NaOH normality)(flow rate, } \ell\text{/hr)}}{\text{current, amps}} \times 2{,}681$$

An overall accuracy of ±1.2% was obtained in the calculation of the current efficiency. The cell was operated continuously for a period of 10 days. Results obtained show a current efficiency average of 82% for the cell.

Example 2: *Comparative* – A cell was operated in accordance with the above procedure with the exception that the expanded metal cathode utilized was a flattened, expanded metal cathode so that turbulence on the surface of the cation-permselective membrane is not induced by directing the evolving cathodic gases toward or away from the membrane. A flattened, expanded metal cathode having openings measuring 3⁄16 inch in height by ½ inch in width was utilized in addition to a flattened, expanded metal electrode having larger ⅜ inch high by 1¼ inch wide openings. Current efficiencies obtained utilizing the cells having flattened, expanded metal electrodes present as cathodes indicate a current efficiency average of 77%.

Energy-Saving Multistage Process

Alkali metal hydroxide and chlorine are produced by *J.N. Andersen; U.S. Patent 4,076,603; February 28, 1978; assigned to Kaiser Aluminum & Chemical Corporation* in a membrane-equipped electrolysis cell system in a staged manner wherein at least one cell of the system is used to generate a dilute caustic solution and wherein the subsequent cell(s) of the system are used to produce a more concentrated caustic solution by electrolysis of fresh brine and the dilute caustic produced in the first stage which is used in lieu of water in the subsequent stage(s). The stage-wise production of caustic of commercially acceptable concentration results in significant energy savings with corresponding increase in the average life of the membranes employed.

Example: A multistage system, such as shown in Figure 7.2 and employing one first-stage and one second-stage electrolysis cell, was employed. In the first stage, 50% of the caustic (NaOH) solution was produced at a concentration of about 16.3%, while from the second stage, caustic solution of 30% by weight NaOH was recovered.

In the first stage, the cell voltage was maintained at 3.57 V and a current efficiency of 95.7% was obtained. Consequently, the energy consumption in the first stage is calculated as follows:

$$\text{Energy consumption/ton of first-stage NaOH} = \frac{607.8\ \text{kAh} \times \text{cell voltage}}{\text{cell current efficiency}} =$$

$$\frac{607.8 \times 3.57}{0.957} = 2{,}267\ \text{kWh}$$

In the second stage, where in lieu of water the NaOH solution produced in the first stage is added to the cathode compartment, the cell voltage employed for the production of caustic of commercial concentration is 4.25 V at a cell current efficiency of 93%. Consequently, the energy consumption of this cell is calculated as follows:

$$\text{Energy consumption/ton of second-stage NaOH} = \frac{607.8\ \text{kAh} \times \text{cell voltage}}{\text{cell current efficiency}} =$$

$$\frac{607.8 \times 4.25}{0.93} = 2{,}778\ \text{kWh}$$

Combined energy consumption = 2,267 + 2,778 = 5,045 KWh, but since in each cell of the multistage system only 50% of the total NaOH is produced, the average energy consumption in the multistage system is

$$\frac{5{,}045}{2} = 2{,}522\ \text{kWh/ton of NaOH end product.}$$

Figure 7.2: Energy Saving Multistage Process

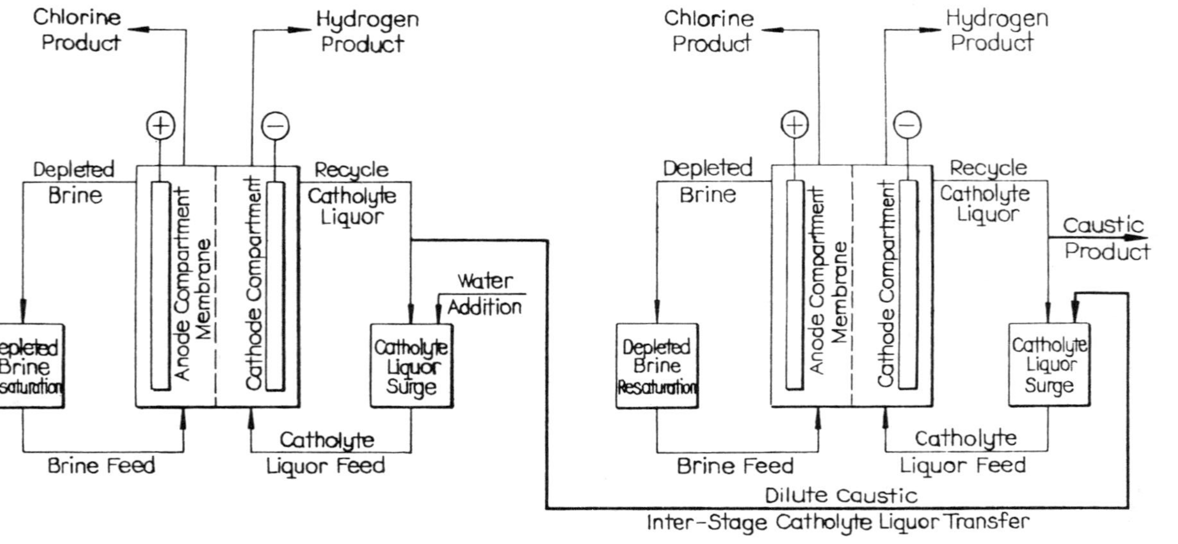

Source: U.S. Patent 4,076,603

In comparing this energy consumption with the consumption in a single-cell system of 2,778 kWh for the same quantity of NaOH of the same concentration, it can be seen that the energy savings provided by the process is 2,778 kWh - 2,522 kWh = 256 kWh, amounting to a savings of approximately 9.2%.

In addition to this significant energy savings, the average membrane life is also considerably extended. Thus, in the first stage, the membrane has a life of approximately 18 months under the process conditions shown above; in the second stage of the system, where the high concentration of caustic is made, the membrane life is about 12 months. The average age of the membrane lives in the first and second stages is

$$\frac{18 + 12}{2} = 15 \text{ months}$$

which, in comparison to the life (12 months) for the single-stage system, is a 25% increase.

Three-Chamber Cell Utilizing Two Types of Membranes

N. Murayama, M. Fukuda, T. Sakagami, S. Suzuki, Y. Kokubu and T. Enoki; U.S. Patent 4,111,780; September 5, 1978; and N. Murayama, K. Nakamura, M. Fukuda, T. Sakagami and S. Suzuki; U.S. Patent 4,076,604; February 28, 1978; both assigned to Kureha Kagaku Kogyo KK, Japan describe an apparatus for the electrolytic treatment of alkali halide solution in a three-chamber type electrolytic bath assembly comprising an anodic chamber, an intermediate chamber and a cathodic chamber arranged one after another in series. Each chamber is separated from its neighboring chamber by means of an anodic ion exchange membrane.

The apparatus is characterized in that the first one of the membranes separating the intermediate and anodic chambers is made of a fluorine-containing resin, while the second membrane separating the intermediate and cathodic chambers includes, as its main ion exchange radical, a pendant type phenolic radical or derivative thereof. The process utilizes a caustic alkali concentration in the intermediate chamber which ranges from about 10 to 20 wt % while the output caustic alkali developed at and delivered from the cathodic chamber amounts to a concentration ranging from about 30 to 50 wt %.

Example: A three-chamber type electrolytic bath assembly, containing an anodic, an intermediate and a cathodic chamber arranged one after another and separated with respective two ion-exchange membranes from each other, was used.

As the first membrane separating the anodic and intermediate chambers, Nafion-315-membrane (to be called membrane No. 1 hereinafter), a composite one including Nafion EW-1100; Nafion EW-1500 and a mesh sheet of ethylene tetrafluoride resin, being laminated one after another, was used.

As the second membrane, separating the intermediate and cathodic chambers, a commercially procurable ion exchange membrane (to be called membrane No. 2 hereinafter), bearing phenolic radicals or derivatives (Maruzen Sekiyu KK) was used.

A hydrolytic treatment of aqueous NaCl brine was then carried out in the following way: The anodic chamber was fed with aqueous saturated NaCl solution and the intermediate chamber was supplied with water. The total quantity of aqueous NaOH solution as produced at the intermediate chamber in the progress of the process was conveyed from the intermediate chamber to the cathodic one in an overflowing way, while high concentration NaOH product solution was taken out from the cathodic chamber. The results are shown in the following table.

Anode	Titanium/ruthenium oxide
Cathode	Steel wire net
Brine	26% aq. NaCl solution
Bath temperature	80°C
Current density	20 A/dm^2
Bath voltage	4.0 V
NaOH concentration,	
in intermediate chamber	19%
in cathodic chamber	39%
Current efficiency	82%
Flow rate of NaOH, aq.	
through communication pipe	1.34 g/Ah
Water penetration rate	
through membrane No. 2	1.49 g/Ah
OH^- penetration rate	
through membrane No. 2	0.22 g/Ah

Carbon Dioxide Injected into Catholyte Compartment

An alkali metal carbonate substantially free of alkali metal chloride is efficiently produced by *K.J. O'Leary, C.J. Hora, Jr., and D.L. DeRespiris; U.S. Patent 4,080,270; March 21, 1978; assigned to Diamond Shamrock Corporation* by electrolyzing an alkali metal chloride in an electrolytic cell having anolyte and catholyte compartments separated by a cation-exchange hydraulically impermeable membrane comprised of a thin film of a fluorinated polymer having pendant sulfonate groups and a cathode spaced apart from the membrane, injecting into the catholyte compartment of the cell carbon dioxide in a quantity sufficient to convert substantially all of the alkali metal hydroxide forming therein to the alkali metal carbonate salt, and utilizing a magnitude of electrolyzing current that reduces alkali metal chloride in the carbonate salt to less than 400 ppm.

Example: An electrolytic cell like that shown in Figure 7.3, generally designated **21**, was divided by a membrane **22** into an anode compartment **23** and a cathode compartment **24** formed by glass cylindrical half-cell members **25** and **26** respectively having front-to-back depths of 4 inches and inside diameters of 2 inches.

In the anode compartment **23**, there was provided a brine inlet tube **27** made of titanium metal to which there was welded a circular anode **28** disposed essentially parallel to the membrane. The anode had a diameter of about 1.95 inches and was made from an expanded mesh of titanium metal bearing a $2TiO_2:RuO_2$ mol ratio coating. Depleted brine was discharged through outlet **29** while chlorine gas was vented off through anolyte standpipe **30**. To insure a good seal, gasket **31** having an inside diameter of about 1.95 inches was positioned between half-cell member **25** and the membrane.

In the cathode compartment, a cathode assembly **32** was positioned in a generally parallel spaced-apart relationship from the membrane by a gasket **33** having an inside diameter of 1.95 inches. The cathode assembly **32** consisted of an apertured stainless steel plate **34** circumscribed by a holding flange **35**. Passing through the holding flange at the bottom of the cell **21** was a carbon dioxide inlet **36** having exit ports **37** directing the carbon dioxide gas generally both upwardly and sidewardly towards the membrane **22**. Catholyte containing carbonate product was discharged through outlet **38** while by-product hydrogen gas was vented off through a catholyte standpipe **39**.

Figure 7.3: Production of Alkali Metal Carbonates

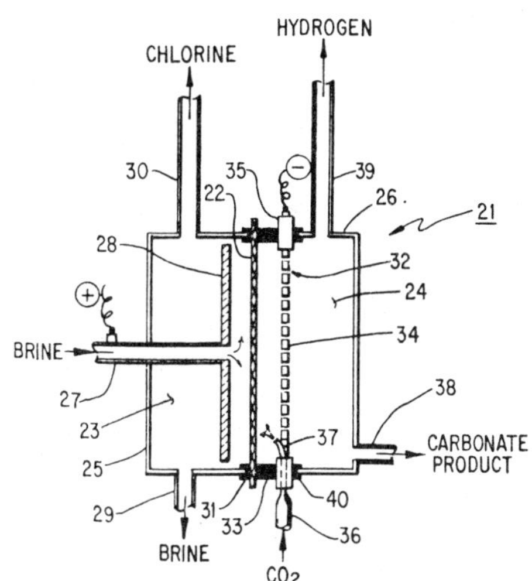

Source: U.S. Patent 4,080,270

The membrane used consisted of a film of a copolymer of tetrafluorethylene and perfluoro[2-(2-fluorosulfonylethoxy)propyl vinyl ether] that was first laminated to a T-12 (Teflon square-woven) fabric and then hydrolyzed and acidified to convert the sulfonyl fluoride groups of the copolymer to sulfonic acid groups. Membrane thickness ranged from 3.5 to 7 mils.

The current density was of about 2 amperes per square inch (Asi) of membrane area exposed to the electrolytes. The brines contained 0.26 cc of 85% H_3PO_4 per liter and were acidified with HCl to a pH of about 2.0.

High current efficiencies are obtained when carbon dioxide is introduced at a rate close to stoichiometric for the production of carbonate, i.e., at a rate producing in the catholyte solids about 90 wt % or more of the carbonate salt and about 10 wt % or less of either the hydroxide or bicarbonate.

Noteworthy is a surprising decline of chloride salt in the catholyte solids with increasing current densities; and the ability of even 3.5 and 5 mil membranes to limit chloride contamination to commercially acceptable levels at about 4 and

2 Asi respectively. It is apparent that the use of the thinner membranes effects a significant reduction in cell voltages without compromising current efficiencies or the ability of the process to produce high-purity commercial-grade carbonate product when a sufficiently high current density is used. Additionally worthy of note is the high concentration of alkali metal carbonate (nearly saturated) produced, which would minimize the energy required for further concentrating or drying of the discharged catholyte to the final products of commerce. Finally, it was observed that voltages increased only slightly upon prolonged running of the cells and seemed to reach a steady state after a few days of continuous cell operation.

High Pressure Process

In carrying out electrolysis of alkali halide in an electrolytic cell having anode and cathode chambers separated by a cation exchange membrane by feeding an aqueous alkali halide solution into the anode chamber to produce halogen gas in the anode chamber and caustic alkali and hydrogen gas in the cathode chamber, the electrolytic cell is pressurized by *S. Ogawa and M. Yoshida; U.S. Patent 4,105,515; August 8, 1978; assigned to Asahi Kasei Kogyo KK, Japan* at higher than atmospheric pressure to obtain various improved results.

Example: Electrolysis is carried out by the system as shown in Figure 7.4. The cation exchange membrane employed has a double layer structure comprising a layer of perfluorosulfonic acid of 6 mils in thickness having molecular weight of 1,200 g per one equivalent of exchange groups and a layer of perfluorocarboxylic acid of 0.2 mils in thickness having molecular weight of 1,200 g per one equivalent of exchange groups.

Figure 7.4: High Pressure Process

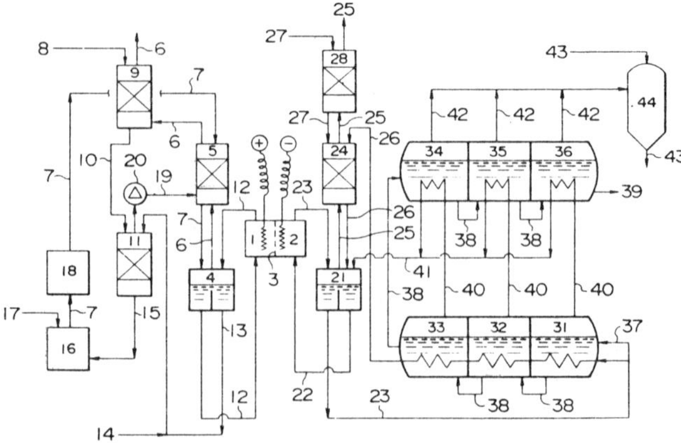

Source: U.S. Patent 4,105,515

There is used an electrolytic cell having anode and cathode chambers divided by this ion-exchange membrane. This electrolytic cell has an effective current passage area of 1.2 m^2. As an anode, a porous anode of a titanium expanded mesh with opening ratio of 60% which has been coated with ruthenium oxide, titanium oxide and zirconium oxide may be employed. As a cathode, a porous cathode of an expanded iron mesh with an opening ratio of 60% is used. The electrolytic cell is a bipolar electrolyzer in which the partition wall is made of explosion-bonded titanium plate and iron plate. Between the cathode mesh and the iron surface of the partition wall is provided an interval of 45 mm and also between the anode mesh and titanium surface of the partition wall an interval of 45 mm. 74 unit cells of this type are connected electrically in series to be assembled in a bipolar electrolyzer for use.

Electrolysis temperature was 102°C, current density was 40 A/dm^2, electrolysis voltage 3.75 V, inner pressure in the cathode chamber 2.9 atm, and inner pressure in the anode chamber 2.8 atm. The concentration of the aqueous caustic soda solution in the cathode chamber was 25% and the concentration of the salt aqueous solution 2.5 N.

When electrolysis is carried out under the above conditions and concentration is effected in a three-stage double-effect evaporator using catholyte as heat source, 25% caustic soda can be concentrated to 43%.

The inner pressures in the anolyte tank **4**, the chlorine gas cooling towers **9** and **5** are pressurized at 1.9 atm.

When chlorine gas and hydrogen gas are discharged out of the system under pressure of about 2 atm, no blower is necessary for carrying the gases to subsequent steps such as hydrochloric acid synthesis tower or chlorine drying tower.

Higher Pressure in Cathode Chamber

According to *M. Seko, S. Ogawa and M. Yoshida; U.S. Patent 4,108,742; August 22, 1978; assigned to Asahi Kasei Kogyo KK, Japan*, in an electrolytic cell having a cation exchange membrane as a diaphragm to partition the cell into cathode and anode chambers, electrolysis of an electrolyte aqueous solution is conducted while generating gas from the anode by keeping the inner pressure in the cathode chamber higher than that in the anode chamber.

Example: 74 units of an electrolytic cell are arranged in series, with a sulfonic acid type fluorine resin membrane interposed between individual cells. A direct current voltage is applied to both ends of the bipolar system electrolytic cell, whereby the current flows in series through the individual cell units. A catholyte and an anolyte are individually charged in series from respective headers through flexible hoses into the individual cell units, and then discharged. The catholyte is charged from a catholyte tank through a catholyte header into the cathode chamber of each cell unit by means of a pump. Subsequently, the catholyte, in the form of a gas-liquid mixture, is discharged as it is, recycled in the catholyte tank and subjected to gas-liquid separation.

Likewise, the anolyte is charged from a catholyte tank through an anolyte header into the anode chamber of each cell unit by means of a pump. Subsequently, the anolyte, in the form of a gas-liquid mixture, is discharged as it is, recycled in the anolyte tank, and subjected to gas-liquid separation.

Electrolysis was conducted, using aqueous sodium chloride as the anolyte and sodium hydroxide as the catholyte. Into each electrolytic cell, both the catholyte and the anolyte were charged at a rate of 600 ℓ/hour. To the anolyte tank, saturated aqueous sodium chloride and hydrochloric acid were added so that the sodium chloride concentration became 2.5 N and the pH became 3 at the outlet of the anolyte chamber. To the catholyte tank, pure water was added so that the sodium hydroxide concentration became 5 N at the outlet of the cathode chamber. Both the cathode chamber and the anode chamber were maintained at an electrolysis temperature of 90°C. To the electrolytic cell a direct current was flowed at a current density of 50 A/dm^2, i.e., a direct current of 14,200 A.

As the result, chlorine gas was generated from the anode, and hydrogen gas from the cathode. The difference between the inner pressure of the cathode chamber and that of the anode chamber was controlled by controlling the inner pressures of the anolyte and catholyte tanks, and the pressure difference between the two chambers was measured by means of a mercury manometer. The relation between the pressure difference between the two chambers and the electrolysis voltage per unit cell was as shown below.

Pressure Difference (meters wc)	Electrolysis Voltage (volts)
-1	4.11
0	3.7-3.9
+0.2*	3.72
+1*	3.65
+2*	3.65
+5*	3.65

*The inner pressure of the cathode chamber was higher than that of the anode chamber.

The effect of the process is clear. The electrolytic cell used in the above was disassembled, but no such phenomenon as burning or like damage of the cation-exchange membrane was observed.

Higher Pressure in Anode Chamber

A process is disclosed by *B.E. Kurtz and R.H. Fitch; U.S. Patent 4,204,920; May 27, 1980; assigned to Allied Chemical Corporation* for producing chlorine and caustic soda in an electrolytic membrane cell which comprises providing a pressure differential between the anode compartment and the cathode compartment sufficient to prevent substantial contact of the membrane with the anode, and reducing fluctuations in the pressure differential by allowing depleted sodium chloride brine and chlorine gas to flow freely from the anode compartment to a brine collection point. Also, the caustic soda and hydrogen gas produced in the cathode compartment are allowed to flow freely from the cathode compartment to a caustic soda collection point.

In Figure 7.5, there is shown unit cell **10** having an anode compartment **15** containing anode **14**, and a cathode compartment **17** containing cathode **16**. The compartments are separated by cationic permselective membrane **12**. Sodium chloride brine is introduced into the anode compartment via header **24** and line **22**, and water or sodium hydroxide solution is introduced into the cathode compartment via header **20** and line **18**.

Upon application of electric current through the electrodes, the sodium chloride in anode compartment **15** is dissociated resulting in the formation of chlorine gas and sodium ions. The sodium ions migrate through membrane **12** into cathode compartment **17** forming sodium hydroxide and hydrogen gas. Depleted brine anolyte and chlorine gas are withdrawn from the anode compartment through line **30** and header **32** to seal pot **42** via dip leg **44** where the liquid and gas separates, chlorine being removed via line **50** and the brine through overflow line **48**. Similarly on the cathode side, sodium hydroxide catholyte and hydrogen gas are withdrawn via line **26**, header **28** and dip leg **36** to seal pot **34** from whence hydrogen is removed via line **38** and catholyte via overflow line **40**.

Figure 7.5: Schematic View of One Unit Cell

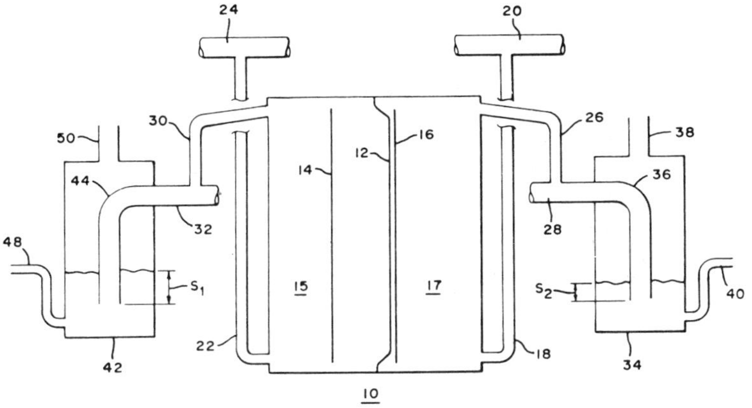

Source: U.S. Patent 4,204,920

By maintaining the submergence of the anolyte dip-leg **44**, designated as S_1, in the proper relationship to the submergence of the catholyte dip-leg **36**, designated as S_2, it is possible to maintain the pressure in anode compartment **15** higher than the pressure in cathode compartment **17**. Adjustments in the submergence of the dip-legs are accomplished by adjusting the heights of the seal leg overflows, **40** and **48**.

Thus, in order to maintain a constant positive pressure differential between anode and cathode compartments, it is only necessary to set the heights of the respective seal leg overflows properly. This pressure differential serves to force the flexible membrane **12** away from the anode **14** and towards the cathode **16**, as shown by the drawing, and may desirably result in the membrane being held securely against the face of the cathode. This serves to prevent contact of the anode with the membrane and also to prevent flexing of the membrane.

Three-Chamber Cell with Membrane and Diaphragm

T. Ueda, T. Nagaya and K. Kawada; U.S. Patent 4,137,136; January 30, 1979; assigned to Asahi Denka Kogyo KK, Japan disclose a method for electrolyzing an alkali metal halide aqueous solution in a three-chamber horizontal type electrolytic cell which comprises a top cathode chamber bounded by a cation exchange membrane, a middle chamber, and a bottom anode chamber bounded by a diaphragm. Halogen gas and dilute alkali metal halide solution are taken out of the anode chamber in such a manner as to make the anode chamber work as a gas chamber.

Example: A cation exchange membrane was prepared by sulfonating styrene-divinylbenzene-ethylvinylbenzene-polybutadiene-dioctyl phthalate-diethylbenzene (1:4:4:1:1:1) copolymer. A neutral diaphragm having a saturated salt solution passage rate of 0.8 ml/cm^2-hr (under a saturated salt solution pressure of 1 cm) was prepared by heat-treating at 300°C a paperlike membrane having a thickness of 0.5 mm obtained from ethylene tetrafluoride, asbestos and titanium oxide (5:3:1) and treating the heat-treated membrane with acetone to make it hydrophilic.

An electrolytic cell was constructed by (a) attaching the above prepared neutral diaphragm onto a meshlike metallic anode horizontally fixed; (b) placing a fluorine resin net having a thickness of 2 mm on the diaphragm as a spacer and placing the above prepared cation-exchange membrane on the spacer, thus constituting a middle chamber; and (c) placing a polypropylene screen having a thickness of 0.5 mm on the cation-exchange membrane and placing an iron mesh cathode on the polypropylene screen.

A saturated salt solution was charged into the middle chamber at a rate of 7 ml/cm^2-hr. The solution was then moved into the anode chamber at a rate of 5 ml/cm^2-hr and was rapidly taken out from the electric cell to keep the anode chamber as a gas chamber.

In the cathode chamber, a 20% sodium hydroxide solution was charged at the beginning of the operation. Water was then added to the cathode chamber in such a manner as to maintain the concentration of 20% sodium hydroxide as the electrolysis proceeds and the alkali metal hydroxide formed in the cathode chamber was taken out of the system.

The operation was carried out at 80°C and at a current density of 20 A/dm^2. The conditions and the efficiency of the operation were as follows: cell voltage, 3.5 V; decomposition rate of the salt solution, 30%; electric current efficiency, 92 to 93%; purity of chlorine gas formed, 98.5%; concentration of chlorine gas in the salt solution in the middle chamber, 0.5 ppm or less.

Even after operating for 8 months, the cation-exchange membrane was not deteriorated and the electrolysis efficiency was not lowered.

Comparative Example: An operation was carried out in the same manner as in the previous example, except that the anode chamber was charged with a salt solution. The decomposition ratio of the salt solution was 30%, but the cell voltage was 5.0 V and the amount of free chlorine gas (Cl_2) in the middle chamber reached 15 ppm. Accordingly, the cation-exchange membrane began to deteriorate after 2 months of operation, and was damaged after 4 months of operation. The electric current efficiency was lowered to 50% and the concentration of sodium hydroxide became less than 20%.

Filter Press Cell with Gas-Liquid Separator

A filter-press type electrolytic cell designed by *K. Sato, Y. Sajima, T. Kuno and H. Ohbe; U.S. Patent 4,149,952; April 17, 1979; assigned to Asahi Glass Co. Ltd., Japan,* comprises alternatively arranged quadrilateral frames and ion-exchange membranes to form anolyte compartments and catholyte compartments. The cell is lightweight and easily prepared at low cost.

Figure 7.6a is a schematic view of an electrolytic cell comprising such frames. A sectional view is shown in Figure 7.6b. In operation, a saturated aqueous solution of sodium chloride is fed from the inlet **8** to the hollow zone **3** corresponding to the lower part of the frame **15** for the anolyte compartment **11** and it is passed through the holes of **7** to the anolyte compartment wherein the electrolysis is conducted to generate Cl_2 gas.

Figure 7.6: Filter Press Cell

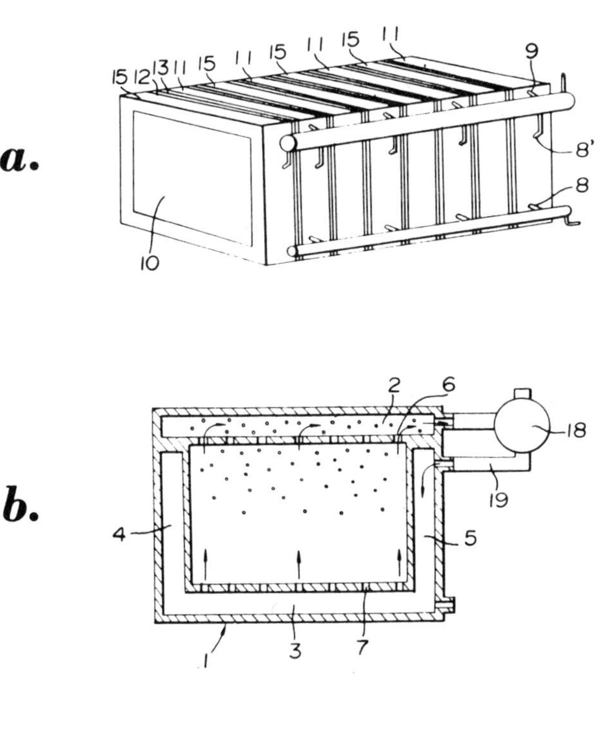

(a) Schematic view
(b) Sectional view of (a)

Source: U.S. Patent 4,149,952

The electrolyzed solution rises in the compartment by gas-lift action and is passed through the holes **6** to the hollow zone **2** corresponding to the upper part of the frame of the anolyte compartment and the solution containing the gas is discharged through the outlet **9** out of the frame.

At the same time, in the frame for the catholyte compartment, water or a dilute aqueous solution of sodium hydroxide is fed from the inlet to the hollow zone corresponding to the lower part of the frame and is passed through the holes to the catholyte compartment, wherein electrolysis produces the aqueous solution of sodium hydroxide and hydrogen gas.

The electrolyzed solution rises in the compartment with the gas and is passed through the holes to the hollow zone corresponding to the upper part of the frame **11** and the solution containing the gas is discharged from the outlet.

The gases discharged from the frames of the anolyte compartment and the frames of the catholyte compartments are respectively fed to the gas-liquid separators **18** wherein the gases are separated. Each part of the separated solutions is flowed down through the pipe **19** to the hollow members **3** in the lower parts of the frames and it is recycled into each of the anolyte compartments or the catholyte compartments.

The gas-liquid separators can be connected to each of the frames and they can be connected to a group of the frames of the same type compartments as the common separators. Thus, the concentration of the solution in the frames of the same type compartments can be uniform.

Pressure-Activated Uniform Spacing

An electrolytic cell designed by *D.D. Justice, B.K. Ahn and R.L. Dotson; U.S. Patent 4,105,514; August 8, 1978; assigned to Olin Corporation*, employing a hydraulically impermeable membrane having a spacing means interposed between the anode and the membrane, is operated by providing a positive pressure differential between the cathode compartment and the anode compartment. The pressure differential is sufficient to maintain contact between the spacer and the membrane to provide uniform spacing between the anode and the membrane.

In addition, this process provides sufficient spacing between the membrane and the cathode to provide efficient release of any gas formed and to prevent gas blinding at the cathode. Employing the positive pressure differential enables the cell to be operated at reduced energy costs when producing, for example, concentrated solutions of sodium hydroxide by careful control of the spacing between the membrane and the electrodes.

Figure 7.7 illustrates schematically a monopolar electrolytic cell **1** having an anode compartment **10** and a cathode compartment **12** separated by a cation permeable separator **14**. Adjustable anode **16** is a foraminous metal screen having threaded flanges **18** which enable the anode to be adjustably secured to anode plate **20**. Spacer **22** separates the anode from cation permeable separator **14**. Adjustable cathode **24** in cathode compartment **12** is a foraminous metal screen having threaded flanges **20** which adjustably secure cathode **24** to cathode plate **26**. Cell **1** has inlets and outlets as shown for the feeding and removal of the anolyte and the removal of the catholyte and the products of electrolysis.

Figure 7.7: Monopolar Electrolytic Cell

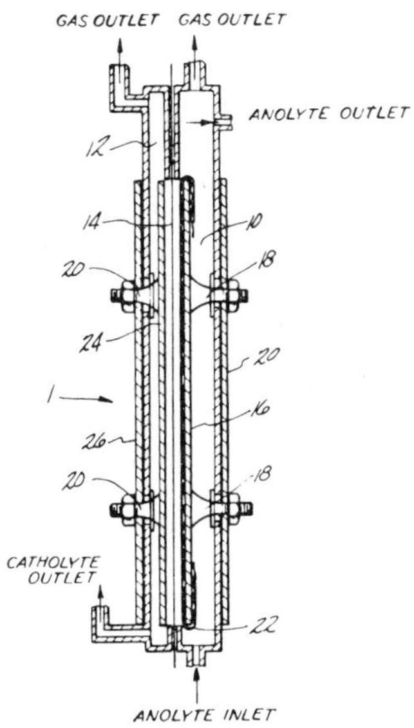

Source: U.S. Patent 4,105,514

Example: A cell of the type of Figure 7.7 was employed where the anode compartment contained a titanium screen coated on one side with an electrochemically active coating of ruthenium dioxide as the anode. The anode was spaced apart from a cation exchange membrane by a plastic net which provided a uniform spacing between the anode and the membrane of 1/16 inch. A perfluorosulfonic acid resin membrane separated the anode compartment from the cathode compartment which contained a steel perforated plate cathode 1/16 inch thick, spaced apart from the membrane a distance of 1/16 inch. The membrane was a homogeneous film 7 mils thick of 1,200 equivalent weight perfluorosulfonic acid resin laminated with a T-12 fabric of polytetrafluoroethylene.

Sodium chloride brine was supplied to the anode compartment at a concentration of 190 to 255 g/ℓ of NaCl, a temperature of 80°C and a pH of about 4.6. The cell was operated until the catholyte liquor became concentrated and it was maintained in the range of from 389 to 473 g/ℓ of NaOH. A vacuum was applied to the gas outlet of the anode compartment. The vacuum and the anolyte level were varied to permit the differential pressure from the anode compartment to the cathode compartment to be varied. As the differential pressure was varied, the cell voltages were recorded and the corresponding voltage coefficients calculated.

The catholyte level was allowed to rise above that of the anolyte level to provide a positive differential pressure from the cathode compartment to the anode compartment. As the pressure varied, the cell voltages were again recorded and the voltage coefficients calculated. The results show that a positive differential pressure from the cathode compartment to the anode compartment results in lower voltage coefficients and thus highly improved cell operation. In contrast, increasing the positive differential pressure from the anode compartment to the cathode compartment results in increasing voltage coefficients.

Electrode-Separator Combination Unit

S.J. Specht and J.O. Adams; U.S. Patent 4,110,191; August 29, 1978; and K.E. Woodard, Jr. and S.J. Specht; U.S. Patents 4,115,237; September 19, 1978; and 4,152,225; May 1, 1979; all assigned to Olin Corporation disclose an electrode-separator combination unit for use in an electrolytic cell having planar interleaved electrodes and a method of assembling such a unit. Electrodes are individually enclosed in a closed envelope of separator material to form individual electrolyte chambers. The separator can be perforated and electrical conductors, fluid supply conduits and fluid outlet conduits can be sealingly passed through the perforations to allow supply of raw materials to the enclosed electrodes and to allow removal of products therefrom. Figures 7.8a through 7.8f illustrate the process.

Figure 7.8: Membrane Cell with Combination Electrode-Separator

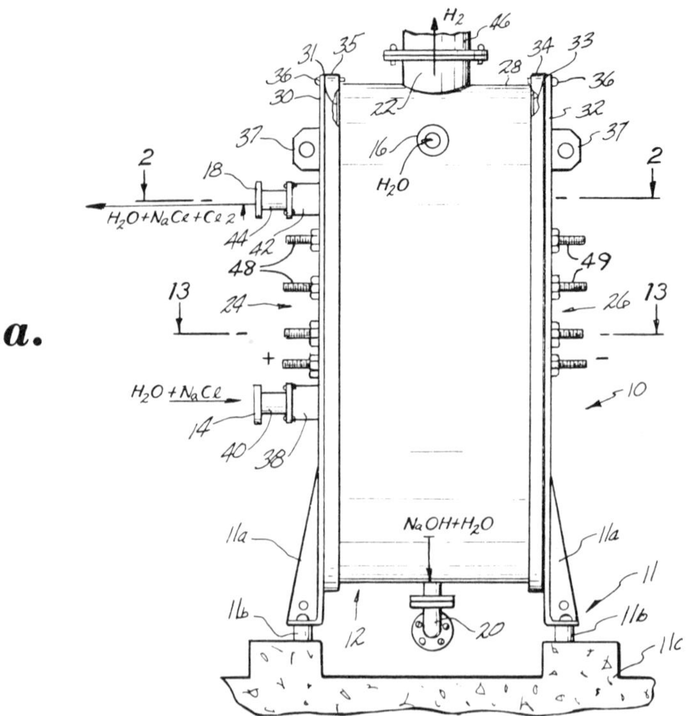

(continued)

Figure 7.8: (continued)

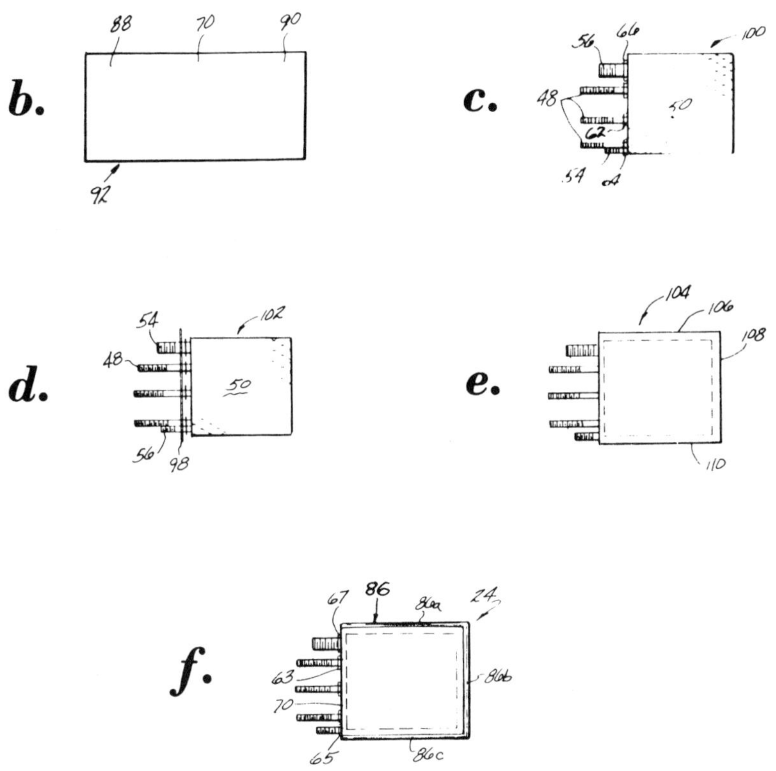

(a) Side elevational view of cell
(b) Top plan view of first stage of assembly of separator
(c)–(f) Side elevational views showing assembly of anode-separator unit

Source: U.S. Patent 4,110,191

Referring to Figure 7.8a, an electrolytic cell **10** is seen which comprises support means **11**, a housing **12**, an anolyte inlet **14**, a catholyte inlet **16**, an anolyte outlet **18**, a catholyte liquid outlet **20**, a catholyte gas outlet **22**, an anode assembly **24** and a cathode assembly **26**. Housing **12** includes a body portion **28**, an anode backplate **30**, an anode backplate gasket **31**, a cathode backplate **32** and a cathode backplate gasket **33**.

Figures 7.8b through 7.8f show the fabrication procedure for assembling the anode-separator assembly or unit **24** of Figure 7.8a. As seen in Figure 7.8b, a rectangular sheet **92** of separator material, for example, a cation exchange membrane of perfluorosulfonic acid resin or other heat sealable impermeable membrane or permeable diaphragm, is the starting point. The sheet can be considered to have a central portion **70** and two side portions **88** and **90**. The central portion can be reinforced by adding an additional layer of separator or other material to the central portion to produce a reinforced sheet for strengthening against damage during assembly or cell operation and because perforations are next made in the central portion at predetermined locations and of predetermined size so as to later receive conductor **48** and pipes **54**, **56** therethrough. Once the perforations are made, the perforated separator sheet is ready for receipt of anode body **100**.

Anode body **100** includes mesh **50**, conductors **48**, pipes **54** and **56** and gaskets **62**, **64** and **66** which are placed around the conductors and pipes **54** and **56**, respectively. After the gaskets are in place, the conductors and pipes are inserted through the perforated separator sheet to produce an unfolded assembly **102**. Side portions **88** and **90** are then folded loosely against opposite sides of mesh **50** to form an unsealed folded assembly **104**, having adjacent edges **106**, **108** and **110**. These edges are then sealed by any suitable means such as heat sealing to encapsulate the mesh and chamber to create a loose fitting anode-separator unit **24** having a U-shaped sealed edge **86**, bordering three sides and the perforated central portion **70** bordering the fourth side, as seen in Figure 7.8f.

Example: A cell of the type illustrated in Figure 7.8a is equipped with a plurality of titanium mesh anodes having portions covered by a coating having ruthenium dioxide as the electroactive component. A fiber glass open fabric coated with polytetrafluoroethylene and having a thickness of 0.035 inch is placed over the mesh anode. The anode mesh and surrounding fabric is enclosed in a perfluorosulfonic acid resin membrane having an equivalent weight of 1,200. The membrane is perforated and heat sealed to form a plurality of individual casings which are placed over the individual anode structures and sealed against the anode plate lining to provide a plurality of self-contained compartments. Intermeshed with the anodes are steel screen cathodes having an open area of about 45%.

The cathodes are spaced apart from the membrane about 0.50 inch to provide an unobstructed hydrogen release area. Sodium chloride brine having a concentration of about 300 g/ℓ of NaCl and at a temperature of 86°C is fed to each of the anode compartments. Sufficient electrical energy is supplied to the cell to provide a current density of 2 kA/m^2 to produce sodium hydroxide liquor in the cathode compartment containing about 400 g/ℓ of NaOH at a cell voltage of 3.5 volts. Hydrogen release from the NaOH liquor is excellent as is the release of chlorine gas from the NaCl brine in the membrane enclosed anodes.

Electrolyte Series Flow

According to *B.R. Ezzell and M.W. Sorenson; U.S. Patent 4,197,179; April 8, 1980; assigned to The Dow Chemical Company*, in an electrolytic chlor-alkali cell, or bank of cells, having a plurality of electrolyte compartments containing electrode pairs (anodes and cathodes) and wherein a hydraulically-impermeable membrane separates the electrolyte compartments into catholyte portions and anolyte portions, the cell or cells being employed to produce chlorine at the anodes and caustic and hydrogen at the cathodes by the electrolysis of an aqueous alkali metal chloride electrolyte, improved operation is attained by flowing anolyte liquor from anolyte portion to anolyte portion, sequentially, while simultaneously, and in the opposite direction, flowing catholyte liquor from catholyte portion to catholyte portion, sequentially. The membrane substantially prevents Cl^- from entering the catholyte liquor from the anolyte, and a high purity caustic, substantially free of salt, is produced.

In Figure 7.9 there are cells **1**, **2**, **3**, **4**, and **5**, each cell comprising a body **51** divided into anolyte portions **20** through **24** and catholyte portions **10** through **14** by a hydraulically-impermeable membrane **50**. Within each anolyte portion there is an anode and within each catholyte portion there is a cathode. The cells are provided with electrical circuitry to provide current for either bipolar or monopolar operation.

Figure 7.9: Electrolyte Series Flow in Chlor-Alkali Cells

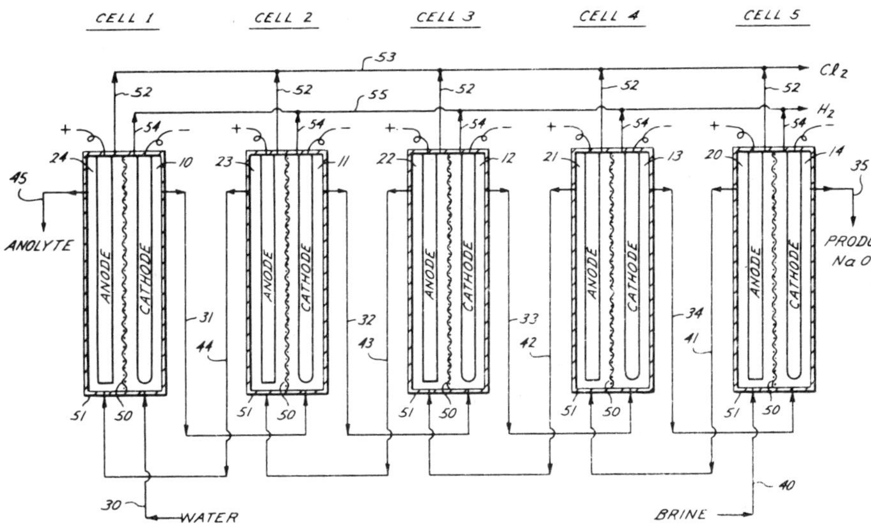

Source: U.S. Patent 4,197,179

During operation the anolyte liquor of each cell is provided by flowing a concentrated aqueous alkali metal chloride solution **40** into the lower part of anolyte portion **20** and out through flow means **41** from the upper part of **20** into the lower part of anolyte portion **21**. In like manner, that anolyte liquor flows sequentially through each anolyte portion **21**, **22**, **23** and **24** through flow means **42**, **43** and **44** until it is removed from the last anolyte portion **24** by flow means **45** as a partially-depleted, or "spent," alkali metal chloride solution.

The catholyte liquor of each cell is provided, in countercurrent manner, by flowing water **30** into the lower part of catholyte portion **10** and out through flow means **31** into the lower part of catholyte portion **11**. The catholyte liquor accrues caustic strength as it flows sequentially through the series of catholyte portions **10**, **11**, **12**, **13** and **14** through flow means **31**, **32**, **33** and **34** and leaves **14** at **35** as a relatively concentrated caustic solution.

Series Catholyte Flow

B.E. Kurtz; U.S. Patent 4,181,587; January 1, 1980; assigned to Allied Chemical Corporation describes a process for producing chlorine and caustic soda involving a bank of electrolytic membrane cells arranged for series catholyte flow. Power efficiency is improved by maintaining at least two of the initial cells in the bank in parallel catholyte flow, combining the catholyte streams from such initial cells and introducing the combined catholyte into the cathode compartment of one or more succeeding cells in the bank.

Curves of current efficiency and power efficiency against concentration of NaOH exiting any cell were fitted and incorporated into a computer program. Various series catholyte flow arrangements were evaluated with the program.

Cells which are in parallel catholyte flow are designated by the assignment of the same integer cell configuration number. Those which are in series are designated by successively higher integer cell configuration numbers. Thus, a 5-cell assembly with the first two cells in parallel and subsequent cells in series would be designated as

Cell Number	Cell Configuration Number
1	1
2	1
3	2
4	3
5	4

It is understood in all cases that the current passing through each cell, and thus the amount of OH^- formed by electrolysis, is the same.

The following tabulation shows the results obtained for a variety of series catholyte flow arrangements, ranked according to overall power efficiency attained, all for a final concentration of 20 wt % NaOH in the catholyte.

Configuration	Overall Power Efficiency (%)
11111	52.7
12345	55.3
11223	55.8
11123	55.9
11112222334	56.0
111222334	56.0
11234	56.2

From these results it is evident that, while simple series catholyte flow (12345) is superior to parallel catholyte flow (11111), modified series catholyte flow, in which cells located at the feed end of the assembly are configured in parallel flow while cells located nearer the product end of the assembly are configured in series flow, is better still. The best of the various 5-cell configurations is 11234, in which the first two cells are in parallel flow and the subsequent three in series flow.

pH Control by Oxidation of Hydrogen

W.A. McRae; U.S. Patent 4,173,524; November 6, 1979; assigned to Ionics Inc. provides an improved method and apparatus for controlling and maintaining the pH of a recirculating anolyte for a membrane-type chlor-alkali electrolysis cell.

Controlling the pH of the anolyte yields several advantages. In a recirculating cell of this type it is important not to contaminate the brine saturated anolyte with unwanted sodium chlorate which will form and accumulate if the hydroxyl ion leakage from the catholyte through the cell membrane into the anolyte is not neutralized. Adding an acid such as HCl from an external source in the prior art manner will increase the cost of and reduce the economic feasibility of the process. Adding a stoichiometric excess of fuel to a catalytic anode for the purpose of creating the acid internally will similarly increase the cost if the resultant pH is below that which is required to efficiently operate the cell, frequently decreasing the amount of chlorine produced substantially.

Further, a lower pH than is necessary may contribute to reduced alkali current efficiency and to the degradation of the cell itself depending upon the construction materials.

Obviously, the reverse of the above is true if the pH is higher than is required, that is, oxygen will be evolved and/or sodium chlorate will form in the recirculating anolyte decreasing cell efficiency.

Referring to Figure 7.10, there is shown a schematic representation of an electrolysis cell **10**. The cell comprises an anolyte compartment **12** and a catholyte compartment **14** separated by a cation permselective membrane **16**.

Although membrane **16** is a cation permselective membrane, some hydroxide ions will still migrate into the anolyte resulting in the formation of sodium chlorate and oxygen unless inhibited by a similar supply of hydrogen ions.

The inhibition may be accomplished by introducing acid directly into the anolyte according to the prior art, or by supplying anode **18** with a substoichiometric amount of fuel, preferably hydrogen, from either an external source **28** or from the catholyte compartment **14**. The quantity of hydrogen so admitted is controlled by valves **30** or **32**. If desired both sources may be employed.

The pH of the anolyte is monitored by a pH meter **34**. The pH may thus be controlled by adjusting the supply of hydrogen by adjusting valves **30** and/or **32**.

Optionally a catalytic cathode may be employed supplied by an external source of oxygen-enriched air or air **36**. The amount of oxygen introduced is controlled by valve **38**. The cathode will catalytically promote the combination of oxygen with water to produce hydroxide ions, the amount of hydrogen evolved around the cathode will thus be reduced and as a result the electrode will be depolarized. Further the amount of hydrogen in the catholyte which is available to the anode will be reduced allowing the reaction to act as an additional control of the pH. The amount of hydrogen removed will depend upon the amount of oxygen available and therefore the setting of valve **38**.

Figure 7.10: Electrolysis Cell Permitting Internal pH Control

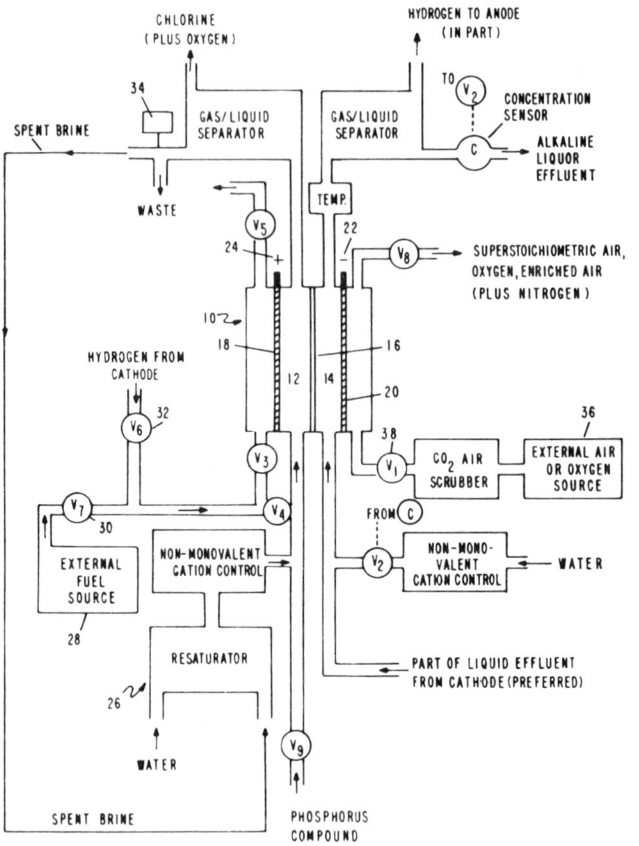

Source: U.S. Patent 4,173,524

Example 1: This example illustrates a preferred operation but without pH control of the anolyte. An electrolyte cell is constructed in accordance with Figure 7.10. The membrane is a perfluorosulfonic acid type, Nafion, and consists of a thin skin having an equivalent weight of about 1,350 laminated to a substrate having an equivalent weight of about 1,100. The membrane is reinforced with a woven polyperfluorocarbon fabric, Teflon. The effective area of the membrane is about 1 dm^2. A perfluorocarboxylic acid membrane, such as that manufactured by the Asahi Chemical Industry Co. may also be used.

The cathode is woven nickel wire mesh; the anode is a woven titanium wire mesh which has been coated on the face adjacent to the membrane with several layers of finely divided ruthenium oxide powder, baked at an elevated temperature to promote adhesion to the mesh as is well known in the art. The electrodes also have apparent areas of about 1 dm^2. The electrodes are spaced from the membrane to permit gas evolution and disengagement. Sodium chloride brine, substantially saturated, is fed to the anode compartment at a rate of about 300 cc per hour. The effluent from the anode compartment is separated into a gas stream and a liquid stream. From about 1 to 10% of the effluent liquid stream is sent to waste; the remainder with additional water is resaturated with salt and used as feed to the anode compartment.

About 5% sodium hydroxide is fed to the cathode compartment. The feed rate is adjusted to produce an effluent from the cathode compartment having a concentration of about 10%. The effluent from the cathode compartment is also separated into a gas stream and a liquid stream. Part of the liquid stream is diluted with water and used as feed to the cathode compartment.

After the flows to the electrode compartments have been established, a direct current of about 25 A is imposed on the cell. After several hours, the voltage of the cell stabilizes at about 4.5 V. The temperature of the effluents from the cell are adjusted to about 80°C by controlling the temperatures of the feeds to the electrodes.

The gas stream separated from the effluent from the anode compartment is analyzed by absorption in cold sodium hydroxide and titration of the latter for available chlorine. The current efficiency for chlorine evolution is found to be about 85%. The pH of the liquid stream separated from the effluent from the anode compartment is found to be substantially greater than 4.

Example 2: This example illustrates the improvements which can be obtained using anolyte pH control. The cell of Example 1 was used. The cell is operated as described in Example 1 except part of the gas separated from the effluent from the cathode compartment is admitted to the brine feed to the anode compartment. The rate of admission of the gas (substantially pure, but humid hydrogen) is adjusted to maintain the pH of the liquid separated from the effluent from the anode compartment in the range of from about 2 to 4. After several hours the voltage of the cell stabilizes at about 4.5 V.

The gas stream separated from the effluent from the anode compartment is analyzed as described in Example 1. The efficiency for chlorine evolution is found to be in the range of about 90 to 95%, higher values being associated with low pHs in the range.

Cathodic Compartment Filled with Conducting Material

O. DeNora and A. Pellegri; U.S. Patent 4,177,116; December 4, 1979; assigned to Oronzio DeNora Impianti Elettrochimici SpA, Italy describe an electrolytic cell with dimensionally stable anodes, nonporous ion-selective membranes separating the anodes from the cathode compartment, and a porous, static bed of loose, conducting cathodic material in the cathode compartment, extending between the conductive walls of the cathode compartment and the membrane and contacting the conductive walls of the cathode and the membranes to carry current between the walls of the cathode compartment and the membranes.

This construction reduces the electrodic gap to substantially the thickness of the membranes and presses the membranes against the anodes. It produces greater uniformity of current density over the entire electrodic area, substantially free from localized differences of current density which tend to cause deterioration of membranes by the creation of localized mechanical and electrical stresses in other types of cells, and provides a method for carrying current from the effective cathodic surface to the walls of the cathode compartment.

The conductive cathodic filling material may be graphite, lead, iron, nickel, cobalt, vanadium, molybdenum, zinc or alloys thereof, intermetallic compounds, compounds of hydridization, carbidization and nitridization of metals, or other materials having good conductivity and resistance to the cathodic conditions.

Bimetallic Electrodes Connected in Series

F.J. Scoville; U.S. Patent 4,196,068; April 1, 1980 has developed apparatus for producing chlorine gas comprising a plurality of separately housed electrolytic cells containing bimetallic electrodes electrically connected in series. Each electrolytic cell is divided into an anode and a cathode compartment separated by an ion permeable membrane which is impermeable to gases and water. Each cell housing contains passageways for simultaneously delivering a pure alkali metal chloride brine solution to the anode compartments and transporting chlorine gas produced at the anode and spent brine solution from the anode compartments.

Passageways are also provided in each cell housing for delivering deionized water to the cathode compartments and conveying hydrogen gas produced at the cathode and an alkali metal hydroxide solution from the cathode chamber. Appropriate piping is provided for delivering brine solution and deionized water to the cell housing and conveying chlorine gas and spent brine and hydrogen gas and an alkali metal hydroxide solution from the cell housing. A positive direct current is supplied to the anode of one end cell of the apparatus and a negative direct current is supplied to the cathode at the other end cell of the apparatus.

Figure 7.11a shows the assembled apparatus consisting of a plurality of electrolytic cells **10** electrically interconnected by wires **11** and joined together at the bottom by feeder lines **12** and **13** which pass through the cell housing and by discharge lines **14** and **15** which lead from the cell housing to collection lines **16** and **17**.

Figure 7.11: Bimetallic Electrodes Connected in Series

a.

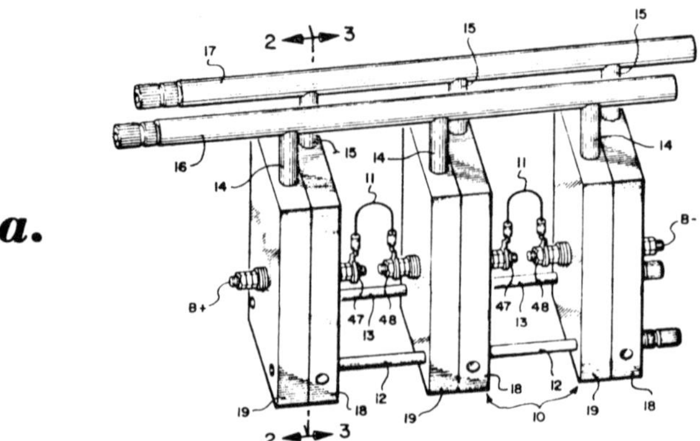

(continued)

Figure 7.11: (continued)

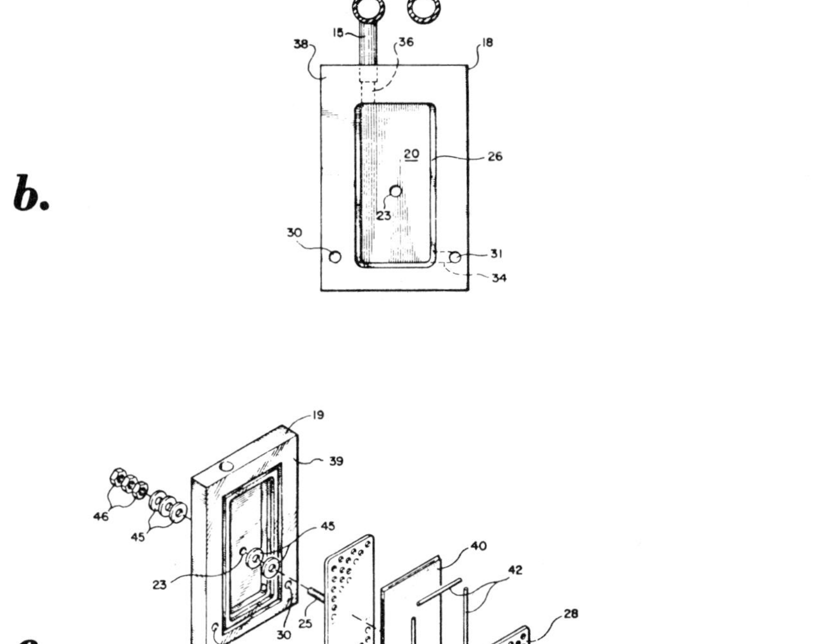

(a) Assembled apparatus
(b) Cross-sectional view along line **3–3** of Figure 7.11a
(c) Exploded view of one complete cell

Source: U.S. Patent 4,196,068

Each electrolytic cell is the same in structure and is shown in detail in Figures 7.11b and 7.11c. The preferred anode is a ruthenium-oxide-covered titanium in sheet form having apertures therein to increase the surface area. The cathodes are an apertured sheet metal preferably made of stainless steel. A preferred membrane **40** is Nafion, a perfluorinated sulfonic acid polymer.

In operation, a pure alkali chloride solution, i.e., preferably sodium chloride, is introduced through brine feeder line **12** into each electrolytic housing and flows through inlets **34** into anode compartments **20**. By application of a direct current source, electrolysis takes place at the anode **28** causing the production of chlorine gas and sodium ions. The chlorine gas and spent salt brine solution exit via outlet **36** into passageway **15** and collection line **17** wherein the chlorine gas and spent salt brine solution pass into a separation unit (not shown) whereby chlorine gas is released and the spent brine is disposed of.

SUPPORT STRUCTURES AND SEALING MEANS

To obtain maximum utility from cells incorporating permselective membranes, a multicell electrolyzer is conventionally employed. In this plural cell design, a number of semiindependent cells are arranged in serial fashion and provided with various means for permitting flow of the fluid medium to be electrolyzed, and means for electrical communication between and among the various cells comprising the electrolyzer.

While such a design takes full advantage of the characteristics of the permselective membranes, precautions must be taken to prohibit fluid and/or gaseous leakage at, for example, points of mechanical connection of the cell components since, obviously, the advantage of the membrane characteristics would otherwise be lost. Thus, the art recognizes the need to provide reliable retaining and restraining structures to achieve a two-fold purpose. Broadly speaking, the individual cell frame members must be maintained in a fluid-tight intimate, face-to-face contact; and, the individual cells should be maintained in a substantially vertical plane.

Support Structure for Plural Cell Electrolyzer

A retaining and restraining support structure for a plural cell electrolyzer described by *R.H. Fitch, B.E. Kurtz and B.B. Smura; U.S. Patent 4,129,495; December 12, 1978; assigned to Allied Chemical Corporation* is comprised of a sleeper assembly for supporting the electrolyzer cell along its longitudinal dimension, and a tension bar assembly for maintaining each of the cells in a substantially vertical plane, while insuring good mechanical connection and fluid communication therebetween, wherein the tension bar assembly includes plural longitudinal, adjustable tension bars and diagonal, adjustable tie bars.

As shown in Figure 7.12, the longitudinal support for electrolyzer **10** is provided by a pair of sleepers **70** which are disposed beneath the electrolyzer. The sleepers also provide appropriate spacing of the electrolyzer from the floor. Appropriate vertical orientation of the individual cells comprising the electrolyzer is achieved by means of diagonal tie bars **82**. Each diagonal tie bar assembly **82** is attached, at one end, to the top of one of the cell end frames **72**, and at the other to an extension **84** of sleeper **70**. However, the lower end of the tie bars might be secured at any convenient anchor point horizontally displaced from the end frame.

The magnitude of the restraining force applied by means of the diagonal tie bars is achieved, as with the longitudinal tension bars **74**, by adjustment of, e.g., threaded nuts cooperating at either end with the tie bar itself. A universal or turnbuckle **86** is similarly provided to aid in adjustment of the magnitude of the restraining force so that the electrolyzer cells **12** are maintained in the necessary vertical orientation. The sense of this force is adjustable to the extent the lower fixture point **88** of tie bar **82** on the extension **84** may be moved inwardly or outwardly.

Figure 7.12: Fragmentary Top Plan View of Plural Cell Electrolyzer

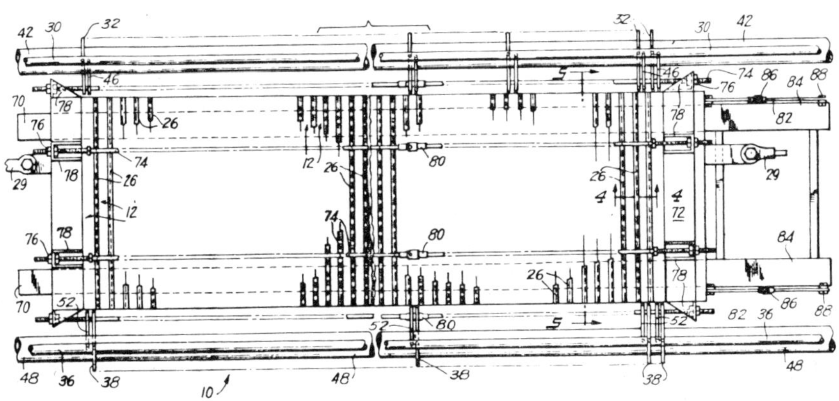

Source: U.S. Patent 4,129,495

The design described above offers many advantages over the prior art structures heretofore employed. Expensive and cumbersome side rail supports and hydraulically or mechanically actuated platens which have been employed to compress the cell frames and members together are eliminated, as are the associated ears or hangers on the cell frames. All of the advantages offered by these more cumbersome and expensive devices are retained, however, inasmuch as the compressive or retaining forces on the cells may be adjustably provided by tension bar assemblies **74** in combination with end frames **72**. Vertical stability is simply provided by means of diagonal tie bar assemblies **82**.

Moreover, the simplicity of the design allows for easy access to interior cells of the electrolyzer. For example, should it be required to remove an internal component, the longitudinal tension bars **74** may be loosened, and those bars on the top of the electrolyzer removed. The flexibility of the various feed and recovery hoses allows individual cells of the electrolyzer to be slightly displaced, and defective or worn out cells removed.

R.H. Fitch, B.E. Kurtz and N.Y. Rothmayer; U.S. Patent 4,153,532; May 8, 1979; assigned to Allied Chemical Corporation describe a cell frame transfer cart for vertical and horizontal displacement of a segment of a plural cell electrolyzer comprised of spaced main and secondary platforms, a vertical displacement member disposed intermediate the platforms, and a support member for supporting the main platform and permitting horizontal displacement of a selected segment of a plural cell electrolyzer when the vertical displacement member is actuated. An assembly including the present transfer cart and a method for disassembly of a plural cell electrolyzer are also described.

The overall dimensions of the transfer cart **40** are such that, prior to actuation of the bladder members **46**, the cart will fit beneath the electrolyzer **10** and between the sleepers **15**, as best viewed in Figure 7.13a. The cart may, accordingly, be freely inserted beneath the electrolyzer to a position subadjacent a segment selected to be removed, as shown in Figure 7.13b. In order to remove a segment of the electrolyzer, it is first necessary to relax the compressive force exerted by the tension bar assemblies **18**.

While Figures 7.13a and 7.13b show all of the tension bar members **18** removed, save for those beneath the electrolyzer, it is possible to remove a portion of the electrolyzer having only first removed the tension bars located along the top side. In either event, however, the individual cells adjacent that section selected for removal must be stabilized as the retaining force normally provided by longitudinal tension bars **18** is no longer present.

To stabilize those sections adjacent that to be removed from the electrolyzer, each of the cell frames **12** is formed with holes **60**, to present at least two sets as shown in Figure 7.13b. A cell grip bar **62**, shown in Figure 7.13c, is also formed with a row of apertures **64**, the spacing between apertures **64** corresponding to the spacing between the holes **60** in adjacent cell **12**.

Figure 7.13: Multicell Electrolyzer with Lifting Device and Transfer Cart

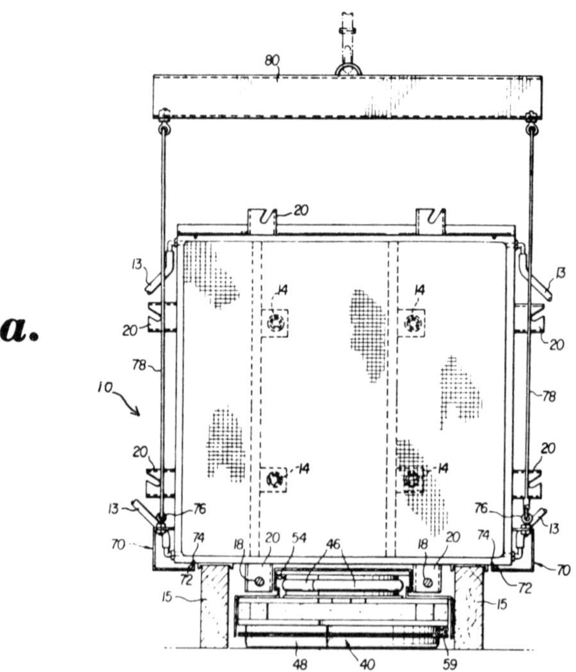

(continued)

Figure 7.13: (continued)

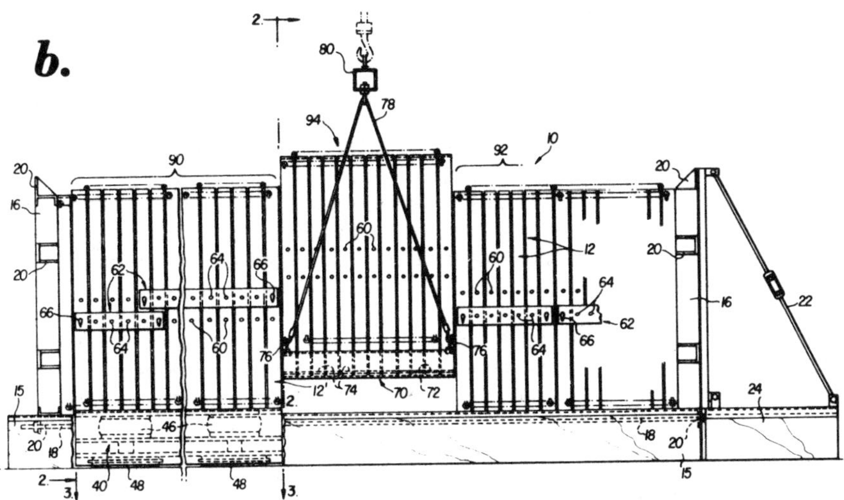

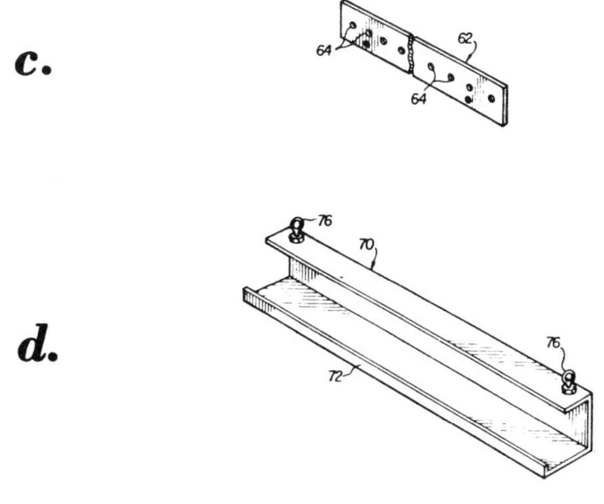

(a) Transverse sectional view showing cells resting on support
(b) Side elevational view
(c) Perspective view of grip bar
(d) Perspective view of lifting member

Source: U.S. Patent 4,153,532

Expandable grip pins **66**, such as Expando-Grip-Pins, are inserted through the apertures **64** into the holes **60** to securely fasten the grip bar **62** to a series of cells **12**. The grip bar **62** may be any convenient length to encompass, for example, 6 or 9 or 12 cells. It is also possible to overlap two grip bars, as shown in Figure 7.13b, to allow for the stabilization of a greater number of cells.

In order to remove a selected segment of the electrolyzer **10**, a pair of lifting bars **70**, shown in Figure 7.13d, are employed. Each lifting bar **70** includes an upturned lip portion **72** which cooperates with notches **74** formed in each of the cell frames, as best viewed in Figure 7.13a. A pair of lifting eyes **76** are located on the lifting bars **70** for attachment to lifting cables **78**. An overhead hoist **80** or similar structure may be utilized to remove the selected segment of the electrolyzer. The lifting bar **70** may be of any desired length, corresponding to the number of cells to be removed.

Implementation of the method for removal of a portion of the electrolyzer **10** is achieved by first placing the cell frame transfer cart **40** beneath the electrolyzer as shown in Figure 7.13b. As all of the tension bar assemblies **18** have been loosened, and at least those on the top side of the electrolyzer **10** removed, it is necessary to modularize the segments **90** and **92** bounding either side of the selected segment **94** which is to be removed, as noted above, whereby the segments **90** and **92** are stabilized.

The expandable bladders **46** are actuated by the controlled admission of air via a line and valve (not shown) thereby causing the secondary platform to displace all of the cells contained in the segment **90** (up to, and including, cell **12'**) off of the sleeper assemblies **15**. The feed and recovery hoses, designated generally as **13** in Figure 7.13a, are disconnected from the cells which constitute the segment **94** selected for removal.

The upturned lip **72** of lifting bar **70** is caused to engage the notch section **74** in each of the cells in segment **94** and, by any convenient lifting means (such as an overhead hoist), the portion **94** may be easily removed from the electrolyzer **10**. As noted above, the length of lifting bar **70** may be appropriately varied to encompass the number of cells selected for removal.

Prior to physical removal of the portion **94**, it is necessary to provide horizontal displacement of the stabilized segment **90** of electrolyzer **10** adjacent that segment selected for removal. For example, in order for an operator to gain access to the internal components of the electrolyzer to manually remove the same, it is necessary to displace the segment **90** sufficiently away from the segment **94** that physical access may be achieved. Horizontal displacement of segment **90** also guards against damage to the cells bounding the line dividing the various segments (e.g., **90**, **94**) during subsequent removal.

When the bladder members **46** are actuated, only the segment **90** will be lifted from the sleeper assemblies **15**, while the cells to the right of **12'** will remain resting on the sleeper supports. Owing to the ease of mobility provided by way of air-bearing members **48**, the segment **90** may be horizontally displaced while supported on the transfer cart **40**. Due to the flexibility of interconnecting hoses, this displacement is readily performed.

Membrane Sealing Means

W.B. Darlington; U.S. Patent 4,175,024; November 20, 1979; assigned to PPG Industries, Inc. describes an electrolytic cell having electrode units of opposite polarity where each of the electrode units have fingered electrodes extending outwardly therefrom toward the electrode unit of opposite polarity. In this way, the electrodes are interleaved between electrodes of opposite polarity.

At least one of the electrode units has a base plate with fingered electrodes mechanically and electrically connected thereto. The fingered electrodes bear a synthetic separator such as a microporous diaphragm or a permionic membrane thereon with lap at the base. The electrolytic cell is characterized by a fingered, interleaved electrode of the electrode unit of opposite polarity compressively bearing upon the lap, whereby to provide an electrolyte-tight seal.

As shown in Figure 7.14, the synthetic separator **61**, that is, the membrane or microporous diaphragm, bears upon cathode finger **43**, and the lap **63** at the open edge of synthetic separator glove bears upon the cathodic back screen **47**. The laps are compressed against the back screen by a bearing surface **71** on the leading edge **37** of the anode blades **33**.

Figure 7.14: Cutaway Top View of a Bipolar Electrolyzer

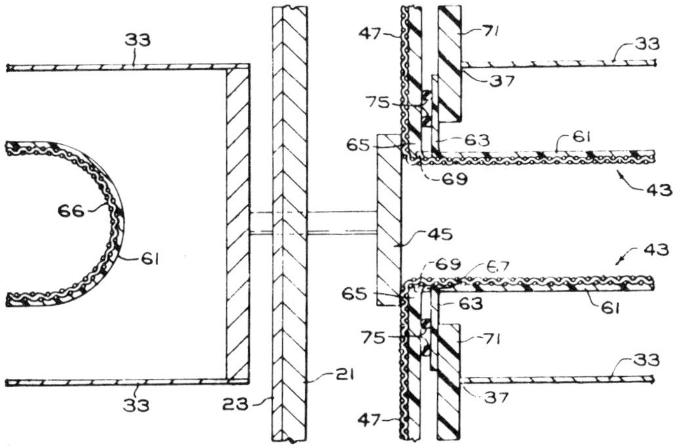

Source: U.S. Patent 4,175,024

The compressive surface **71** may be joined to the anode by mechanical means and may be a ceramic, a polymeric, or a noncatalytic metallic member. For example, the bearing or compressive surface may be fabricated of an undercoated valve metal. The compressive surfaces may be T- or L- shaped members of the anode blade, as a T welded onto or bent into the anode blade or an L welded onto or bent into the anode balde.

The compressive surface can be formed by a pair of anode blades forming one anode and bent in the form of a U or V, or welded to a U or V, or welded parallel to each other with a cross member at the leading edge.

In the exemplification thus described, the U or V formed by the loading edges **37**, or the cross member joining the leading edges **37** bears on the separator **61**.

In one exemplification, the laps **63** overlap and are sealed, for example, between cathode fingers **43** by the bearing surface **71**.

In an alternative exemplification, there is a liner **65** on the back screen **47** with the laps **63** laying on the liner **65**. In a particularly preferred embodiment of this exemplification, the liner **65** has lips **67** to provide both a further seal and a compressible surface for the compressive means **71** to bear upon.

In a still further alternative exemplification, there is a liner **65** on the back screen **47** with the liner **65** laying atop the laps **63** and being held in compression by compressive means **71**, thereby to provide a seal.

In a still further embodiment of the seal of the exemplification described above, there may be sealing means, e.g., O rings or gaskets **75** interposed between a pair of overlapping laps **63** or between the liner **65** and an overlap **63**.

The synthetic separator **61** can be in the form of separate envelopes for each finger with joints at this top **65**, bottom **66**, and leading edge (not shown), and lap **63** at the trailing edge **69**. Thereby the seals or seams can be welded, sewn, heated, or produced by chemical reaction prior to placing the individual gloves or envelopes on individual electrodes. Alternatively the individual fingers may be unitary envelopes, produced, for example, by casting, blow molding or injection molding.

MEMBRANE MATERIALS

A large portion of the chlorine and alkali metal hydroxide produced throughout the world is manufactured in diaphragm-type electrolytic cells wherein the opposed anode and cathode are separated by a fluid permeable diaphragm, usually of asbestos, defining separate anode and cathode compartments.

Through the years, substitution of a membrane material for the diaphragm has been proposed. These membranes are substantially impervious to hydraulic flow. In operation, an alkali metal chloride solution is introduced into the anode compartment wherein chlorine is liberated. Then, in the case of a cation permselective membrane, alkali metal ions are transported across the membrane into the cathode compartment.

The concentration of the relatively pure alkali metal hydroxide produced in the cathode compartment is determined by the amount of water added to this compartment, generally from a source exterior the cell. While operation of a membrane cell has many theoretical advantages, its commercial application to the production, for example, of chlorine and caustic has been hindered owing to the low current efficiencies obtained and the often erratic operating characteristics of the cells.

Polyamine-Modified Membranes

C.J. Hora and A.D. Babinsky; U.S. Patent 4,081,349; March 28, 1978; assigned to Diamond Shamrock Corporation provide an efficient, relatively supple, cation-exchange membrane for use in the production of chlorine and alkali metal hydroxide.

A cation-exchange membrane is derived from a fluorinated polymer containing sulfonyl groups on pendent side chains thereof, one surface of which has been treated with a polyamine whereby a majority of the sulfonyl groups to a depth of at least 10 μ has been converted to the form $SO_2NRR'NRR''$ wherein R is H, Na, or K; R' is C_{3-6} alkylene, $Z_2N(R'')$, or $ZN(R'')Z_2N(R'')$; Z is C_{2-6} alkylene; and R'' is R or $-SO_2-$.

Example: The membranes to be treated are like copolymers formed from 7 parts of tetrafluoroethylene and 1 part of $CF_2{=}CFOCF_2(CF_3)OCF_2CF_2SO_2F$ having an initial thickness of 208 μ and an equivalent weight of 1,100 (grams of polymer per equivalent of proton). Each membrane is sealed across the bottom of a treating vessel into which is introduced a solution of 18 parts by volume of 1,3-diaminopropane and 1 part by volume of water. Treatment continues for the length of time indicated in the table below at a temperature held between 60° and 65°C, followed by washing, removal from the vessel, and saponification in a solution containing 600 ml water, 400 ml dimethylsulfoxide, and 13 wt % NaOH maintained at 85° to 90°C for 70 min.

Each treated membrane is removed, rinsed, and installed wet in a membrane cell spaced 1.6 mm from a nickel cathode and 3.2 mm from an opposed expanded tiatnium metal anode bearing a $2TiO_2:RuO_2$ mol ratio coating in its surface. Water is added to the cathode compartment, saturated brine (315 g/ℓ) at a temperature of 85°C and a pH of 2.0 to the anode compartment and operation is commenced at 31 A/dm². On attaining equilibrium, the operating conditions of the cell are determined as noted below.

Membrane	Time of Treatment (min)	Depth of Amination (microns)	Voltage	Current Efficiency (%)	NaOH (g/ℓ)
1	0	0	4.1	73	466
2	17	70	4.3	82	536
3	20	123	4.4	86	572
4	30	185	4.4	88	572
5	60	208	4.2	90	590
6*	0	0	4.3	75	500
7*	600	206	4.8	88	536

*1,200 equivalent weight copolymer.

Improved operation is evident from the table. In addition, the membranes remain supple and resistant to mechanical failure.

Heat Treatment of Polyamine-Modified Membranes

S.F. Burkhardt and D.E. Maloney; U.S. Patent 4,168,216; September 18, 1979; assigned to Diamond Shamrock Corporation and E.I. Du Pont de Nemours and Company find that the useful life of these membranes can be extended by heating them for at least 15 min at temperatures in the range of 100° to 160°C.

In the example that follows, repeated reference will be made to two treating baths. Treating Bath A consists of 11 to 14 wt % potassium hydroxide, 30 to 36 wt % dimethylsulfoxide, and the balance water.

Treating Bath B consists of 94 to 96 wt % ethylenediamine and 4 to 6 wt % water as determined with Karl Fischer reagent.

Example: A film of copolymer of tetrafluoroethylene and perfluoro-3,6-dioxa-4-methyl-7-octenesulfonyl fluoride having an equivalent weight of 1,150 and having a thickness of 0.15 mm (6 mils) was hydrolyzed on one surface to form potassium sulfonate groups to a depth of 0.013 mm (0.5 mil) by treating one side of the film with Treating Bath A at 59°C for 8.5 min.

The film was then reinforced according to U.S. Patent 3,925,135 by embedding into the nonhydrolyzed side thereof, by vacuum lamination, a fabric of polytetrafluoroethylene fibers which is a warp knit with weft insertion having 4.7 200 denier threads per centimeter (12 threads per inch) in the warp and 5.9 400 denier threads per centimeter (15 threads per inch) in the fill (fabric designation T-25).

The lamination was carried out with the fabric against the vacuum roll which was heated to 235°C, the film toward the curved stationary plate which is heated to 362°C, the vacuum in the vacuum roll being 78,000 Pa (23 inches of mercury), at a speed of 30.5 cm/min (1 ft/min).

The resulting laminate was then immersed in Treating Bath B at 35°C for 11 min, which resulted in a sulfonamide layer about 0.025 mm (1 mil) thick on the nonhydrolyzed side of the membrane and a second sulfonamide layer about 0.025 mm (1 mil) thick just beneath the hydrolyzed surface. The laminate was then immersed in Treating Bath A at 95°C for 35 min to hydrolyze the remaining sulfonyl fluoride groups to potassium sulfonate groups, and the resulting membrane was washed with acetic acid to leach excess potassium hydroxide from it.

As a control, one piece of the membrane was installed in a chlor-alkali cell with the exposed sulfonamide surface layer facing the cathode, and such that the active membrane area was 45.2 cm^2 (7 in^2). Saturated sodium chloride solution (brine) was electrolyzed with a current of 14 A (2 A/in^2) at 80°C, using a brine flow rate of 0.4 ml/A/min and addition to the brine of HCl at a rate of 0.14 mg/A/min, and with water slowly added to the cathode compartment to maintain production of caustic at a goal concentration of about 28 to 30 wt %. The performance was as follows.

Days	Current Efficiency (%)	Voltage	NaOH (% by wt)
1	89	4.0	28
5	85	3.9	27
16	85	3.8	27
23	83	3.7	28
28	83	3.8	27

A second piece of the same membrane was heated in a vacuum oven at 110°C for 24 hr under a vacuum of about 95,000 Pa (28 inches of mercury). It was tested in a chlor-alkali cell as described above, except that the rate of HCl addition to the brine was 2.1 mg/A/min. The performance was as follows.

Days	Current Efficiency (%)	Voltage	NaOH (% by wt)
2	89	4.35	29
4	87.5	4.0	29
7	87.5	3.95	29
14	87	3.8	29
23	87	3.8	29

N-Monosubstituted Sulfonamido Groups in First of Two Layers

An ion exchange film developed by *P.R. Resnick and W.G. Grot; U.S. Patents 4,085,071; April 18, 1978 and 4,113,585; September 12, 1978; both assigned to E.I. Du Pont de Neumours and Company* comprises a fluorine-containing polymer with (a) ion exchange sites present in a first layer as N-monosubstituted sulfonamido groups or salt thereof formed through reaction of a primary amine, and (b) ion exchange sites present in a second layer of the film other than as the N-monosubstituted group or salt thereof.

Example 1: In this example a film is employed of a copolymer of tetrafluoroethylene and

$$CF_2{=}CFOCF_2\underset{\displaystyle CF_3}{\underset{|}{C}}FOCF_2CF_2SO_2F$$

and is referred to as precursor polymer containing pendant sulfonyl fluoride groups. The equivalent weight of the polymer is given and, illustratively, at a mol ratio of tetrafluoroethylene to the other monomer of 7:1, an equivalent weight of 1,146 would be obtained. Equivalent weight is the weight of the polymer in grams containing one equivalent of potential ion exchange capacity.

To a stoppered Erlenmeyer flask was added 50 cc of a 40% aqueous solution of methylamine and the precursor fluorinated polymer containing pendant sulfonyl fluoride groups (EW 1151). Stirring took place at room temperature for 16 hr followed by washing with water. Infrared spectra by attenuated total reflectance (ATR) indicated the conversion of sulfonyl fluoride groups to $-SO_3^-$ and $-SO_2NHCH_3$.

Example 2: A 5 mil film of precursor fluorinated polymer containing pendant sulfonyl fluoride groups (EW 1108) was placed in a Pyrex baking dish. Ethylenediamine (99% purity) was poured on top of the film so as to contact only the top surface. The surface of the liquid was covered with a second, similar film to minimize the exposure to moisture. After 15 min at room temperature, the amine was poured off, the film rinsed first with diglyme, then benzene and finally with warm water at about 40°C. Staining of a cross section of the film with Sevron Red indicated reaction to a depth of 0.7 mil.

The remaining pendant sulfonyl fluoride groups were converted to $-SO_3K$ groups by immersing the film in a solution of 15% potassium hydroxide and 30% dimethyl sulfoxide in water for 6 hr at 60°C.

The film was clamped in a chlor-alkali cell with the amine treated side toward the cathode. The chlor-alkali electrolysis cell was constructed of two identical half-cell housings made from Teflon TFE resin into which were mounted in the respective cell housings a dimensionally stable anode and a perforated stainless steel cathode. The clamped film gave an active area of the electrodes and membrane of 4 x 4 inches. The electrolytes, saturated brine and sodium hydroxide, were circulated through respective cell halves with a temperature maintained at 85°C by heaters installed in the circulatory lines. Fresh brine was pumped into the anode section of the cell and distilled water was pumped into the cathode section of the cell.

In operation of the cell a current efficiency of 91% was realized at a cell voltage of 3.6 V. A sodium hydroxide concentration of 14% was obtained.

Thin Secondary Layer

T. Sata, S. Murakami and Y. Murata; U.S. Patents 4,169,023; September 25, 1979; and 4,166,014; August 28, 1979; both assigned to Tokuyama Soda Kabushiki Kaisha, Japan describe an electrolytic diaphragm consisting essentially of a main layer composed of a polymeric membranous material uniformly containing cation exchange groups and fluorine atoms chemically bonded thereto and a secondary layer having a smaller thickness than the main layer and composed of at least one of an electrically neutral layer and a layer containing an anion exchange group, the secondary layer being in intimate contact with the main layer at at least one surface layer portion or interior of the main layer.

Example: An oxidation-resistant cation exchange membrane (Nafion XR-480, a perfluoro-type cation exchange membrane containing a sulfonic group bonded to the pendant chain by E.I. Du Pont) was immersed in 1 N hydrochloric acid to convert the sulfonic group to an acid type. The membrane was then dried for 4 hr in an electric dryer at 80°C, and then immersed in thionyl chloride and refluxed for 48 hr to convert the sulfonic group to a sulfonyl chloride group (base membrane).

The resulting base membrane was thoroughly washed with carbon tetrachloride, immersed in an 8% ethyl alcohol solution of linear polyethyleneimine [obtained by ring-opening polymerization of 2-oxazoline and subsequent hydrolysis of the resulting product by the method described in *Polymer Journal*, 3, 35 (1972)] for 24 hr at 40°C, washed with water, and then immersed in 4.0 N sodium hydroxide to convert the sulfonyl chloride group in the membrane again to a sulfonic group.

Consequently, in the surface layer portion of the membrane, ethyl sulfonate and the linear polyethylene were bonded to the membrane. The surface layer was cut out in a thickness of about 10 μ using a microtome, and analyzed for S and N. It was found that the ratio of S:N was 1:6. The membrane was then immersed in acetic anhydride for 25 hr at 30°C to convert the remaining primary and secondary amino groups in the polyethyleneimine bonded to the surface layer portion to the carboxamide, thereby to form a neutral layer consisting of sulfonamide, carboxamide and sulfonate on the surface layer portion of the cation exchange membrane.

Using this membrane, a saturated sodium chloride solution was flowed to the anode compartment with the flow rate in the anode compartment being 6.0 cm/sec, and pure water was flowed to the cathode from outside the electrolytic cell so that 4.0 N sodium hydroxide could be obtained steadily. The cathodic solution was flowed at a flow rate of 6.0 cm/sec in the cathode compartment. The temperature of the solution was maintained at 70°C during the electrolysis. As a result, the current efficiency for obtaining sodium hydroxide was 96%, and the amount of sodium chloride in the 4.0 N sodium hydroxide was 0.0002 N.

Separately, Nafion XR-480 was used for the electrolysis under the same conditions without prior treatment. As a result, the current efficiency for obtaining sodium hydroxide was 82%, and the amount of sodium chloride in 4.0 N sodium hydroxide was 0.002 N. The electric resistance of this membrane was 4.5 Ω-cm^2 both in direct and alternate currents. The treated membrane also had an electric resistance of 4.5 Ω-cm^2 both in direct and alternate currents.

The transport number measured on the basis of the potential of the membrane was 0.85 for the untreated membrane, and 0.88 for the treated membrane.

Hydrophilic Membranes Formed by Reacting Fluoropolymers with Sulfur- or Phosphorus-Containing Compounds

According to *J.C. Fang; U.S. Patents 4,153,520; May 8, 1979; 4,169,024; September 25, 1979; and 4,189,369; February 19, 1980; all assigned to E.I. Du Pont de Nemours and Company* certain fluoropolymers, when chemically modified by reacting them with sulfur- or phosphorus-containing compounds, become hydrophilic materials useful for making ion-exchange membranes, especially diaphragms for electrolytic cells, particularly chlor-alkali cells used in the production of chlorine, hydrogen and sodium hydroxide from brine.

Example: A piece of Teflon fluorocarbon resin cloth (T-162-42 by Stern & Stern Textiles, Inc.) was preshrunk by baking it for 10 min at 270°C.

A 15% solids aqueous dispersion of tetrafluoroethylene/bromotrifluoroethylene 86/14 copolymer was sprayed on both sides of the cloth to an overall thickness of 100 μ (dry) and the cloth was then baked for 20 min at 270°C.

The resulting coated fabric was then placed in a vessel containing a solution of 11 parts potassium sulfide, 6.4 parts sulfur, and 3.0 parts cesium fluoride in 300 parts of dimethylacetamide.

The vessel was heated on a steam bath for 4 hr, while the solution containing the fabric was stirred. The coated fabric was then removed from the vessel, washed with distilled water, submerged in saturated chlorine water and kept there for 16 hr at 20°C.

The fabric was then rinsed with distilled water, dried and placed in the diaphragm position of a laboratory chlor-alkali cell containing saturated brine, where, in operation, it required a voltage of 3.0 to 3.1 to achieve a current density of 0.204 A/cm^2 of diaphragm area.

The flow rate through the diaphragm was found to have been 1.87 cm^3 of brine per square centimeter per minute. The fabric coating had a water solubility of less than 1%.

Sulfonyl/Carboxylic Acid Copolymer Blends

C.J. Molnar, E.H. Price and P.R. Resnick; U.S. Patent 4,176,215; November 27, 1979; assigned to E.I. Du Pont de Nemours and Company disclose ion-exchange films, membranes and laminar structures incorporating a layer of a blend of a first fluorinated polymer which contains sulfonyl groups in ionizable form and a second fluorinated polymer which contains carboxylic acid functional groups, which when used to separate the anode and cathode compartments of an electrolysis cell, permit operation at high current efficiency and low power consumption.

They can be made by synthesis of precursor polymers in melt-fabricable form, blending, fabrication of the film, membrane or laminar structure, and hydrolysis of the functional groups to ionizable form, such as the free acid form or alkali metal salt thereof.

Example: A first copolymer of 61 wt % of tetrafluoroethylene and 39 wt % of perfluoro(3,6-dioxa-4-methyl-7-octenesulfonyl fluoride)(having an equivalent weight of 1,148 after hydrolysis) was prepared according to the procedure of U.S. Patent 3,282,875. A second copolymer of 63 wt % of tetrafluoroethylene and 37 wt % of:

$$CF_2{=}CF{-}O{-}CF_2{-}\underset{\displaystyle CF_3}{\underset{|}{C}F}{-}O{-}CF_2{-}CF_2{-}COOCH_3$$

(having an equivalent weight of 1,133 after hydrolysis) was also prepared.

A 50:50 blend of the above two copolymers was made by mixing equal weights (2.80 g each) of them and grinding to a powder in a Spex freezer mill at the temperature of liquid nitrogen. The mixture was pressed to a film in a press at 240°C and 13,000±500 kg, with a 60 sec warm-up period, a 60 sec press period, and slow cool down.

The film was cut up in small pieces and again pressed into film under the same conditions. The film was ground to powder in the freezer mill, and the powder pressed a third time under the same conditions into a film 125 to 150 μm (5 to 6 mils) thick. The film was hydrolyzed to an ion exchange film having sulfonic acid and carboxylic acid groups in the potassium salt form by placing it in a bath made up of 392 g of potassium hydroxide (86 wt % KOH), 1,778 ml of water, and 848 ml of dimethylsulfoxide (the bath containing about 13 wt % potassium hydroxide) at 90°C for 1 hr, and then washing in distilled water for 30 min.

The film was used to separate the anode and cathode compartments of a laboratory-size chloralkali cell, and 26 wt % aqueous sodium chloride solution placed in the anode compartment was electrolyzed at a current density of 2.0 A/in^2. The following results were obtained.

Days	Current Efficiency (%)	NaOH (%)	Power Consumption	..(kWh/ton caustic)..
1	88.0	37.1	4.20	2,830
3	94.25	34.7	4.10	2,650
4	93.0	34.7	4.13	2,690
9	91.68	34.5	4.18	2,760
10	93.43	32.9	4.11	2,670

(continued)

Days	Current Efficiency (%)	NaOH (%)	Power Consumption	..(kWh/ton caustic)..
15	92.55	33.4	4.17	2,720
25	88.43	33.7	4.26	2,930
36	88.74	33.6	4.26	2,930
39	81.34	41.0	4.54	3,410
40	83.84	36.7	4.43	3,210
45	83.48	37.1	4.38	3,190
50	87.94	24.4	3.90	2,700
60	91.91	30.6	4.08	2,700
70	86.09	29.1	4.08	2,880
79	74.24	43.1	4.40	3,530
80	87.52	31.6	4.22	2,930
91	84.24	31.5	4.26	3,050
101	81.45	35.3	4.33	3,230

Fluoropolymer with Pendant Carboxylic Acid Groups

M. Seko, Y. Yamakoshi, H. Miyauchi, M. Fukumoto, K. Kimoto, I. Watanabe, T. Hane and S. Tsushima; U.S. Patent 4,151,053; April 24, 1979; assigned to Asahi Kasei Kogyo Kabushiki Kaisha, Japan disclose a cation exchange membrane comprising a fluorocarbon polymer characterized by the presence of pendant carboxylic acid groups of the formula $-OCF_2COOH-$ and derivatives thereof.

Example: Tetrafluoroethylene and perfluoro(3,6-dioxa-4-methyl-7-octene sulfonyl fluoride) were copolymerized in 1,1,2-trichloro-1,2,2-trifluoroethane in the presence of perfluoropropionyl peroxide as the initiator. The polymerization temperature was held at 45°C and the pressure maintained at 5 atm during the copolymerization. The exchange capacity of the resultant polymer, when measured after saponification, was 0.95 mg equivalent per gram of dry resin.

This copolymer was molded with heating into a film 0.3 mm in thickness. It was then saponified in a mixture of 2.5 N caustic soda per 50% methanol at 60°C for 16 hr, converted to the H form in 1 N hydrochloric acid, and heated at 120°C under reflux for 20 hr in a 1:1 mixture of phosphorus pentachloride and phosphorus oxychloride to be converted into the sulfonyl chloride form.

At the end of the reaction, the copolymer membrane was washed with carbon tetrachloride and then subjected to measurement of the attenuated total reflection spectrum (hereinafter referred to as ATR), which showed a strong absorption band at 1,420 cm^{-1} characteristic of sulfonyl chloride. In a crystal violet solution, the membrane was not stained.

Between frames made of acrylic resin, two sheets of this membrane were fastened in position by means of packings made of polytetrafluoroethylene. The frames were immersed in an aqueous 57% hydroiodic acid solution so that one surface of each membrane would undergo reaction at 80°C for 24 hr. The ATR of the membrane was then measured. In the spectrum, the absorption band at 1,420 cm^{-1} characteristic of the sulfonyl chloride group vanished and an absorption band at 1,780 cm^{-1} characteristic of the carboxylic acid group appeared instead.

In the crystal violet solution, a layer of a thickness of about 15 μ on one surface of the membrane was stained. The cation exchange groups existing on the surface were found to be carboxylic acid groups (100%) by measurement of ATR.

By saponifying this membrane in an aqueous solution of 2.5 N caustic soda per 50% methanol at 60°C for 16 hr, there was obtained a homogeneous and strong cation exchange membrane.

The diaphragm was equilibrated in an aqueous 2.5 N caustic soda solution at 90°C for 16 hr, incorporated in an electrolytic cell in such a way that the treated surface fell on the cathode side. It was utilized as the membrane in the electrolysis of sodium chloride and its current efficiency measured. The result was 95%.

Fluoropolymer with Pendant Carboxylic and/or Sulfonic Acid Groups

Cation exchange membranes characterized by carboxylic or carboxylic and sulfonic acid groups pendant from a fluorocarbon polymer are utilized for the electrolysis of aqueous sodium chloride by *M. Seko; U.S. Patent 4,178,218; December 11, 1979; assigned to Asahi Kasei Kogyo Kabushiki Kaisha, Japan.*

Example: A copolymer of perfluoro[2-(2-fluorosulfonylethoxy)propylvinyl ether] with tetrafluoroethylene was molded according to a conventional polymer-molding process into a membrane 0.12 mm in thickness. The membrane was hydrolyzed to prepare a perfluorosulfonic acid type cation exchange membrane having an exchange capacity of 0.91 meq/g of dry resin.

This cation exchange membrane was heated at 100°C for 3 hr in a solution containing 15% acrylic acid, 15% divinylbenzene, 55% styrene and 0.01% of benzoyl peroxide to impregnate the membrane with the monomer mixture, which was then polymerized at 110°C.

The thus obtained cation exchange membrane, which was a polymer mixture comprising a perfluorosulfonic acid type polymer and a crosslinked acrylic acid polymer, contained about 0.81 meq/g dry resin of exchange groups in terms of sulfonic acid groups and 0.23 meq/g dry resin of carboxylic acid groups. The thickness of the cation-exchange membrane was 0.14 mm.

This cation exchange membrane, which had an effective area of 100 dm^2, was used to divide an electrolytic cell into a cathode chamber and an anode chamber. 50 units of such an electrolytic cell were arranged in series so that the respective adjacent electrodes formed a bipolar system comprising 50 electrolytic cells.

Using the thus prepared electrolytic cell assembly, electrolysis was conducted by charging 305 g/ℓ of an aqueous sodium chloride solution to each cell through the inlet of the anode chamber, and an aqueous sodium hydroxide solution was recycled while being controlled at a concentration of 20% by adding water to the outlet of the cathode chamber. The electrolysis was carried out while applying in series a current of 5,000 A to the chambers.

In this case, the amount of the solution charged to the anode chamber was controlled to 11,515 kg/hr, the amount of the water added to the outlet of the cathode chamber was controlled to 1,063 kg/hr, and the aqueous sodium hydroxide solution at the outlet of the cathode chamber was recycled. As the result, the amount of chlorine generated in the anode chamber was 314.5 kg/hr, the amount of 20% sodium hydroxide recovered from the cathode chamber was 1,521.8 kg/hr, and the amount of hydrogen generated from the cathode chamber was 9,325 g/hr. The current efficiency of the sodium hydroxide recovered from the outlet of the cathode chamber was 95.1%.

Example: *Comparative* – A copolymer of perfluoro[2-(2-fluorosulfonylethoxy)-propylvinyl ether] with tetrafluoroethylene was molded into a membrane 0.12 mm in thickness, which was then hydrolyzed to prepare a cation exchange membrane containing 0.90 meq/g dry resin of sulfonic acid groups.

This membrane was utilized in the same manner as above, but the current efficiency while producing sodium hydroxide of 35.1% concentration was only 55.7%, and the amount of NaCl in NaOH was 2,000 ppm.

Fluoropolymer with Sulfonic Acid Groups plus Less Acidic Groups

According to *M. Seko, Y. Yamakoshi, H. Miyauchi, K. Kimoto and Y. Masuda; U.S. Patent 4,123,336; October 31, 1978; assigned to Asahi Kasei Kogyo Kabushiki Kaisha, Japan*, an aqueous alkali metal halide solution is electrolyzed in an electrolytic cell using a cation exchange membrane which is a fluorocarbon polymer containing sulfonic acid groups and at least one cation exchange group which is less acidic than the sulfonic acid group with higher proportion of the latter in surface stratum on the cathode side of the membrane than in the entire membrane, while controlling proton concentration in the anolyte at not higher than critical proton concentration at which no substantial amount of protons in the anolyte penetrate into the membrane.

The cation exchange groups with lower acidity than sulfonic acid group may include carboxylic acid groups, phosphoric acid groups, phosphite groups, sulfonamide groups, N-monosubstituted sulfonamide groups, alcoholic or phenolic hydroxyl groups, thiol groups and sulfinic acid groups. Among them, carboxylic acid groups and phosphoric acid groups are preferable from the standpoint of their characteristics and stabilities. In particular, carboxylic acid groups are most preferred.

Example: Tetrafluoroethylene and perfluoro(3,6-dioxa-4-methyl-7-octene sulfonyl fluoride) were copolymerized in an emulsion in the presence of ammonium persulfate as the initiator and ammonium perfluorooctoate as the emulsifier at 70°C under the pressure of 4 atm of tetrafluoroethylene.

The exchange capacity of the resultant polymer, when measured after washing with water and saponification, was 0.83 mg equivalent per gram of dry resin.

This copolymer was molded with heating into a film of 0.3 mm in thickness. It was then saponified in a mixture of 2.5 N caustic soda per 50% methanol at 60°C for 16 hr, converted to the H form in 1 N hydrochloric acid at 90°C for 16 hr, and heated at 120°C under reflux for 40 hr in a 1:1 mixture of phosphorus pentachloride and phosphorus oxychloride to be converted into the sulfonyl chloride form. At the end of the reaction, the copolymer membrane was washed under reflux with carbon tetrachloride for 4 hr at 40°C and then measured for attenuated total reflection spectrum (hereinafter referred to as ATR), which showed a strong absorption band at 1,420 cm^{-1} characteristic of sulfonyl chloride but no absorption of sulfonic acid group at 1,060 cm^{-1}. In a crystal violet solution, the membrane was not stained.

Between frames made of acrylic resin, two sheets of this membrane were fastened in position by means of packings made of polytetrafluoroethylene. The frames were immersed in an aqueous 57% hydroiodic acid solution so that one surface of each membrane would undergo reaction at 80°C for 30 hr. After washing with water at 60°C for 30 min, the treated surface of the membrane was subjected to measurement of ATR. In the spectrum, the absorption band at 1,420 cm^{-1} characteristic of sulfonyl chloride group vanished and an absorption band at 1,780 cm^{-1} characteristic of carboxylic acid group appeared instead. In a crystal violet solution, a layer of about 25 μ on one surface of the membrane was stained.

This membrane was saponified in a mixture of 2.5 N caustic soda/50% aqueous methanolic solution at 60°C for 16 hr and the treated surface was again subjected to measurement of ATR, whereby the absorption of carboxylic acid group was found to be shifted to 1,690 cm^{-1}. The specific conductivity of this membrane, when measured in 0.1 N aqueous caustic soda solution after being treated with oxidizing agent in a mixture of 2.5 N caustic soda/2.5% aqueous sodium hypochlorite solution at 90°C for 16 hr, was 10.0×10^{-3} mho/cm.

The membrane was stained again after the above treatment with oxidizing agent. From observation of the cross section stained, the thickness of the surface stratum containing carboxylic acid groups was found to be 7 μ.

The specific conductivity of the membrane was determined by initial conversion to a complete Na form, keeping the membrane in a constantly renewed bath of an aqueous 0.1 N caustic soda solution at normal temperature for 10 hr until equilibrium and subjecting it to an alternating current of 1,000 cycles while under an aqueous 0.1 N caustic soda solution at 25°C for measurement of the electric resistance of the membrane.

The aforementioned Na form cation exchange membrane was equilibrated in an aqueous 5.0 N caustic soda solution at 90°C for 16 hr, incorporated in an electrolytic cell having a dimensionally stable metal electrode as an anode and an iron plate as the cathode in such a way that the treated surface fell on the cathode side.

While the concentration of sodium chloride on the anode side was kept at 4 N and the alkali concentration on the cathode side at 8 N in the absence of hydrochloric acid added in the anolyte, current was passed at 90°C at a current density of 50 A/dm^2 for 5 hr. The current efficiency determined from the amount of sodium hydroxide formed was 91% with voltage of 3.7 V. The thickness of the desalted layer, measured by the method described above was 1×10^{-2} cm. Concentration of HCl corresponding to $d = 1 \times 10^{-2}$ cm and current efficiency = 91% was found to be 0.022 N.

Electrolysis was continued under the same conditions described above except that the proton concentration in the anolyte was maintained by addition of 0.013 N HCl which is lower than the critical value as determined above. During passage of current, which continued for 300 hr, the current efficiency was found to be stable at 91% at a voltage of 3.7 V. At the end of this period, the membrane was taken out and inspected by microscope, No unusual changes in the membrane were observed.

Heat-Treated Copolymers of Tetrafluoroethylene and Sulfonylfluoride Perfluorovinyl Ether

According to *M. Krumpelt and S.T. Hirozawa; U.S. Patents 4,089,759; May 16,*

1978, and 4,127,457; November 28, 1978; both assigned to BASF Wyandotte Corporation, membranes for use in chlor-alkali cells, made of a copolymer of tetrafluoroethylene and sulfonylfluoride perfluorovinyl ether, have their selectivity improved, with resulting substantial decrease in consumption of electric power per mol of sodium hydroxide produced by being heat-treated at 100° to 275°C for several hours to four minutes. The current efficiency is substantially increased, and the power consumption, per unit of sodium hydroxide produced, is usually decreased by about 10% or more.

Example 1: A 0.125 mm (5 mil) thick piece of polytetrafluoroethylene-reinforced membrane material, made of a copolymer of tetrafluoroethylene with sulfonated perfluorovinyl ether and having an equivalent weight number of about 1,100, was boiled briefly in a 1 N aqueous solution of hydrochloric acid and then removed.

In this state, it could have been inserted directly into a chlor-alkali cell. The piece was wiped dry, sandwiched between two sheets of polytetrafluoroethylene, and placed into a hydraulic press that had been preheated to 225°C. A pressure of 6.83 kg/cm^2 (7 $tons/ft^2$) was then applied for a period of 5 min, and the membrane was then allowed to cool in the press after the pressure had been released. This took about 15 min. The membrane was removed from between the sheets of polytetrafluoroethylene and inserted into a chlor-alkali cell having dimensionally stable anodes and steel cathodes.

The cell was then operated at a cell current of 25 A. Saturated brine having a pH of 4 was fed to the anode compartment at a rate of about 200 ml/hr, and 80 ml/hr of water were fed to the cathode compartment, which produced an 18 wt % aqueous solution of sodium hydroxide. The cell operated at 3.85 V and with a current efficiency of 78%. The energy consumption was 132 Wh/mol of sodium hydroxide.

For comparison, a similar membrane was inserted into a similar chlor-alkali cell, immediately after having been boiled briefly in hydrochloric acid. This chlor-alkali cell was operated under substantially the same conditions, exhibiting a cell voltage of 3.35 V, a current efficiency of 59%, and an energy consumption of 152 Wh/mol of sodium hydroxide. The thermal treatment increased the current efficiency from 59 to 78%, and it lowered the energy consumption from 152 to 132 Wh/mol.

Examples 2 through 4: A saturated solution of sodium chloride was introduced into the anode compartment of a two-compartment electrolytic cell containing a ruthenium oxide coated titanium mesh anode and a steel mesh cathode separated from the anode by a cation active selectively permeable diaphragm of 116 cm^2 effective area having a total film thickness of 0.2 mm and being composed of a 0.1 mm layer of a copolymer of tetrafluoroethylene and sulfonated perfluorovinyl ether having an equivalent weight of about 1,100 and a 0.05 mm layer having an equivalent weight of 1,500, the polymers prepared according to U.S. Patent 3,282,875.

The membrane was utilized without heat conditioning to improve selectivity. The cathode compartment was initially filled with dilute aqueous sodium hydroxide at a concentration of 80 g/ℓ and water added subsequently to maintain a sodium hydroxide concentration of 19%.

Chlorine gas evolved from the anode compartment was vented through a pipe and hydrogen evolved at the cathode was separately vented from the cathode compartment. A pipe for removal of caustic liquor was located in the cathode compartment. A temperature of about 80°C was maintained in the cell which was operated at a current density of about 1.4 A/in^2 of membrane.

Samples of the anolyte liquor were taken at intervals and analyzed for sodium chloride and sodium chlorate. Current efficiencies for sodium hydroxide, sodium chlorate and oxygen were calculated for each level of salt conversion (i.e., 40, 53 and 93%) and sodium chlorate formation. The data from this run are set forth below.

Example	Salt Conversion (%)	Rate of Chlorate Formation (mols/hr)	. . . Current Efficiencies . . . . NaOH (%)	$NaClO_3$ (%)	O_2 (%)
2	40	24.0×10^{-3}	76.3	19.2	5.3
3	53	20.9×10^{-3}	75.6	15.4	6.6
4	93	9.2×10^{-3}	75.6	6.2	14.5

Examples 5 through 8: Following the procedure of Examples 2 through 4, a saturated solution of sodium chloride was subjected to electrolysis in an electrolytic cell. The selectively permeable membrane utilized in the cell was subjected to a heat treatment prior to use at a temperature of 200°C for a period of 2 hr in order to provide improved selectivity, exhibit higher current efficiency and lower energy consumption per unit of product. The conditions of electrolysis were similar to those described in Examples 2 through 4. The results are set forth below.

Example	Salt Conversion (%)	Rate of Chlorate Formation (mols/hour)	. . . Current Efficiencies . . . . NaOH (%)	$NaClO_3$ (%)	O_2 (%)
5	24	2.03×10^{-3}	90.7	1.3	5.6
6	46	1.01×10^{-3}	92.2	0.7	6.1
7	47	1.03×10^{-3}	91.2	0.7	7.2
8	85	0.48×10^{-3}	89.9	0.3	9.5

These data indicate that the rate of chlorate formation in the electrolysis of a sodium chloride brine can be substantially reduced by operating the chlor-alkali cell at a salt conversion percentage in the anolyte compartment of about 60 to 80%. The data also indicate that the rate of chlorate formation can be substantially reduced when a selectively permeable membrane composed of a copolymer of tetrafluoroethylene and sulfonate perfluorovinyl ether is subjected to a heat-treatment step prior to its use in order to increase selectivity of the membrane.

Heat plus Diene-Derivative Treatment

T. Seita, T. Satoh and A. Shimizu; U.S. Patent 4,189,361; February 19, 1980; assigned to Toyo Soda Manufacturing Co., Ltd., Japan prepare a cation exchange membrane by treating one surface of a membrane of a perfluorocarbon polymer, having sulfonyl halide groups on branched chains, with a diamine or a polyamine and heating the treated polymer membrane at a temperature from 170°C to a deterioration temperature of the membrane if necessary, hydrolyzing it and immersing a diene derivative having a carboxyl group or a group which can be converted to carboxyl group into the membrane to polymerize partially the diene derivative, and dipping the membrane into an organic solvent which is water-miscible and has a boiling point of higher than 120°C in 760 mm Hg and holding the treated membrane between smooth plates and heating the membrane at 80° to 180°C and, if the diene derivative having a group which can be converted to carboxyl group is used, converting the groups into carboxyl groups.

Example: A membrane made of a copolymer of $CF_2{=}CF_2$ and

$$CF_2{=}CF{-}O{-}CF_2{-}\underset{\displaystyle CF_3}{\underset{|}{C}F}{-}O{-}CF_2{-}CF_2{-}SO_2F$$

(EW = 1,100 and thickness of 7 mil) was used. Ethylenediamine was contacted with one surface of the membrane and then, the surface was washed and dried. According to a coloring test of a sectional part of the membrane, it was found to react in a depth of 1.1 mil.

The membrane was reinforced with polytetrafluoroethylene fabric in the non-treated side by a heat bonding and it was heated at 180° to 200°C and hydrolyzed to obtain a cation exchange membrane. The cation exchange membrane in acid type was used.

Then, an ether solution of butadiene-1-carboxylic acid (20 wt %) was prepared and the membrane was immersed in the solution at room temperature for 2 days. The membrane was taken out and the surfaces were wiped off and the membrane was held between glass plates and heated at 110°C for 2 hr to partially polymerize the monomer. Then, the membrane was immersed in ethylene glycol for 7 hr and the membrane was held between smooth plates made of a glass plate, a rubber plate and a polyester sheet and heated at 110°C for 6 hr and immersed into 0.5 N NaOH for 2 days.

An electrolytic cell having effective area of 30 x 30 cm^2 was prepared by partitioning an anode compartment and a cathode compartment with the resulting cation exchange membrane (the treated surface was faced to the cathode side). An electrolysis was carried out by feeding an aqueous solution of sodium chloride having a concentration of 300 g/ℓ into the anode compartment at a rate of 2,600 cc/hr and feeding water into the cathode compartment so as to maintain 20 wt % of the concentration of sodium hydroxide and passing through a current of 270 A.

In the normal operation, a current efficiency for sodium hydroxide obtained from the cathode compartment was 95% and a cell voltage was 3.7 V. On the other hand, when the nontreated cation exchange membrane was used, a current efficiency was 88% and a cell voltage was 3.9 V.

Pretreating by Hydrolyzing and Stretching

D.T. Tokawa, B.J. Mentz, J.D. Eng, E.H. Cook, Jr. and G.R. Marks; U.S. Patent 4,124,477; November 7, 1978; assigned to Hooker Chemicals & Plastics Corp. describe an electrolytic cell suitable for use in electrolyzing ionizable chemical compounds, particularly alkali metal halide brines and hydrohalic acids, which comprises a cell body having an anode compartment containing a porous anode and a cathode compartment containing a cathode, the compartments being separated from each other by a prestretched, taut membrane barrier which is substantially impervious to gases and liquids and which is selected from a hydrolyzed copolymer of a perfluorinated hydrocarbon and a sulfonated perfluorinated hydrocarbon and a sulfonated perfluorovinyl ether, and a sulfostyrenated perfluorinated ethylene propylene polymer, the barrier being pretreated by hydrolyzing and stretching prior to insertion in the cell.

Such cells can be operated at constant low voltage, and are not subject to erratic operating voltages which are due, in part at least, to the accumulation of gases between the anode and diaphragm.

Example 1: A section of Teflon fabric reinforced membrane of the hydrolyzed copolymer of perfluorinated hydrocarbon and a fluorosulfonated perfluorovinyl ether type was soaked in boiling water by immersion for 2 hr. The hot, hydrolyzed membrane was found to be soft and pliable. The membrane was then stretched on a wooden frame. Stretching was accomplished by initially stretching the membrane diagonally and securing the ends of the membrane to the frame at the corners with clamps and subsequently stretching the sides and securing the sides of the membrane to the sides of the frame with clamps.

The corner clamps were then removed and the stretched, taut membrane was left to dry. A ½" wide strip of Devcon 2 Ton Epoxy resin cement was spread around the inside of an anode flange. The wooden frame with the pretreated, taut membrane was then positioned on the cemented surface of the anode so that the membrane is in contact with the cement and the wooden frame clamped to the anode frame.

After 2 hr at room temperature the clamps and wooden frame were removed. The result was a pretreated, taut membrane positioned on an anode member. The excess membrane material around the edges of the anode member was trimmed and the anode member and membrane were ready for use in an electrolytic cell.

Example 2: A conventional two compartment cell was utilized in this example. In this cell, the anode compartment containing a ruthenium oxide coated titanium clad steel mesh anode was fed with an acidified concentrated brine solution which was circulated continuously during the electrolysis. The cathode compartment was filled initially with dilute aqueous caustic soda which during the electrolysis was fed continuously to the cathode compartment as make-up.

The anode and cathode compartments were separated by a membrane composed of a hydrolyzed copolymer of perfluoroethylene and a sulfonated perfluorovinyl ether supported on a Teflon cloth as described above. The membrane was pretreated and stretched as described in Example 1. The membrane was about 7 mils in thickness, and was tautly positioned on the front face of the anode.

The cell was operated over a period of 30 hr by applying a decomposition voltage of 2 A/in^2 of anode surface. During this run a constant voltage of 3.88 was observed. The other operating conditions during this run were: 155 g/ℓ caustic concentration, 90° to 94°C catholyte temperature, 88° to 91°C anolyte temperature, 292 to 309 g/ℓ anolyte salt concentration, and 3.8 to 4.6 anolyte pH.

The caustic soda liquor produced in the cathode compartment contained less than 1.0% sodium chloride.

The chlorine evolved from the anode compartment was free from hydrogen, and the hydrogen evolved from the cathode compartment was free from chlorine.

Radiation-Grafted Films

Polymers and membranes are produced by *V.F. D'Agostino, J.Y. Lee, and E.H. Cook, Jr.; U.S. Patent 4,107,005; August 15, 1978; assigned to Hooker Chemicals & Plastics Corporation and RAI Research Corporation* by radiation techniques to provide improved products.

For example, α,β,β-trifluorostyrene in an inert organic solvent is grafted onto an inert film, such as tetrafluoroethylene-hexafluoropropylene copolymer, by irradiation, i.e., with Co-60 γ-radiation at a dose of several megarads. The grafted film is then sulfonated, preferably in a chlorosulfonic acid bath. The resulting film is useful as a membrane or diaphragm in various electrochemical cells such as chlor-alkali or fuel cells.

A preferred starting material is α,β,β-trifluorostyrene (TFS). This material may be made in accordance with several well known methods, such as those described in U.S. Patents 3,489,807, 3,449,449 and 2,651,627.

It should be noted that pure TFS is not required for this process. As the grafting will generally be done from solution due to economic considerations, commerical TFS containing other materials may be used. It is necessary, however, that the diluent be inert and a nonsolvent for the base film. This means that the diluent must not either homopolymerize, nor copolymerize with the TFS under the radiation conditions of this process, nor adversely affect the radiation grafting.

The general procedure for the graft polymerization of TFS onto the base film, used in each of the following embodiments, is as follows. The film base, covered with an interlayer of paper and wound into a roll, is placed in a chamber. A solution of trifluorostyrene in a suitable solvent is then added to completely immerse the film base roll. A vacuum and nitrogen flush is applied and reapplied several times to remove any dissolved oxygen from the solution and chamber.

The chamber, under a nitrogen atmosphere, is sealed. The temperature is kept sufficiently low during this procedure to prevent vaporization of the solvent. The film and trifluorostyrene solution, in the sealed chamber, are irradiated, i.e., for about 2 weeks at a suitable dose rate. The grafted film is then removed from the chamber and washed and dried.

Generally the dose rate and temperature are not critical. It is possible to utilize dose rates in the range of 5,000 rads/hr up to 300,000 rads/hr. Preferably the dose rate is kept below about 100,000 rads/hr. Too great a total dose may cause degradation of the base film.

Generally a total dose within the range of 0.1 to 10 Mrad is useful, while a range of 0.7 to 5 Mrad is preferable. Higher dose rates also yield lower percentage grafts for the same total dose. The temperature may also be widely varied within the range of -78° to 60°C; room temperature is preferred for convenience of operations.

Microporous Layer Joined to Membrane

A selective composite diaphragm is provided by *P. Bouy, J. Bachot and J.-L. Bourgeois; U.S. Patent 4,135,996; January 23, 1979; assigned to Rhone-Poulenc Industries, France* comprising an ion-exchange membrane having at least one microporous layer joined to it. The microporous layer is of high, homogeneous porosity and is preferably at least as thick as the membrane. The diaphragm is particularly useful for electrolysis of sodium chloride, enabling pure concentrated alkaline solution to be obtained with a high yield.

The nature of the microporous layers may vary greatly and depends particularly on the type of electrolysis carried out. It is necessary in particular for them to have good physical properties in order to resist the stresses to which they are subjected, particularly during handling, and good chemical resistance vis-a-vis the constituents of the electrolytic bath.

The material forming these layers may be based on a wide variety of organic polymers, particularly fluorocarbon or polyvinyl polymers, or organic or inorganic gels, such as those of silica, alumina, titanium oxide or thorium, silica aluminate or various phosphates. Layers of porous diaphragms based on asbestos and polytetrafluoroethylene are generally preferred.

Example 1: A cation exchange membrane consisting of a sulfonated fluorocarbon polymer comprising active sulfonated groups (Nafion XR 315 by Du Pont) is joined to a microporous sheet 0.9 mm thick with 80% porosity and pores of an average diameter of 0.3 μ (hydraulic diameter). The sheet is obtained from a mixture of the following composition: 1.8 parts of polytetrafluoroethylene latex (Soreflon) with 60% dry extract; 0.4 part asbestos (fiber of Asarco 5R4); 5 parts of calcium carbonate, calibrite 14 grade (Omya); and 0.04 part of dodecylbenzenesulfonic acid.

The mixture is put into sheet form by calendering, then dried and sintered under the conditions described in Belgian Patent 831,963.

The resultant sheet is then applied to the ion exchange membrane. For this purpose it is kept in an electric press on the preheated membrane for 6 min at 220°C then for another 4 min at a pressure of 4 kg/cm^2.

The composite unit formed is then immersed for 48 hr in a 20 wt % aqueous solution of acetic acid, degassed under reduced pressure in water and kept moist until it is used.

It is then mounted vertically in an electrolytic cell with a cross section of 1 cm^2 between the two electrodes. The microporous sheet faces towards the anode, which consists of a titanium grid covered with a platinum-iridium alloy. The cathode is an iron grid. The distance between the two electrodes is 7 mm.

The anode compartment thus formed is supplied with a brine containing 295 g/ℓ of NaCl at a rate of 270 ml/hr; the cathode compartment has been previously filled with water which is made conductive with a small quantity of soda (20 g/ℓ).

A current of 25 A is passed between the electrodes. In the normal working state (after 48 hr) the potential difference between the electrodes is 4.5 V.

The temperature of the bath is 85°C. A 420 g/ℓ soda solution is collected, containing less than 0.1 g/ℓ of NaCl and no chlorate that can be revealed by chemical analysis. The yield is 73%.

Example 2: *Comparative* – In the same cell as that in Example 1, but fitted solely with the ion exchange membrane instead of the composite diaphragm, the anode compartment is supplied with a sodium chloride brine containing 295 g/ℓ at a rate of 700 ml/hr.

A current strength of 25 A is maintained. After 48 hr the potential difference between electrodes is 5.5 V and a soda solution containing 600 g/ℓ is obtained; the yield is then 50%. The situation does not remain static but develops, with the voltage rising to 6 V while the soda concentration increases. It becomes impossible to continue the experiment.

Porous Layer on One or Both Surfaces of Membrane

K. Moeglich; U.S. Patent 4,124,458; November 7, 1978; assigned to Innova, Inc. provides a porous layer on one or both surfaces of a mass-transfer membrane to enhance or modify the passage of ions. The membrane may be a film of any of a variety of polymeric materials, and the porous layer may be a surface layer on the membrane or a separate porous structure placed in contact therewith.

Example 1: *No Current Using Plain Plastic Film* – An electrolytic cell, equipped with two electrodes, was divided into two chambers by a membrane consisting of a film of polyethylene 0.0013 cm thick (½ mil). The anolyte was deionized water; the catholyte was dilute phosphoric and sulfuric acid with a pH of 1.5.

A potential of 150 V dc was applied to the electrodes. During and after 8 hr of voltage applied, no current was measured, indicating a membrane resistance in excess of several megaohms. The pH of the deionized water did not change. This test shows that under normal conditions, a film of polyethylene plastic will not serve as a membrane.

Example 2: *Current Flow Using Film Plus Paper* – The same cell as in Example 1 was used, with the same film material, except that the film was covered on both sides with filter paper. The electrolytes were the same as in Example 1. With a potential between the electrodes of 30 V dc, the current increased to 0.5 A within 8 hr.

After 350 A-min, the deionized water in the anode compartment changed in pH from 7 to 0, while the acid in the cathode compartment changed in pH from 1.3 to 7.5, showing a transfer of the negative phosphate and sulfate ions through the membrane. Each chamber held 200 ml of electrolyte. The anode was platinum, the cathode was graphite, and the active membrane area was 13 cm^2.

Example 3: *Current Efficiency* – Polyethylene film ½ mil thick was used between layers of filter paper as the membrane. The anolyte was 400 ml of 5% KOH; the catholyte was 400 ml of deionized water. The potassium ions were transferred across the membrane into the catholyte where purified KOH was built up to a concentration of 25% before the test was discontinued.

The initial low current efficiency shown in Figure 7.15 is due to the conditioning process of the new membrane. For comparison purposes, in this same figure is shown the current efficiency of a new ion exchange membrane Nafion by DuPont, designed for sodium hydroxide use. It may be noted from this figure that even without cell optimization, the membrane of this process is superior to Nafion at the higher levels of hydroxide concentration.

Figure 7.15: Current Efficiency vs Percent Hydroxide Concentration

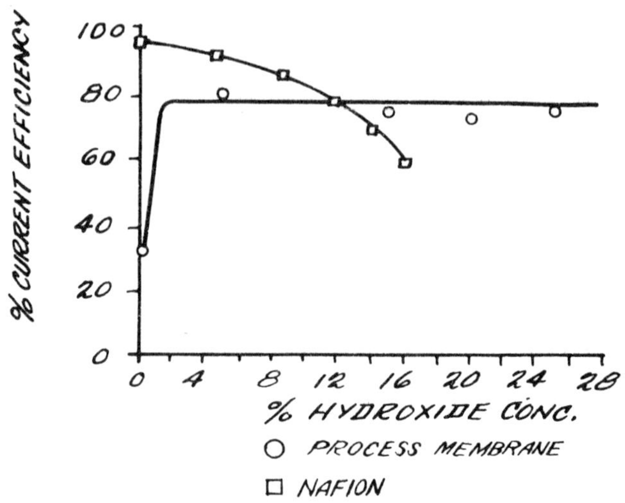

Source: U.S. Patent 4,124,458

Two-Layer Ceramic Membrane

P.M. Spaziante and A. Nidola; U.S. Patent 4,119,503; October 10, 1978; assigned to Oronzio de Nora Impianti Elettrochimici SpA, Italy have developed two-layer ceramic membranes for electrolysis cells comprising on the anodic side a layer of at least one oxide selected from the group consisting of Sb_2O_5, Bi_2O_5, MoO_3, WO_3 and V_2O_5 and on the cathodic side a layer of at least one oxide selcted from the group consisting of ZrO_2, Nb_2O_5, Ta_2O_5 and TiO_2, electrodes provided with a two-layer ceramic membrane applied thereto, an electrolysis cell provided with a two-layer ceramic oxide membrane and an electrolysis process wherein a two layer ceramic membrane is in the electrodic gap.

Example: An expanded iron cathode base which had been sandblasted was precoated on its backside with a thin layer (1 to 2 mm) of zinc by plasma jet to close the voids and avoid deposition of oxides thereon. The opposite surface of the cathode which will face the anode was provided by plasma jet with a first layer of zirconium oxide and a second layer of antimony pentoxide with a thickness of less than 150 μ for each oxide layer. The zinc layer was then removed by soaking the cathode in a 5% nitric acid solution at 20°C for 1 to 2 min and washing with distilled water to remove traces of acid. The resulting membrane had a zirconium oxide surface on the cathodic side where it is stable to the alkaline conditions of the anolyte and an antimony pentoxide surface on the anodic side where it is stable to the acid chlorinated conditions of the anolyte.

The isoelectric point for the zirconium oxide surface was at least a pH of 5 and for the antimony pentoxide was very low at a pH of less than 2.3. The interface between the two oxide layers is a mixture of the two oxides.

The ceramic oxide coated electrode was placed in an electrolysis cell to electrolyze a sodium chloride solution of 230 to 300 g/ℓ of NaCl at 90°C and a current density of 2,000 A/m^2. The catholyte flow rate was 0.3 to 0.1 ℓ/hr with a head of 100 mm (H_2O). After 3 days of operation the catholyte composition was NaOH 130 g/ℓ, NaCl 50 g/ℓ and the faraday efficiency was 90%.

REJUVENATION OF MEMBRANES

Treatment with Purified Brine

Chlor-alkali cells of the membrane type are highly susceptible to multivalent metallic impurities in the brine feeds. As the impurity concentration builds up on and/or in the membrane material, cell voltage increases and current efficiency is lost. By the process developed by *M.M. Dorio and A.J. Stacey; U.S. Patent 4,116,781; September 26, 1978; assigned to Diamond Shamrock Corporation* the membrane is rejuvenated and the current efficiency of the cell increased by intermittently replacing the normal brine feed with a brine which is of relatively high purity with respect to the multivalent metallic impurities found in the normal brine feed.

Generally, the pure brine feed should have little or no hardness (calcium and magnesium level) or, if present, the hardness should be completely complexed by a phosphate treatment. The other multivalent metallic impurities should be present in the pure brine at less than 2 mg/ℓ in total and preferably less than 1 mg/ℓ thereof, while each individual other metallic contaminant should not exceed 0.5 mg/ℓ and preferably it would be below 0.2 mg/ℓ.

The means for purifying the intermittently fed pure brine to the desired degree can be any method known to those skilled in the art. The procedure will vary based on the quantity and type of impurities present in the normally found brine feed at a given locale. Generally, however, a purified brine feed may be conveniently obtained by treating the brine with an ion exchange resin so as to remove such metallic impurities to the desired degree.

Often it is most economical to remove the majority of impurities present in a brine by means of conventional purification methods such as precipitation and filtration and to follow this rough purification procedure by an ion exchange resin technique.

Particularly effective are the ion exchange resins formed from polymers containing a functional group. Typically, these include such compounds as ethylene diamine tetraacetic acid, trimethylenediamine tetraacetic acid, and iminediacetic acid.

Typical commercially available ion exchange resins useful for this purpose include Diamond Shamrock's Duolite C-433 and Dow Chemical's Dowex A-1.

Example 1: A completely fluorinated copolymer containing pendant sulfonic acid groups was chosen as the membrane material for this and the following examples. The membrane material is available as Nafion by E.I. Du Pont de Nemours and Company. The particular membrane chosen was one formed from the polymerization of seven parts of tetrafluoroethylene and one part of:

$$CF_2{=}CFOCF_2CF(CF_3)OCF_2CF_2SO_2F,$$

had an equivalent weight of 1,200 (grams of polymer per equivalent of proton), was reinforced by a T-12 woven PTFE fabric laminated to the membrane, and had a nominal thickness of 7 mils. This membrane was then treated on one side in an asymmetric vessel at a temperature of 26°C with an ethylene diamine/water solution (18:1 volume ratio) for 9 min, water washed, and removed from the treating vessel.

Tests of the membrane showed that the ethylene diamine (EDA) treatment penetrated approximately 1.5 mils into the surface of the membrane. The membranes were then saponified at about 90°C for 70 min in a solution of 600 ml water, 400 ml dimethylsulfoxide and 13 wt % NaOH. The membrane, after water rinsing, is immersed in an aqueous 150 g/ℓ NaOH solution for 4 hr at room temperature, following which it is removed, blotted dry with absorbent toweling, and allowed to air dry for 24 hr again at room temperature.

Without further treatment, the membrane was installed in an electrolytic cell between an expanded steel mesh cathode and an expanded titanium metal anode bearing a $2TiO_2:RuO_2$ mol ratio coating on its surface. The cell was a 25 in² cell and the electrode gap was approximately ¼". A 350 g/ℓ NaOH solution was initially added to the cathode compartment while saturated brine at temperature of 25°C and a pH of about 0.8 was flowed through the anode compartment.

The brine fed during electrolysis was a so-called pure brine (multivalent metallic impurities other than hardness contaminants totaled less than 2 mg/ℓ and the effective hardness concentration was effectively reduced below 0.5 mg/ℓ by acid treatment.) An analysis of this pure brine feed is as follows.

Al	0.13 mg/ℓ
Be	<0.07
Ca	1.4
Co	<0.004
Cr	<0.005
Cu	<0.2
Fe	<0.2
Hg	<0.02
Mg	0.3
Mo	<0.02
Ni	<0.01
Si	<0.2
Ti	<0.5
V	<0.01
Zn	<0.1
NaCl	~280 g/ℓ
pH	<1
H_3PO_4	375 mg/ℓ

As can be noted from the above analysis, phosphoric acid at an addition rate of 375 mg/ℓ of brine was added to complex the hardness ions.

Utilizing a continuous brine feed, electrolysis was commenced at a current density of 2.0 A/in^2. After continued operation for over 90 days, cell operation was found to range from 4.4 to 4.6 V while cathode current efficiencies remained above 90% throughout the test period.

Example 2: The experiment of Example 1 was repeated utilizing an impure brine feed. (Pure brine feed of Example 1 was contaminated so that it contained 20 mg/ℓ $CaCO_3$ and 20 mg/ℓ Hg.) Current efficiency decayed from an initial efficiency of 90+% to 78% in 10 to 12 days. Voltages were around 4.3 V during the run.

Example 3: This example is intended to illustrate the recoverability of current efficiency effected by substitution of a pure brine feed in a contaminated membrane cell.

The cell and operating conditions of Example 1 were utilized while the cell was operated initially on the impure brine of Example 2. At startup, the current efficiency was approximately 95% and decayed gradually until after 21 days on line the current efficiency had decayed to 83%.

At that time, the brine being fed to the cell was switched to the pure brine of Example 1 and over a period of 10 days continued operation of the cell the current efficiency gradually recovered to 93%. The voltage over the entire test period remained constant in the 4.9 to 5.1 V range.

Treatment with Alkaline Brine

A method is provided by *S.J. Specht; U.S. Patent 4,118,308; October 3, 1978; assigned to Olin Corporation* for renewing the permeability of a porous diaphragm comprised of a synthetic fluorocarbon resin having ion exchange properties which comprises feeding an alkaline solution of an alkali metal chloride having a pH of from 9 to 12 to the anode compartment. A first portion of the alkaline brine passes through the porous diaphragm and a second portion contacts the porous diaphragm and is then removed from the anode compartment.

Figure 7.16 represents a schematic view of an electrolytic diaphragm cell used for employing the process.

Diaphragm cell **2** is divided into an anode compartment **4** and a cathode compartment **6** by porous diaphragm **8**. Anode **10** is located in anode compartment **4** and cathode **12** is positioned in cathode compartment **6**.

An alkaline alkali metal chloride brine is fed through inlet **14** into anode compartment **4**. The alkaline brine contacts porous diaphragm **8**, with a portion of the alkaline brine passing through porous diaphragm **8** into cathode compartment **6**, the remainder of the alkaline brine being continuously removed through brine outlet **20**.

Diaphragm cell **2** has outlets **22, 24** and **26** for Cl_2, catholyte liquor and H_2 which are produced during the electrolysis process.

Figure 7.16: Schematic View of A Diaphragm Cell

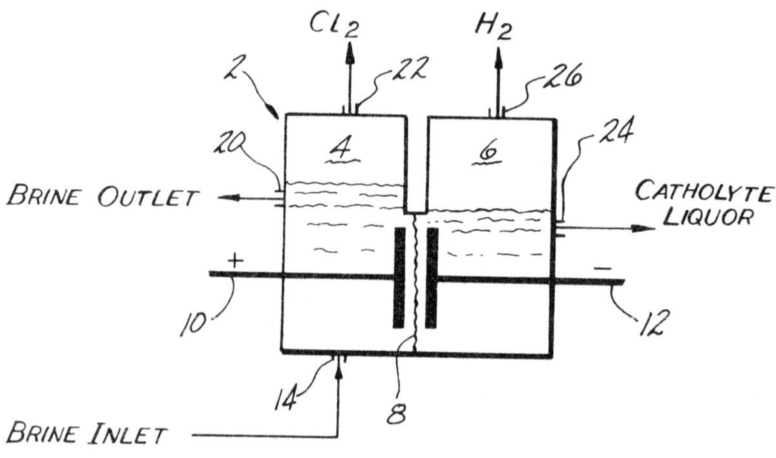

Source: U.S. Patent 4,118,308

A diaphragm cell of the type illustrated in Figure 7.16 was operated employing a porous diaphragm comprised of a perfluorosulfonic acid resin film 7 mils thick having an equivalent weight of 1,100 laminated to a fabric of polytetrafluoroethylene known as Nafion (E.I. Du Pont de Nemours and Company).

After the cell had operated for a period of about 75 days, the catholyte flow rate was reduced to about 10% of the desired flow rate. The current to the cell was shut off and the diaphragm contacted with a sodium chloride brine having a pH of about 10.5 and a calcium content of about 1.5 ppm. The alkaline brine, at a temperature of about 25°C and having a NaCl concentration of about 300 g/ℓ, was fed to the anolyte compartment at a rate of about 0.47 ℓ/min/m² of diaphragm surface area.

A differential pressure between the anode compartment and the cathode compartment of about 30 cm of brine was maintained. The ratio of the first portion of the alkaline brine fed to the anode compartment which initially passed through the porous diaphragm to the second portion which circulated through the compartment and contacted the surface of the porous diaphragm was in the range of 0.05:0.95-0.25:0.75.

This second portion was removed from the anode compartment through outlet **20** and discarded. The catholyte flow rate was monitored and after a period of 29 hr had returned to the desired flow rate. Feeding alkaline brine through the anode compartment was discontinued, the current turned on and the electrolysis process resumed.

Acid Treatment

Deposits of alkali metal ions such as calcium and magnesium on chlor-alkali cathodes and diaphragms are removed by *W.E. Britton and M. Krumpelt; U.S. Patent*

4,204,921; May 27, 1980; assigned to BASF Wyandotte Corporation by feeding an acid such as hydrochloric acid into the cathode feed in an amount sufficient to reduce the pH below 8 for a time sufficient to remove these deposits and restore the cell to normal operation. The use of the process allows the use of relatively high hardness water as the cathode feed solution during normal operation of a membrane chlor-alkali cell.

Example 1: A laboratory scale chlor-alkali membrane cell housing a 3" x 6" steel cathode and a 3" x 6" dimensionally stable anode comprised of ruthenium dioxide and titanium separated by a sulfonated perfluorocarbon ion exchange membrane was utilized for the purpose of electrolyzing sodium chloride brine to produce chlorine and reducing water to produce caustic and hydrogen.

This cell was operated for 413 days using the tap water with typically 30 ppm hardness as a cathode feed solution. The feed rate was 125 ml/hr yielding a sodium hydroxide of 5.2 M concentration. The initial potential of the cell had been 3.3 V but had increased to 3.75 V at the end of 413 days. The cathode had a cell potential of 1.48 V versus a saturated calomel reference. A normal HCl solution was added to the cathode feed solution for 2½ hr at the rate of 100 ml/hr during which time the pH fell from 14 to 1.

This condition was maintained for 30 min and a yellow discoloration was noted in the solution. The colored component proved to be ferric hydroxide (20 ppm analysis by thioglycolic acid method). After returning to normal water feed, the cell voltage was 3.38 V and the half cell potential was down to 1.41 V. After disassembling the cell, it was noted that much of the coating had been removed.

Example 2: A chlor-alkali membrane cell similar to that of Example 1 was operated for 85 days with demineralized water as the cathode feed. Then, in order to increase the amount of magnesium ion for testing purposes, the cathode feed solution was spiked with 100 ppm of magnesium chloride for 25 days.

During this addition, the voltage rose from 3.66 V at 5 N caustic to 4.0 V. At this time, the cathode was fed a 5 M hydrochloric acid solution and the effluent was monitored for iron concentration and magnesium concentration as a function of time. The pH stayed between 3 and 8 and the acid flow was maintained for 4 hr. The data are recorded in the table below. The cell voltage dropped from 4.0 to 3.7 V after the acid addition.

Time from Beginning	.... Effluent Values		
of Acid Feed (hr)	ph	Iron (ppm)	Mg (ppm)
0	14	0	0
0.5	14	0	0
1.0	14	0	0
1.5	14	0	0
2.0	14	0	0
2.5	8	0	60
3.0	4	0	80,000
3.5	3	2	7,000
4.0	3	5	100

From the foregoing description, it is evident that a method of rejuvenating membrane type chlor-alkali cells, and in particular, rejuvenating the operative cathode surfaces thereof is provided. It is also seen that the rejuvenation may be accomplished without shutting down the cell.

Conversion of Carboxylate Group to Acid or Ester

According to *T. Asawa and T. Gunjima; U.S. Patent 4,115,240; September 19, 1978; assigned to Asahi Glass Company Ltd., Japan* electrochemical properties of a cation exchange membrane of a carboxylic acid type fluorinated polymer which is used for an electrolysis of an aqueous solution of an alkali metal chloride are recovered by converting ion exchange groups of $\left(COO\right)_m M$, wherein M represents an alkali metal or an alkaline earth metal, and m represents a valence of M, to the corresponding acid or ester groups of –COOR wherein R represents hydrogen or a C_{1-5} alkyl group and heat-treating the fluorinated polymer having the groups of –COOR.

Example: Tetrafluoroethylene and

$$CF_2{=}CFO{-}CF_2{=}\underset{\displaystyle CF_3}{\underset{|}{C}}F{-}O(CF_2)_3{-}COOCH_3$$

were copolymerized with a catalyst of azobisisobutyronitrile in trichlorotrifluoroethane to obtain a fluorinated copolymer having an ion-exchange capacity of 1.17 meq/g polymer and T_Q of 190°C. The resulting fluorinated polymer was press-molded to form a film having a thickness of 200 μ. The membrane was hydrolyzed in an aqueous methanol solution of sodium hydroxide whereby a carboxylic acid type fluorinated cation exchange membrane was obtained.

A two-compartment type electrolytic cell was prepared by partitioning an anolyte and a catholyte with the cation exchange membrane and using an anode of titanium coated with rhodium and a cathode made of stainless steel with a space between the electrodes of 2.2 cm and an effective area of 25 cm^2. The electrolysis of sodium chloride was carried out under the following conditions.

The anode compartment was filled with 4 N NaCl aqueous solution and the cathode compartment was filled with 12 N NaOH aqueous solution. The electrolysis was carried out by feeding 4 N NaCl aqueous solution at a rate of 150 cc/hr into an anode compartment and feeding water into a cathode compartment so as to result in 14.4 N NAOH aqueous solution with a current density of 20 A/dm^2 at 85°C.

The aqueous solution of sodium chloride was overflowed from the anode compartment and the current efficiency was measured from the amount of NaOH which was produced by the electrolysis; current efficiency was 91%. The electrolysis was continued for 360 days; current efficiency had dropped to 85%.

The electrolytic cell was disassembled and the membrane having deteriorated properties was taken out and was treated in a 1 N HCl solution containing 20% of dimethyl sulfoxide at 90°C for 16 hr to convert the ion-exchange groups to acid type groups. The membrane was pulverized in a hammer mill to obtain a powder (100 μ) of acid type cation exchange resin and the powder was press-molded at 210°C under the pressure of 50 kg/cm^2 for 5 min to obtain a cation exchange membrane having a thickness of 200 μ, and it was used for the electrolysis of an aqueous solution of sodium chloride in the same condition.

The powder of the acid type exchange resin was also treated in methanol containing 1% HCl at 60°C for 16 hr to convert the ion exchange groups to methyl ester groups and the powder was press-molded at 280°C under the pressure of 60 kg/cm^2 for 5 min to obtain a cation exchange membrane having a thickness of 200 μ, and it was also used for the electrolysis of an aqueous solution of sodium chloride. In both cases, the resulting membrane had a current efficiency of 90%.

Cell Design Allowing for Easy Membrane Removal, Reactivation and Reuse

E.D. Creamer and M. Krumpelt; U.S. Patent 4,175,025; November 20, 1979; assigned to BASF Wyandotte Corporation have designed an electrolysis cell assembly of the filter press type having plastic frames and a membrane wherein the membrane is formed to fit between adjacent frames with one or more recesses provided on one or both of the adjacent frames together with sealing gasket means which are formed for effectively controlling electrode spacing and for retaining the membrane in position without membrane damage. The construction not only provides proper sealing and protection of the membrane during use, but also allows for the membrane to be removed, reactivated, and reused.

In the assembly, the membrane preferably extends considerably beyond the periphery of the frames when in assembled position. In this way, the membrane may be removed and reactivated, and still returned to the frame in the same location or in another position even though the membrane will have shrunk somewhat. In addition, it is possible to reactivate the membrane a plurality of times and still have it fit properly in the assembly.

There is shown in Figure 7.17, a pair of adjacent filter press frames **12** and **14** constructed from plastic such as filled polypropylene. The frames are of typical construction and contain the various units of an electrolytic filter press cell which may be monopolar or bipolar. The cell assembly also contains a membrane **16** which is typically larger than the frames **12** and **14** and extends beyond the periphery thereof at **18**.

The frame **14** has a recess **20** cut therein and formed with its inner surfaces **22** and outer surfaces **24** rounded. A gasket **26** is provided which is formed to fit into the recess **20** and secure the membrane **16** therein. Gasket **26** is formed to protrude from the recess and has a flat surface **28** formed to bear against the adjacent frame member **12**. The gasket is an elastomer such as EPDM (a polymer from ethylene propylene dimonomer), Hypalon or Neoprene.

Example: An EPDM Nordell 70 Shore A durometer gasket having cross-sectional dimensions of ⅜" x ⅜" was used to caulk a Nafion 313 membrane. The gasket was forced into a ⅜" wide by 5/16" deep groove with the membrane held therebetween as in the embodiment of Figure 7.17.

The groove overall perimeter was rectangular at 7' x 4' with the corners thereof rounded. This system sealed with a compressive force of 200 lb ram pressure per inch of gasket.

Figure 7.17: Cross-Sectional View of a Portion of Adjacent Filter Press Cell Frames

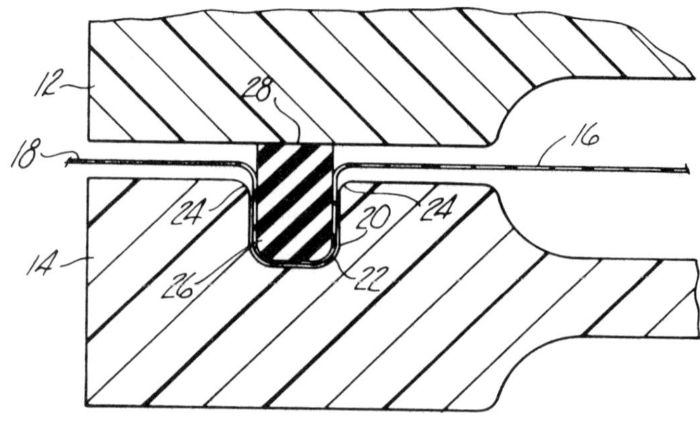

Source: U.S. Patent 4,175,025

Periodic Change in Anode Polarity

Electrolysis of impure saline solutions containing dissolved manganese causes deposits to form on the anodes of the electrolysis cell which rapidly reduces the efficiency of the cell.

The efficiency of the cell is rejuvenated periodically by *J.E. Bennett; U.S. Patent 4,087,337; May 2, 1978; assigned to Diamond Shamrock Corporation* by changing the polarity of the anode for 1 to 10 min at an amperage of 2 to 50 mA/in^2.

In Figure 7.18, cell **1** is connected electrically with rectifier **2** via connecting bus **3** and **4**. Rectifier **2** supplies the power required for normal operation of cell **1**. A secondary rectifier **5** is operatively connected to cell **1** to effect the brief low level current reversal through the cell when rectifier **2** is deactivated.

As shown in the figure, secondary rectifier **5** is connected to connector **3** and to connector **4** through connector **7** and resistor **6**, respectively. The resistor is sized so as to effectively prevent any more than minor amounts of current flow through the secondary rectifier **5** when the electrolytic cells are in normal operation. When it is desired to remove the anodic manganese deposits in cell **1** so as to improve current efficiency, main rectifier **2** is first deactivated. This is followed by the activation of the secondary rectifier **5** in order to supply the desired reverse current through the cell for the appropriate time period to remove the manganese build-up on the anodes in the cell.

During this time the diodes in the primary or normal rectifier **2** have sufficient resistance to effectively prevent the flow of current from the secondary rectifier **5** through rectifier **2** and force the current through the cell **1** in the reverse direction of normal operation. If no anodic manganese deposits are present in the cell, or if the reduction cycle has been activated long enough to effectively remove any anodic manganese deposits present, the resistance of cell **1** will then be higher allowing a significant portion of the current from secondary rectifier **5** to pass through primary rectifier **2**.

The circuit shown by Figure 7.18, therefore, allows a low-level current to pass through the cell **1** in a direction opposite to that of normal operation when anodic manganese deposits are available for reduction, but effectively limits the passage of low level reverse current when manganese deposits are not present.

Figure 7.18: Cell Equipped with Secondary Rectifier for Low Level Reverse Current

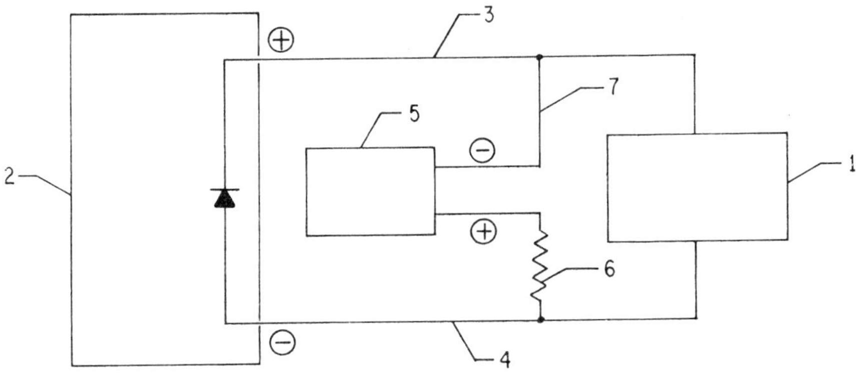

Source: U.S. Patent 4,087,337

This further prevents unnecessary reverse current from passing through the cell which may over a period of time damage the dimensionally stable anode used in such systems. At no time is it necessary or desirable to disconnect the main conducting bus **3** and **4** from the circuit.

Example: A laboratory cell provided with two Hastalloy C-276 cathodes and two anodes provided with ruthenium dioxide-titanium dioxide coating with an active area of 11.25 in^2 was tested for efficiency by batch electrolysis of a solution containing 28 g/ℓ of sodium chloride and 0.004 g/ℓ of $Na_2Cr_2O_7$. The electrolysis was conducted for 30 min during which 8.08 g/ℓ of chlorine were generated at a current efficiency of 63.6%.

The cell was then operated using the method shown by Figure 7.18 with an aqueous solution of 28 g/ℓ of sodium chloride and 0.2 ppm of Mn^{+2} at 1.0 A/in^2. The high concentration of Mn^{+2} in this test has the effect of accelerating the loss of efficiency caused by the deposition of manganese on the anode surface.

Normal electrolysis was effected for 30 min and the current was then reversed at 8 mA/in^2 for 5 min. Following current reversal the cell was returned to normal operation and this sequence was repeated for 338 cycles. At the end of the last sequence, the cell was again tested for efficiency by the 30 min batch electrolysis test described previously. During the test 8.17 g/ℓ of chlorine were generated at a current efficiency of 64.3% indicating no loss of efficiency due to anodic manganese deposits.

The cell was then returned to electrolysis in the 28 g/ℓ sodium chloride, 0.2 ppm Mn^{+2} solution. Electrolysis was again conducted at 1.0 A/in^2, using current reversal at 8 mA/in^2 at 30 min intervals as before. A batch efficiency test was again conducted after 2,500 cycles. During this 30 min test 9.23 g/ℓ of chlorine were generated at a current efficiency of 73.4% again indicating no loss of efficiency due to anodic manganese deposits.

During this accelerated test the Hastalloy C-276 cathodes lost only 1% of their original weight. This is considered satisfactory since under normal conditions of electrolysis 2,500 cycles represents several years of lifetime.

MOLTEN SALT METHODS AND OTHER CHLORINE TECHNOLOGY

NONAQUEOUS ELECTROLYTIC METHODS

Downs Cell

A considerable proportion of the elemental alkali metals which are manufactured for commerce is produced by the electrolysis of molten halogen salts of the metals, especially eutectic mixtures of such salts with other salts which are substantially inert.

Previously, the riser/cooler on Downs-type electrolytic cells has been comprised of a vertical tube rising from the upper end of the inverted trough collector through which the molten metal rises by difference in gravity into the bottom of a molten product receiver or, alternatively, up to and over an open weir located at some predetermined higher level in the side of the riser/cooler so that the molten product spills over into the receiver.

Though such arrangements have been satisfactory during normal cell operations, they have nevertheless presented serious problems of safety. More particularly, the prior art arrangement of open weirs or pipes between the riser and the receiver results in contamination of the inert gas-purged vapor space of the receiver whenever hydrogen or chlorine is produced inadvertently on the cathode side of the cell. This, of course, increases the amount of inert gas which must be used for operation of the cell and reduces the margin of safety in operation of the cell.

F.J. Ross; U.S. Patent 4,092,228; May 30, 1978; assigned to E.I. Du Pont de Nemours and Company has found that the abovementioned disadvantages of the prior art cells can be effectively overcome by modification of those means by which material is transferred between the riser/cooler and the receiver.

The operating characteristics of a Downs-type electrolytic cell are improved by incorporating a sealed weir between the riser/cooler and the molten metal receiver and by providing means for venting inert gas from the receiver through the vapor space of the riser/cooler.

A Downs-type electrolytic cell is shown in Figure 8.1, having a steel shell **1** lined with ceramic brick **3** containing a bath of molten salt **5** which has been charged to the cell in particulate solid form through loading port **6**. An anodic bus bar **7** is connected electrically with each of a plurality of cylindrical graphite anodes **9**. Surrounding each of the cylindrical anodes is a concentric steel cathode **11**, which is connected electrically through cathode arm **12** with bus bar **14**.

Figure 8.1: Vertical Cutaway Section of a Downs Cell

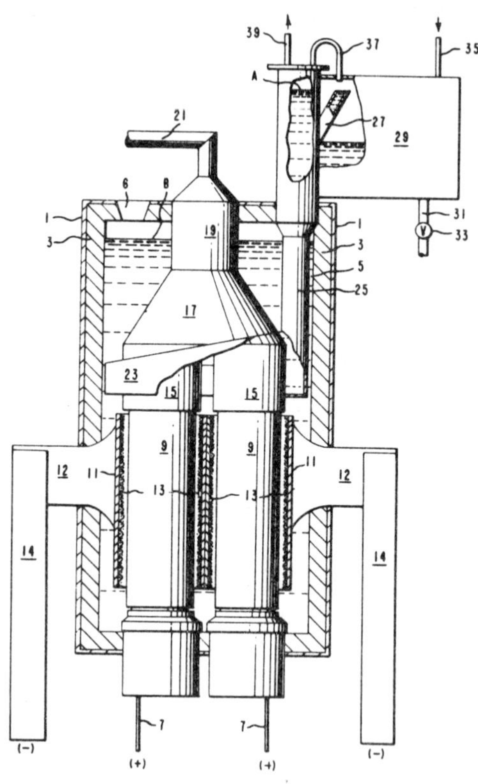

Source: U.S. Patent 4,092,228

In the annulus between the anode and cathode is positioned a steel screen diaphragm **13** which prevents mixing of the anode and cathode products. The cell shown contains four symmetrically-positioned electrode assemblies of which only the front two appear in the view of the drawing. Gas formed at the surface of the anode passes upward between the diaphragm and the anode into collector ring **15** and then into gas collector cone **17**.

The gas is removed from the cell via gas dome **19** through gas outlet line **21** to storage or other disposition. It is customary to maintain a slight vacuum on the exit line in order to prevent seepage of halogen gas into the work areas where the cells are located. Molten metallic alkali metal formed at the cathode rises upward by the force of gravity difference into molten alkali metal collector **23**, which is in the shape of an inverted sloping trough. The molten alkali metal flows along the top of the metal collector and passes into the lower end of riser/cooler **25** in which it is cooled as it rises toward the top of the riser/cooler. The difference between the level of molten metal in the riser/cooler **A** and the level of molten salt in the main part of the cell **B** will be mainly a function of the difference in the specific gravities of the molten metal and the molten salt bath.

In the side of the riser/cooler is a sealed weir conduit **27**, the top of which is positioned at a level corresponding to the level of metal required for the riser/cooler. Overflow through the sealed weir spills over into molten metal receiver **29** in which the metal can be cooled still further. The still molten alkali is removed periodically from the receiver for molding, storage, shipping, etc. via alkali metal outlet **31** through valve **33**.

Cell with Solid Electrolyte Tubes Permeable to Selected Cations

C.H. Lemke; U.S. Patent 4,089,770; May 16, 1978; assigned to E.I. Du Pont de Nemours and Company describes an electrolytic cell for the electrochemical separation of selected metals from electrodissociatable compounds in the molten state utilizing as electrode separator a plurality of solid electrolyte tubes which, under the influence of an electrical potential, are permeable to the flow of selected cations, but impermeable to fluids and the flow of anions and other cations.

In Figure 8.2, a preferred form of the cell is shown comprising in combination an enclosed shell having a topwall **1**, upper and lower sidewalls (**3a** and **3b**, respectively) and a bottom wall **5**. The topwall member is constructed of transparent material, such as glass, to permit viewing into upper collection zone **100**, which is formed by an upper horizontal fluid-tight partition **7** positioned below the top of the cell and extending between the upper sides of the cell **3a**.

The upper horizontal fluid-tight partition functions as a tube sheet having joined thereto and suspended therefrom a plurality of cylindrical tubes **9**, closed at the lower end and made of solid electrolyte material which is permeable to the flow of monovalent cations, such as Na^+, but impermeable to the flow of fluids, anions and polyvalent cations. The tubes are positioned and supported on the upper horizontal partition by means of open riser **11** which is joined in a fluid-tight manner to the partition.

Though the tubes are closed at their lower ends, they are in fluid communication with the upper collection zone at their upper ends in such manner that monovalent metal formed at the inner surface of the tubes is collected in the tubes and rises within the tubes to overflow onto the top surface of upper partition **7**. Monovalent metal flowing onto the top of the upper partition is removed from the cell via collecting channels **13** through outlet line **15**.

During normal operation of the cell, an inert atmosphere is maintained in the upper collection zone by maintaining a small flow of inert gas which is provided via inert gas inlet line **17**. The upper collection zone is also equipped through wall **3a** with access means comprising a glove assembly **19** and access port **20** by which certain maintenance functions can be carried out within the upper collection zone **100** without having to remove the top member **1**. In particular, when a tube fails, it is removed from the tube sheet using glove assembly **19**. Access port **20**, which during normal operation is sealed by means of a flange and bolted cover, is then opened and the failed tube is removed therethrough.

The replacement tube can then be inserted into the metal collection zone via the open access port. The access port is then resealed and the replacement tube is placed into operating position using the glove assembly. During this operation, it will usually be preferred to purge the chlorine collection zone with inert gas which is supplied via a second inert gas inlet **22**. In place of the bolted flange and cover used here, an air lock assembly might also be used.

Figure 8.2: Vertical Section of Separation Cell

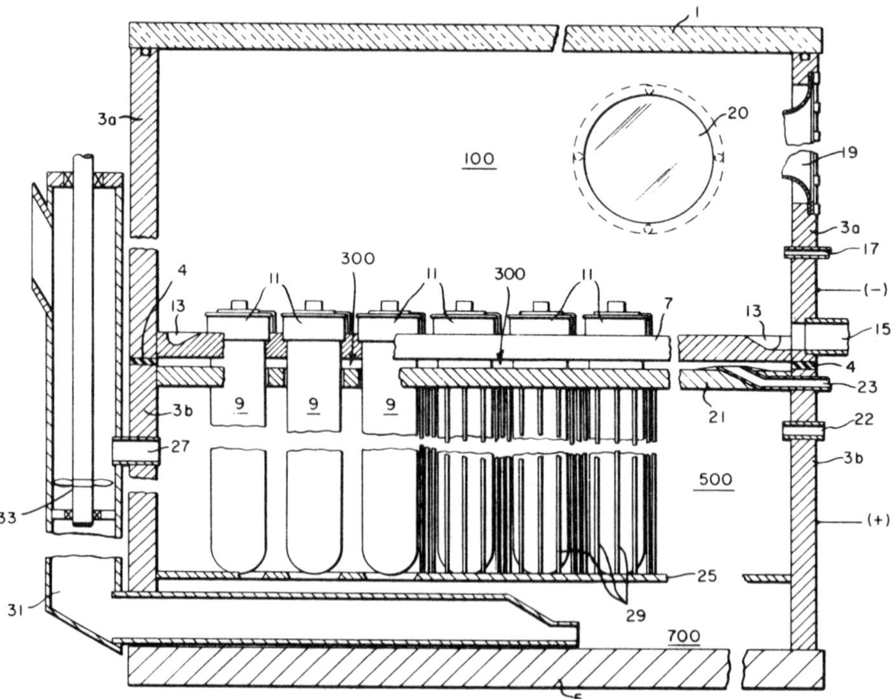

Source: U.S. Patent 4,089,770

In the cell illustrated, the open ends of the solid electrolyte tubes **9** (or inert extensions thereof) protrude above the surface of the tube sheet **7** and are supported atop the tube sheet by riser **11** above the desired liquid level on the sheet. By this arrangement, when a tube is broken, molten metal in the metal collection zone will drain off in its usual path and will not drain into the electrolyte circulation zone through any opening left by the fractured tube.

The upper horizontal partition **7** as well as the upper sidewalls of the cell **3a** are constructed of electrically conductive material and together function as negative current collector for the cell. The upper part of the cell is insulated electrically from the lower part of the cell by means of insulating gasket **4** placed between the abutting edges of the upper and lower cell sidewalls.

An intermediate horizontal partition **21** extending between the lower sides of the cell **3b** is positioned below the upper horizontal partition **7**, thus forming a lower second collection zone **300** in which gas formed outside the solid electrolyte tubes **9** is collected. Gas within zone **300** is removed from the cell through gas outlet line **23**. The intermediate horizontal partition is perforated in such a manner that an annular space is formed between the edge of the perforations and the outer surfaces of the solid electrolyte tubes **9** near the upper end thereof.

Positioned near the closed lower end of the solid electrolyte tubes is a lower horizontal partition **25** which, with the intermediate partition **21**, forms an electrolyte circulation zone **500** surrounding the solid electrolyte tubes **9**. The lower horizontal partition **25** is also provided with perforations through which molten electrolyte flows into the zone and around the solid electrolyte tubes. Molten electrolyte is discharged from circulation zone **500** through liquid electrolyte discharge line **27**.

Extending between the intermediate and lower horizontal partitions **21**, **25** in close proximity with each solid electrolyte tube **9** is a positive pole assembly comprised of a plurality of metal rods **29** positioned in circular array around the solid electrolyte tube **9**. In the cell illustrated, both the intermediate partition **21** and the lower sidewall **3b** are constructed of electrically conductive material and together function as positive current collector for the cell.

The lower horizontal partition **25** separates the circulation zone of the cell **500** from a molten salt inlet zone **700**. Feed materials are passed to the cell through feed line **31**. A positive flow of salt feed and recirculation of molten salt is maintained by operation of impeller assembly **33**, which is located within the salt feed line **31**.

When the abovedescribed cell is assembled and the appropriate feed, product and electrical connections are made, startup of the cell is quite easy. This is illustrated by the following procedure for starting up and operating the cell for the manufacture of sodium from an approximately equimolar mixture of NaCl and $AlCl_3$.

Appropriate quantities of granular NaCl and $AlCl_3$ are fed to a solids blender, such as a ribbon mixer, to form a uniform mixture of the two materials. The thusly mixed granular salts are then placed in a suitably heated melt tank in which they are melted by heating to 200° to 250°C, which is well above the solidus of the bath. The molten salt feed mixture is pumped to the inlet of the cell and the circulation zone is filled up to the level of the electrolyte discharge line. Circulation of the feed through the cell is then established.

After bath circulation is started, the space within the molten metal collection zone is purged with inert gas and the solid electrolyte tubes are then filled with molten sodium to a level sufficient to provide electrical contact with the upper horizontal partition.

The cell is then started merely by turning on the power to the cell which can be done either gradually or fully at once. Operation of the cell is then continued with either continuous or bath addition of granular NaCl to the cell at a rate to maintain the NaCl composition of the molten salt bath at the desired level.

The cell, when making sodium at 200°C, operates at a voltage of 6 as compared to about 7 for conventional Downs cells making sodium at 600°C. Average current (coulombic) efficiency for this process cell is essentially 100% compared to a range of 80 to 90% for Downs cells. Power consumption of the process at the same productivity is about 30% lower than the Downs cell.

Chlorine from Ammonium Chloride

B.M. Abraham; U.S. Patent 4,172,017; October 23, 1979 describes a process for producing chlorine from ammonium chloride which comprises vaporizing ammonium chloride, contacting the vapors with a metal selected from the group consisting of tin, zinc and cadmium, with tin preferred, whereby a metal chloride is formed and then electrolyzing the metal chloride in the presence of a fused salt electrolyte.

The ammonium chloride is preferably obtained from the Solvay soda ash process and the process for producing chlorine is integrated with the Solvay process. The integrated process produces chlorine at a substantial energy savings and without calcium chloride being produced as a by-product.

It is preferred that the process occur continuously. Ammonium chloride which is recovered from the Solvay process is dried, preheated with heat derived from the ammonia and hydrogen off-gases and fed to the bottom of a vertical reaction vessel. The ammonium chloride is vaporized and the vapors of hydrogen chloride and ammonia are countercurrently contacted with a descending spray of molten tin. The liquid stannous chloride formed is recovered from the bottom of the reaction vessel and gravity fed to electrolysis cells. The stannous chloride is then subjected to a fused salt electrolysis. The molten tin formed is recycled to the chlorinator while chlorine is recovered and fed to a storage vessel.

The ammonia generated by the reaction of molten tin with ammonium chloride vapors is recycled to the Solvay scrubber for reuse in the Solvay process and the hydrogen generated by the same reaction is burned to obtain a fuel credit. The heat value of hydrogen per gram mol of chlorine produced is 242 kJ. Hydrogen could lower the effective cell voltage required by 0.33 volts based on an 80% boiler efficiency and a 33% generating efficiency.

The process according to this method has several advantages over the Solvay process and the electrolysis of brine. First of all, there is a reduced electrical requirement since the actual cell voltage needed to produce chlorine lies between 1.5 to 1.7 volts, including ohmic losses. With the hydrogen credit voltage, the cell voltage would be reduced to 1.2 to 1.4 volts. Secondly, cell maintenance is simplified. Although chlorine does react with the graphite used as the electrode to produce a number of chlorinated carbon compounds, the rate is so slow that this mechanism of anode attrition is negligible. There is no need for a dimensionally stable anode as in the electrolysis of brine inasmuch as there is no oxygen in the system. Thirdly, pollution of the environment is eliminated due to the fact that calcium chloride is not produced as in the Solvay process. There is no added financial burden attached to the disposal of calcium chloride. The lime produced from the decomposition is marketed instead of being reacted with ammonium chloride. Finally, the process permits the production of chlorine independently from that of sodium hydroxide.

R.J. Botton; U.S. Patent 4,110,423; August 29, 1978; assigned to Societe Rhone-Progil, France discloses a reaction mass for use in the preparation of chlorine and ammonia by contact with ammonium chloride, in which the reaction mass contains iron oxide and an alkali metal chloride, with the ratio of iron oxide and alkali metal chloride being such that the ratio of the chlorine chemically combined with the iron to the chlorine chemically combined with the alkali metal is less than 1.

Example: An apparatus similar to that shown as Figure 2 in French Patent 1,395,701 is utilized. This apparatus comprises a sublimater, supplied with solid ammonium chloride, a chlorination reactor, a dechlorination reactor, and a reduction reactor; these three cylindrical reactors, 100 mm in diameter and 100 cm in height, contain fluidized beds and are separated by fluidized bed type separators.

Three different masses are prepared, on a basis of iron-bearing bauxite containing approximately 42% iron. The ore is crushed, then calcined at 900°C for 1 hr, and then finally ground to produce particles having dimensions from 10 to 150 μ, the dimensions of 20% of the particles being less than 40 μ. Sodium chloride and cuprous chloride are crushed very finely. The following three masses are then prepared by simple mixing of the following.

	NaCl	Bauxite	CuCl
		(%)	
Mass number 1	8.5	91.5	0
Mass number 2	14	86	0
Mass number 3	14	85.5	0.5

The apparatus is supplied with mass No. 1, the height of the fluidized mass being fixed at 60 cm for all the reactors. Taking account of the flow rates of the reagents, the contact periods are from 4 to 10 sec. The mass in each reactor and each separator is brought into the fluidized state by very moderate flows of inert gas. The reactors for dechlorination and reduction are heated to approximately 500°C, and the reactor for chlorination to 450°C.

Hydrogen is admitted at a flow rate of 1,400 ℓ/hr, such that the temperature of the reduction reactor becomes adjusted to 560°C and that of the chlorination reactor to 440°C. Some of the mass and of the ammonium chloride are introduced into the sublimator, and 950 ℓ/hr of oxygen is introduced into the dechlorination reactor. The temperature in the reactor is regulated to 550°C by regulating the injection of 450 ℓ/hr of air into the reduction reactor. In the separators, a flow of 200 ℓ/hr of nitrogen is utilized.

92% of the ammonia which was introduced in the form of ammonium chloride is recovered. At the outlet from the dechlorination reactor a mixture is obtained of chlorine and oxygen, the chlorine content of which relative to the oxygen is 62%. At the outlet from the reduction reactor, a mixture of nitrogen, steam and hydrogen is obtained. The quantity of hydrogen amounts to 11% of that introduced into the reactor. The mass leaving the chlorination reactor contains 3% of chlorine fixed to the iron. The amount of chlorine fixed to the sodium is 4%.

Mass No. 2 is processed under the same conditions as mass No. 1 and the same results are obtained.

The same experience is obtained with mass No. 3, and at the outlet of the chlorination stage, 95% of the ammonia which was introduced in the form of ammonium chloride is recovered; at the outlet from the dechlorination reactor, a mixture of chlorine and oxygen having a chlorine content relative to oxygen of 68% is obtained; at the outlet from the reduction reactor there is a mixture of nitrogen, steam and hydrogen. The quantity of hydrogen represents 6% of that introduced into the reactor.

Oxidation of Magnesium Chloride

M.H. Feilchenfeld; U.S. Patent 4,073,875; February 14, 1978; assigned to Yissum Research Development Co. of the Hebrew Univ. and Israel Chemicals Ltd., both of Israel has developed a method for the production of chlorine and magnesium oxide. According to the process substantially anhydrous magnesium chloride, containing at least 95% $MgCl_2$, is admixed with particulate magnesium oxide and contacted with gas comprising oxygen at a temperature from 700° to 1100°C.

One of the advantages of the process is the utilization of the exothermicity of the oxidation of the molten magnesium chloride, the reaction being self-sustained. According to a preferred embodiment, the particulate magnesium oxide is preheated at a temperature of above 720°C before it is admixed with the molten magnesium chloride and serves as the heat transfer agent for the subsequent oxidation of the magnesium chloride.

The process may be carried out in a stack reactor or in a fluidized bed reactor. The oxidation is effected with substantially dry, pure oxygen or air.

The magnesium oxide obtained in the oxidation reaction may be recycled to be utilized as a support for the magnesium chloride oxidation. The process can be conducted either batchwise or continuously.

Example: 5 g magnesium oxide were heated in a quartz reactor to above 900°C in a stream of argon, thereby eliminating substantially all absorbed moisture. The reactor was then cooled to 850°C, 1 g of anhydrous magnesium chloride was then added through an opening at the reactor top and the contents of the reactor were mixed by applying a vibrator. The argon was then replaced by a stream of oxygen and the exit gases were analyzed. The results are summarized in the following table.

Time (min)	Yield of Chlorine (%)	Conversion of Magnesium Chloride (%)	Selectivity ($Cl_2/Cl_2 + 2Cl$)
1.5	4.2	5.5	0.759
3.5	7.6	9.5	0.792
5.5	11.1	13.4	0.831

(continued)

Time (min)	Yield of Chlorine (%)	Conversion of Magnesium Chloride (%)	Selectivity (Cl_2/Cl_2 + 2Cl)
11.5	18.0	21.1	0.850
15.5	21.6	25.3	0.855
20.5	25.3	29.7	0.854
30.5	45.4	52.6	0.874
45.5	79.3	88.7	0.895
65.5	80.4	90.2	0.891

The exit gas at the maximum rate of the reaction contained 35.5 mol % of elementary chlorine and 5.8 mol % of hydrogen chloride, the balance being oxygen.

Iron Chloride in Oxidation Furnace

In the process developed by *S. Fukushima; U.S. Patent 4,073,874; February 14, 1978; assigned to Mitsubishi Kinzoku KK, Japan* for recovery of chlorine from iron chloride by causing a reaction between an oxidizing gas and a gas containing iron chloride as the predominant constituent within an oxidation furnace, the initial meeting of the two reacting gases is caused to take place in an unobstructed space within the furnace, the oxygen being injected into the furnace in directions and at a velocity such that the resulting turbulent flow due to the initial collision of the two gases will not reach the furnace wall, whereby depositing of Fe_2O_3 on the furnace wall is reduced to a minimum.

Referring to Figure 8.3, wherein one preferred embodiment of the oxidation furnace is illustrated, the oxidation furnace comprises: a closed furnace structure **1** of vertical cylindrical type; an iron chloride conduit **2** for injecting an iron chloride gas, this iron chloride conduit being mounted at the top end center of the furnace structure coaxially therewith; a plurality of oxygen nozzles **3** for injecting an oxidizing gas provided around the furnace structure and supplied with the oxidizing gas from an oxidizing gas supplying header or pipe **4** by way of respective valves **5**; an exhaust pipe **6** connected to a lower part, preferably a part close to the lower end of the vertically straight side wall, thereof; and an extraction port **7** for extracting Fe_2O_3 at the bottom of the furnace structure **1**.

Figure 8.3: Oxidation Furnace

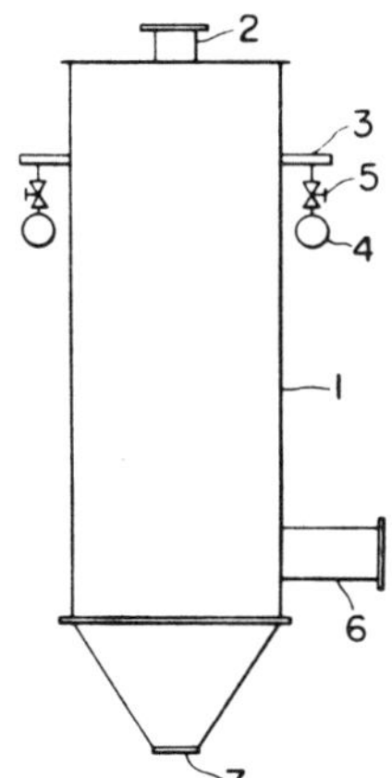

Source: U.S. Patent 4,073,874

Example: This example illustrates a case wherein an oxidation furnace as shown in Figure 8.3 is used. The furnace used had a cylindrical furnace structure **1** of an inner diameter of 0.8 m and an overall furnace length from the top of the furnace to the Fe_2O_3 extraction outlet **7** at the bottom thereof of 3.7 m. The diameter of the inlet **2** for the gas containing iron chloride at the top of the furnace was 0.12 m. Three oxygen injection nozzles **3** each of 0.025 m inner diameter were installed with 120° spacing in the furnace side wall at a level 0.3 m below the inlet and with centerlines perpendicularly intersecting the central axis of the furnace at one point thereon.

The temperature in the inlet was maintained at 850°C, and a termperature gradient of decreasing temperature with distance in the downstream direction was established with the temperature being maintained at 600°C in the gas exhaust outlet **6**.

Through the inlet, a gaseous mixture heated to 850°C and composed of 44.8% of $FeCl_3$, 33.2% of CO_2, 0.3% of Cl_2, 0.8% of O_2, and 20.9% of N_2 was blown into the furnace at a flow rate of 660 ℓ/min (atmospheric pressure and 850°C). Simultaneously, oxygen preheated to 300°C was injected into the furnace uniformly through the three nozzles at a flow rate of 120 ℓ/min (atmospheric pressure and 300°C) and caused to react continuously with the above-described gaseous mixture for 72 hr.

As a result, the reaction product gas taken out through the exhaust pipe **6** was found to have an average composition of 2.2% of $FeCl_3$, 51.5% of Cl_2, 27.4% of CO_2, 3.0% of O_2, and 15.9% of N_2, and an average reaction rate of the iron chloride of 94.0% was obtained. The Fe_2O_3 of the reaction product was found to be adhering in a soft state to the side wall of the furnace downstream from the mixing zone with an average thickness of the order of merely 20 mm. With respect to the mixing zone, an extremely thin layer of powdery iron oxide and absolutely no scale formation was observed. Approximately 250 kg of Fe_2O_3 powder was taken out through the Fe_2O_3 extraction outlet **7** and, in addition, Fe_2O_3 was discharged as dust through the exhaust pipe **6**.

Iron Chloride in Fluidized Bed Reactor

An improved process is provided by *D.J. Haack and J.W. Reeves; U.S. Patent 4,144,316; March 13, 1979 and J.W. Reeves, R.W. Sylvester and D.F. Wells; U.S. Patent 4,174,381; November 13, 1979; both assigned to E.I. Du Pont de Nemours and Company* for producing chlorine and iron oxide in a fluidized-bed reactor by treating ferric chloride in the vapor phase with an excess of oxygen at a temperature of 550° to 800°C in the presence of a catalyst made from sodium chloride and iron oxide.

A carbonaceous fuel is fed to the reactor bed to supply supplemental heat. The improvement comprises using a fluidized-bed reactor in which a portion of the bed material is continuously recycled and the carbonaceous fuel is supplied in a pulverized solid form in an amount equal to between 1 and 9 wt % of the iron chloride feed. The fuel has a stable ignition temperature in air of no higher than 500°C and contains hydrogen amounting to between 0.5 and 2.5 wt % of the fuel. The process is particularly useful for avoiding potential pollution problems associated with the disposal of iron chloride by-product from ilmenite chlorination processes while recovering valuable chlorine.

As shown in Figure 8.4, oxygen is fed to line **1**. The oxygen, which is not heated, is supplied at a pressure of about 100 psig (6.8 atm). The size of line **1** is such that the oxygen feed is maintained at a sufficiently high velocity to permit transport of iron chloride, sodium chloride, pulverized fuel and recycle material fed to line **1** from pressurized storage vessels **2**, **3** and **4** which are pressurized with a gas, for example N_2, and recycle line **28**, respectively.

The materials in the storage vessels are maintained dry. The oxygen and any gas which may exit from the pressurized storage vessels and the solid materials conveyed with the gas through line **1** enter the reactor, which comprises the equipment designated by numerals **51, 52, 53** and **54**, through the bottom of reactor vessel **51**.

Figure 8.4: Schematic Diagram of Continuous Process for Production of Chlorine from Iron Chloride

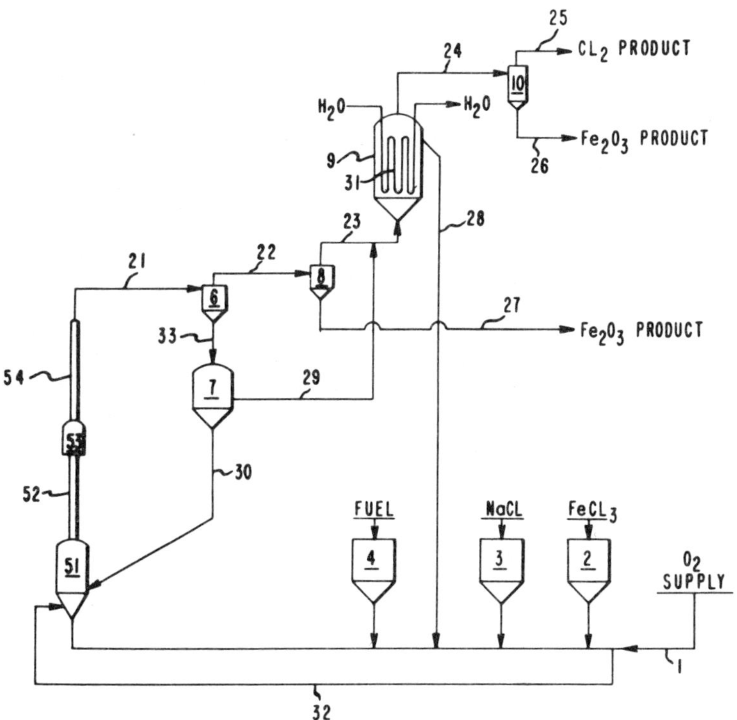

Source: U.S. Patent 4,144,316

The reactor is divided into several sections. The first section, reactor vessel **51**, into which the materials from line **1** enter, is ceramic-lined to an internal diameter of 2½ ft (76 cm) and has a height of 9.3 ft (2.85 m), including a height of 2 ft (0.61 m) for the conical bottom and 1.25 ft (0.38 m) for the hemispherical head. Four supplemental oxygen inlets, supplied through line **32**, are located at about the midheight of the bottom conical portion of reactor vessel **51**. The nozzles are spaced 90° apart and inject oxygen radially into the center of the cone. Approximately 5 to 15% of the total oxygen fed to the reactor vessel is supplied through these nozzles. The fluidized bed particles, reactants and products formed flow concurrently from reactor vessel **51** through ceramic-lined pipe **52**, which measures 9" (22.9 cm) in inside diameter and about 6' (1.83 m) in length, into enlarged, ceramic-lined section **53**, which measures about 2' (61 cm) in inside diameter and 4' (1.22 m) in length.

In pipe **52**, because of the higher velocity of the stream, the solids concentration is lower than in reactor vessel **52**. The enlarged section **53** serves to reduce the velocity of the particles and acts as a mixer prior to the entry of the stream into ceramic-lined pipe **54**, which is of the same diameter as pipe **52** and measures 28' (8.54 m) in length.

Within reactor vessel **51**, the materials fed from line **1** are heated to temperatures of 550° to 800°C; ferric chloride is vaporized; the carbon is burned and sodium chloride and ferric oxide form catalytic bed particles; this ferric chloride and oxygen react in reactor vessel **51** as well as in reactor sections **52**, **53** and **54** to form chlorine and ferric oxide product.

The stream exiting from reactor vessels **51**, **52**, **53** and **54** enters cyclone separator **6** from line **21**. The difference in elevation between the top of reactor vessel **51** and the inlet to cyclone separator **6** is approximately 70' (21.3 m). In cyclone separator **6**, coarse iron oxide particles are separated from the stream and deposited via line **33** in hot-solids storage tank **7** from which they are recycled via line **30** to the bottom of the cylindrical section of reactor vessel **51**.

The gaseous stream and fine iron oxide particles exiting cyclone separator **6** are transported via line **22** to cyclone separator **8** which operates at a higher separation efficiency than cyclone separator **6**, and removes most of the remaining solids from the gaseous product stream. This gaseous product stream is then fed via line **23** to the bottom of fluid-bed condenser **9**. The condenser is provided with water-cooled internal coils **31** which reduce the temperature in the condenser to about 150°C.

In condenser **9**, unreacted gaseous iron chloride is condensed onto a bed of iron oxide particles, which were fed to condenser **9** from hot-solids storage tank **7** via line **29** through line **23**. The unreacted ferric chloride and iron oxide particles are returned to the reactor via line **28** through line **1**.

The cooled pressurized gaseous product leaving fluid bed condenser **9** is fed via line **24** to final cyclone separator **10** to remove any remaining entrained solids. The gaseous product is primarily chlorine which can be recycled directly to an ilmenite chlorination process or can be collected for other uses.

Part of the iron oxide product is obtained from line **27**; the remainder from line **26**. It is possible to operate with cyclone separator **8** removed from the system, in which case, iron oxide product could be removed from a tap in reactor recycle line **30**.

The following startup procedure has been found satisfactory for the above-described system. The reactor system **51**, **52**, **53** and **54**, the first cyclone separator **6**, the hot solids storage tank **7**, the iron-oxide recycle line **30** to the reactor and the interconnecting piping **21, 33** are heated to temperatures in the range of 350° to 500°C with air, which is preheated to about 1000°C and supplied to the equipment through the oxygen and feed materials inlet line **1**.

Iron oxide particles are fed to and circulated through the system during the initial heat-up to provide the inventory needed for the fluidized bed. When the temperature of the system has reached the 350° to 500°C range, the air is replaced with unheated oxygen and pulverized carbonaceous fuel is fed (from storage vessel **4**) into the reactor system where it burns and further heats the equipment and iron oxide particles to the desired operating temperature range of 550° to 800°C.

The sodium chloride is fed (from storage vessel **3**) to the reactor system to combine with the recirculating iron oxide fluidized-bed particles to form the catalyst. The amount of sodium chloride fed is sufficient to provide a sodium chloride concentration in the range of 0.1 to 10%, preferably 0.4 to 1.0 wt % of the bed particles. At this point, the system is ready for establishing the desired steady-state operating conditions and material flows. The following ranges of operating conditions are suitable.

Reactor temperature	500°-800°C
Reactor inlet pressure	50-150 psig (3-7 atm)
Ferric chloride feed	3,000-15,000 lb/hr (1,360-6,820 kg/hr)
Excess oxygen feed	3-70%
Carbonaceous fuel feed	150-400 lb/hr (68-180 kg/hr)
Sodium chloride feed	50-300 lb/hr (23-136 kg/hr)
Iron oxide recycle to reactor	15,000-60,000 lb/hr (6,800-27,200 kg/hr)

For these conditions, the conversion of iron chlorides to iron oxide generally exceeds 90%. When the reactor outlet temperature is greater than 600°C, conversions of 95% or more are usually obtained.

Combination of Reduction and Oxidation Reactions to Recover Chlorine from Iron Oxide

J.H.W. Turner and C.E.E. Shackleton; U.S. Patent 4,140,746; February 20, 1979; assigned to Mineral Process Licensing Corporation BV, Netherlands disclose the recovery of chlorine values from iron chloride by-produced from the chlorination of a titaniferous material containing more than 5 wt % iron oxide, and particularly from the carbochlorination of ilmenite which, for example, can be the first stage in the so-called chloride route to form titanium dioxide pigment.

The iron chloride, which may be ferric chloride or ferrous chloride, is subjected to a combination of reduction and oxidation reactions. In the reduction reaction, ferric chloride is dechlorinated to ferrous chloride by a reducing agent suitable for producing a chloride compound for recycle to the chlorination process or for sale and in the oxidation reaction ferrous chloride is oxidized to ferric oxide and ferric chloride, the ferric chloride being recycled to the reduction reaction.

By this method the chlorine values are recovered from by-product iron chloride by a route which avoids the difficult reaction between ferric chloride and oxygen to produce chlorine and ferric oxide.

Example: *Stage A* – Ferric chloride obtained from the carbochlorination of ilmenite with an analysis of minimum 96% $FeCl_3$ and maximum 1% $FeCl_2$, was charged in solid powder form together with reagent grade flowers of sulfur to a reaction flask. Care was taken during transfer to the flask to minimize the contact of the reactants with moist air as ferric chloride especially is very hygroscopic.

The reactants within the sealed flask were then thoroughly mixed by shaking together since it had been found in previous experimental work that preliminary mixing was of primary importance in achieving a good reaction. The charge weight of ferric chloride was 100 g, and the sulfur was 95% stoichiometric approximately according to the following equation.

$$2FeCl_3 + 2S \longrightarrow 2FeCl_2 + S_2Cl_2$$

The reaction flask was connected to a condenser and to a nitrogen line. The reactor and condenser were flushed with nitrogen to displace oxygen and moisture before initiating the reaction. The reaction was initiated by heating the flask reactor with a mantle furnace to a temperature above the melting point of sulfur. Thereafter, the bed temperature was increased slowly over the reaction period up to a final temperature of 400°C. The total reaction period was 2 hr. The solid reactor product had the following analysis: $FeCl_3$, 0.3%; S, 2.5% and $FeCl_2$, 97%.

The physical nature of the ferrous chloride product was a mixture of lumps and fines, the lumps consisting of many small crystals which were readily broken down into fine material. The gaseous sulfur monochloride product was condensed in the water-cooled condenser and collected in a flask. It contained no ferric ions and had a liquid gravity of 1.677.

Stage B – 450 g of ferric oxide in powder form and 75 g of ferrous chloride (manufactured in Stage A above) were thoroughly mixed together. The material was then charged to a vertical stirred bed reactor to rest on a sintered disc at the reactor's base. The material was heated up to 550°C, using external heaters around the reactor cylinder. Throughout the heat-up, nitrogen was introduced through the sintered disc to purge the material and the reactor system of any moisture or oxygen.

During heat-up, HCl gas was identified at the exit of the condenser in line after the reactor, indicating that there was some loss of chloride during the heat-up, although the losses were small in proportion to the chloride input to the reactor as can be seen from the results below. When bed temperature stabilized at 550°C, the nitrogen flow was discontinued and oxygen (dried with $CaCl_2$) was introduced through the sintered disc.

The flow rate was 500 ml/min. Total running time was 30 min, although the reaction was apparently completed sooner. $FeCl_3$ was evolved during the first 5 min of the reaction period, and was collected in the condenser. The weight of ferric chloride was 57 g which was 93% of the theoretical amount, calculated from $6FeCl_2 + 1\frac{1}{2}O_2 \longrightarrow Fe_2O_3 + 4FeCl_3$. 0.7 g of chlorine were collected in a KI bubbler downstream of the condenser. The iron oxide residue weight was 464.4 g which was close to the theoretical weight. The residue contained no iron chloride.

Stage C – The ferric chloride produced in Stage B was recycled to Stage A.

The S_2Cl_2 produced may either be recycled directly to the chlorinator, as would be appropriate in the case of chlorination in the presence of sulfur, or its chloride content may be processed into a form suitable for recycle to a carbochlorinator, e.g., by reacting sulfur monochloride with carbon bisulfide to form carbon tetrachloride and sulfur or by heating sulfur monochloride above its dissociation temperature (444°C) and cooling the resulting gas rapidly to produce chlorine and chlorine polysulfides.

Oxidation of Ferric Chloride from Selective Chlorination of Titaniferous Material

According to *J.P. Bonsack and G.R. Walker; U.S. Patent 4,094,954; June 13, 1978; assigned to SCM Corporation*, ferric chloride from the selective chlorination of titaniferous material such as ilmenite is partially oxidized to obtain a chlorine-rich stream and by-product iron oxide. Unreacted ferric chloride is separated as solid particles from the stream and at least a portion of the particles are returned to the chlorination operation for absorbing heat generated in such operation.

Example: Referring to Figure 8.5, the following is set forth as a basis of design for ilmenite ore beneficiation on commercial scale. Fluidized bed **22**, acting as a selective chlorination zone, is ilmenite ore and coke. It is maintained at 1000°C and approximately 1 atm total pressure in chlorinator **21**. The ore is Australian ilmenite (54% TiO_2, 30% total Fe, passing 40 mesh and retained on a 200 mesh U.S. Standard sieve).

Level **29** represents the top of bed **22**. A mixture of beneficiated ilmenite ore (about 92% TiO_2 and 3% Fe with a small proportion of unreacted coke) continuously overflows from the bed through line **26**, which line is purged with a small flow of nitrogen gas not shown. Chlorine, in a proportion of 750 parts per hour per 1,300 parts per hour of fresh ore and 180 parts per hour of fresh petroleum coke feed (+40 mesh) from lock hopper **37**, enters inlet **25** and passes through porous support plate **23** for fluidizing bed **22**.

Vapors from bed **22** (mainly ferric chloride) ascend into oxidizer **28**, the oxidation zone for this operation. Substantially pure oxygen gas is passed in through inlet **27** in a proportion of 200 parts per hour. The oxidation zone is at 900°C from the oxidation inlet to just below its top. The superficial vapor velocity therein is maintained slightly in excess of 1 ft/sec for suppressing back-flow of oxygen into the chlorination zone and maintaining entrainment of all the iron oxide formed.

Such superficial vapor velocity is calculated on the basis of an instantaneous reaction to equilibrium of the oxygen and ferric chloride to chlorine gas and solid iron oxide at 900°C and 1 atm total pressure while neglecting solids and assuming plug flow of the cross section of the narrowed oxidation zone. Residence time in this zone is about 6 sec whereupon about 80% of the ferric chloride ascending from the chlorination zone is converted into solid iron oxide and chlorine gas.

Figure 8.5: Flow Sheet for Commercial Process for Beneficiating Ilmenite

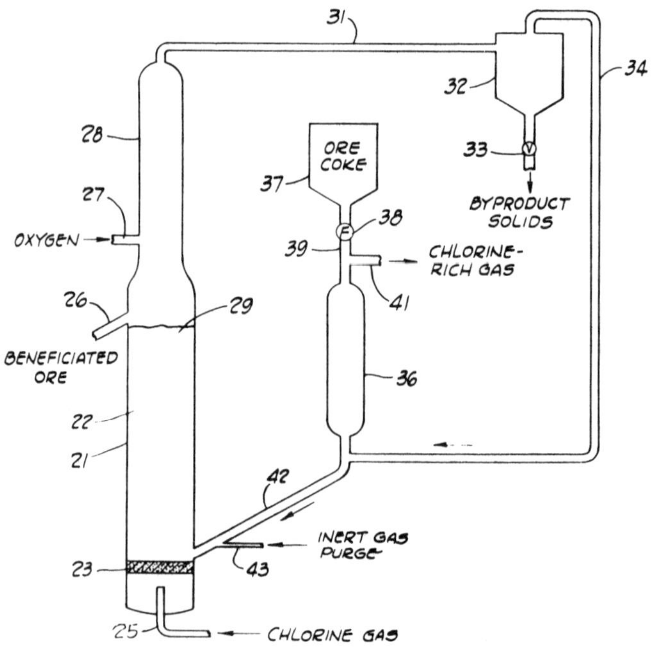

Source: U.S. Patent 4,094,954

At the very top of the oxidation zone the exit vapors are quenched to about 600°C with injection of liquid chlorine, not shown, for suppressing further oxidation. Exit vapors in line **31** are further cooled indirectly by means not shown. They discharge into separator **32** wherein iron oxide is separated at 270° to 300°C and removed from the system through rotary valve **33**. Separator **32** is a hot cyclone separator followed in series by an electrostatic precipitator/bag house type iron oxide collector. Ferric chloride vapor borne in a chlorine-rich stream discharges through outlet line **34**.

These vapors pass through line **34** and upwardly through vessel **36** countercurrent to a downward flow of fresh cold coke and ore feed. Such feed discharges through lock hopper **37**, screw feeder **38**, line **39** and into vessel **36**. In vessel **36** these solids mix thoroughly with the chlorine-rich vapors from line **34** and cool such vapors to a temperature (e.g., below 200°C) whereby substantially all of the ferric chloride precipitates, mostly on the ore and coke feed. Chlorine-rich product gas thus stripped of ferric chloride is withdrawn through outlet **41**. The feed solids enriched by iron chloride fall downwardly into dense feed leg **42**. They flow continuously from there, with a small nitrogen gas purge entering line **43**, into the bottom of bed **22**.

Oxidation of Halides in Conical Combustion Chambers

R. Nowak and G. Holland; U.S. Patent 4,120,941; October 17, 1978; assigned to Halomet AG, Switzerland describe a process of oxidizing halides, including mixtures thereof, with oxygen, oxygen-containing gases and/or oxygen-liberating substances in a combustion chamber, whereby the vapor pressure of the halides at introduction to the combustion or oxidation chamber is greater than the pressure within the combustion chamber. Recovery of halides at greater purity with substantial decrease of by-product deposits on operating equipment is realized.

Example 1: Iron chloride (Fe_2Cl_6) was burned with industrial oxygen (about 98 vol % O_2 and about 2 vol % N_2) in a cylindrical chamber whose inner walls were lined with sillimanite by means of a double nozzle from an oil burner as follows.

Fe_2Cl_6, liquid	325°C
Industrial O_2	21°C
Temperature at hottest part of reaction zone	750°C
Fe_2O_3 and chlorine-containing exhaust gas	740°C

The vapor pressure of the molten Fe_2Cl_6 in a closed container was slightly below 1 atm, and the delivery pressure was generated with compressed air in the container. The pressure in the combustion chamber was exactly 1 atm. The exhaust gas of the combustion had the following composition in percent by volume.

Cl_2	92.43
Fe_2Cl_6	3.33
O_2	2.22
N_2	2.02

After 40 min combustion, the double nozzle was incrusted with Fe_2O_3 to such an extent that the combustion operation came to a stop. At the inner walls of the combustion chamber deposits had formed which threatened to clog the chamber if operation were continued.

Example 2: The Fe_2Cl_6 was heated in a closed pressure vessel to 440°C and its vapor pressure rose to 3 atm. It was possible to carry on combustion with this superheated melt for as long as desired without incrustations forming at the burner nozzle. The combustion chamber did become increasingly incrusted with deposits of Fe_2O_3 with time. The exhaust gas composition remained substantially the same throughout the reaction and the temperature in the hottest part or section of the reaction zone was 860°C.

Example 3: The cylindrical sillimanite chamber was replaced by three conically widened chambers made up of copper sheet as shown in Figure 8.6. These chambers **10a**, **10b** and **10c**, are water cooled, water at 15°C being fed into inlet **11** and the warm water taken off at outlet **12**.

The Fe_2Cl_6 was again supplied to the burner nozzle as a superheated melt of 440°C and 3 atm vapor pressure at inlet **13**. In this third test the Fe_2O_3 passed out of the apparatus at outlet **14** along with the exhaust gas at a temperature of 450°C. O_2 is fed into chamber **10a** at inlet **15**. The composition of the gas was as follows in percent by volume.

Cl_2	96.41
Fe_2Cl_6	0.65
O_2	0.98
N_2	1.96

A thermodynamic calculation showed that the gas equilibrium during the cooling of the reaction products from 860° to 600°C switched over continuously, considerably increasing the chlorine yield and substantially reducing the portion of iron chloride and oxygen.

Figure 8.6: Oxidation of Halides in Conical Combustion Chamber

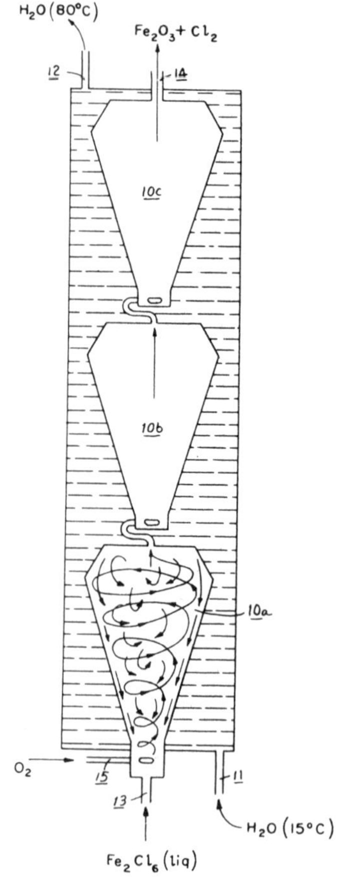

Source: U.S. Patent 4,120,941

CHLORINE FROM HYDROCHLORIC ACID

The production of halogen by the electrolysis of aqueous hydrogen halides such as hydrochloric acid has generated substantial commercial interest because of the large quantity of excess hydrochloric acid available throughout the world as a by-product of organic chlorination operations and other industrial processes. The demand for hydrochloric acid as such, on the other hand, has not kept pace with the amount of such low-grade hydrochloric acid available.

Disposing of such excess by-product HCl presents a troublesome environmental problem to the chemical industry since these large quantities of hydrochloric acid must be disposed of without polluting the environment. Thus, direct electrolysis of hydrochloric acid in water solutions is of great interest to industry since it provides a method for disposing of large quantities of excess hydrochloric acid, while at the same time producing chlorine for which there is a large and increasing industrial demand.

Catalytic Electrodes Bonded to a Solid Polymer Electrolyte

In a process developed by *R.M. Dempsey, T.G. Coker and A.B. La Conti; U.S. Patent 4,210,501; July 1, 1980; assigned to General Electric Company* catalytic electrodes in the form of fluorocarbon bonded, thermally stabilized, reduced oxides of platinum group metals are bonded to at least one surface of the membrane and an aqueous hydrochloric acid solution is brought in contact with the bonded anode to generate chlorine at the anode and hydrogen at the cathode.

The perfluorocarbon-polytetrafluoroethylene, also known as Teflon (Dupont), bonded, graphite electrode includes reduced oxides of platinum group metals such as ruthenium, iridium, ruthenium-iridium, etc., in order to minimize chlorine overvoltage at the anode. The reduced oxides of ruthenium are stabilized to produce an effective, long-lived anode which is stable in acids and has very low chlorine overvoltage.

Figure 8.7 illustrates diagrammatically the reactions taking place in various portions of the cell during HCl electrolysis, and is useful in understanding the electrolysis process and the manner in which the cell functions. An aqueous solution of hydrochloric acid is brought into the anode compartment which is separated from the cathode compartment by means of the cation membrane **13**.

The bonded graphite electrodes containing reduced oxides of Ru stabilized by reduced oxides of iridium or titanium, etc., are as shown, pressed into the surfaces of membrane **13**. Current collectors **15** and **16** are pressed against the surface of the catalytic electrodes and are connected, respectively, to the negative and positive terminals of the power source to provide the electrolyzing voltage across the electrodes.

The hydrochloric acid brought into the anode chamber is electrolyzed at anode **24** to produce gaseous chlorine and hydrogen ions (H^+). The H^+ ions are transported, across membrane **13**, to cathode **14** along with some water and some hydrochloric acid. The hydrogen ions are discharged at the cathode electrode which is also bonded to and embedded in the surface of the membrane. Cathode **14** may, for example, also consist of a fluorocarbon bonded graphite with thermally stabilized, reduced oxides of platinum group metals and valve metals, viz., Ru, Ir, Ti, Ta, etc. The reaction in various portions of the cell is as follows.

Anode reaction: $2Cl^- \longrightarrow Cl_2\uparrow + 2e^-$

Membrane transport: $2H^+(H_2O, HCl)$

Cathode reaction: $2H^+ + 2e^- \longrightarrow H_2\uparrow$

Overall reaction: $2HCl \longrightarrow H_2 + Cl_2$

In this arrangement, the catalytic sites in the electrodes are in direct contact with the cation membrane and the ion exchanging acid radicals attached to the polymer backbone (whether $SO_3H{\cdot}H_2O$ sulfonic acid radicals or $COOH{\cdot}H_2O$ carboxylic acid radicals).

As a result, there is no IR drop to speak of in the anolyte or the catholyte fluid chambers (usually referred to as electrolyte IR drop) and this is one of the principal advantages of this process. Furthermore, because the chlorine and hydrogen are generated right at the electrode and membrane interfaces, there is no IR drop due to the so-called bubble effect which is a gas mass transport loss. That is, in prior art systems, gas formation occurs between the catalytic electrode (spaced away from the membrane) and the membrane. This layer or film of gas at least partially blocks ion transport between the catalytic electrode and the membrane and introduces a further IR drop.

Figure 8.7: Schematic Illustration of Cell for HCl Electrolysis

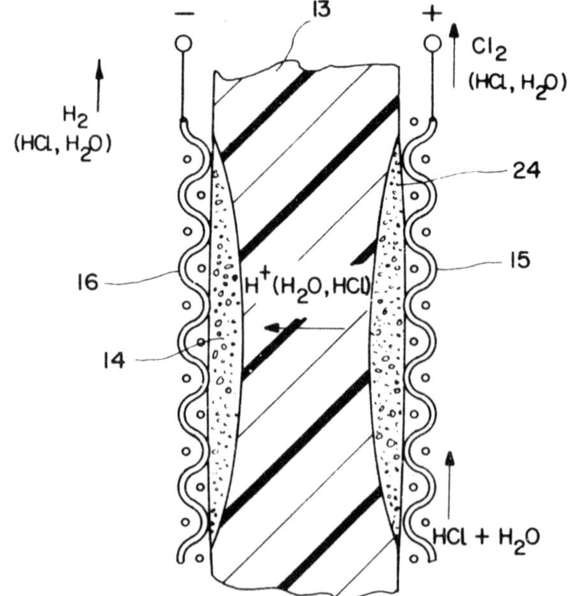

Source: U.S. Patent 4,210,501

Catalytic Molten Salt Mixtures

A process for oxidizing hydrogen halides by means of a catalytically active molten salt is disclosed by *C.A. Rohrmann; U.S. Patent 4,107,280; August 15, 1978; assigned to Battelle Memorial Institute*. The subject hydrogen halide is contacted with a molten salt containing an oxygen compound of vanadium and alkali metal sulfates and pyrosulfates to produce an effluent gas stream rich in the elemental halogen. The reduced vanadium which remains after this contacting is regenerated to the active higher valence state by contacting the spent molten salt with a stream of oxygen-bearing gas.

Referring to Figure 8.8, the general process for regenerating either chlorine, bromine or iodine from their respective halides is shown. A waste or by-product hydrogen halide gas is fed to a fused salt contactor **10** via line **12**. The contactor contains a molten salt mixture including a dissolved oxygen compound of vanadium.

Figure 8.8: Flow Diagram for Oxidation of Hydrogen Halides Using a Catalytically Active Molten Salt

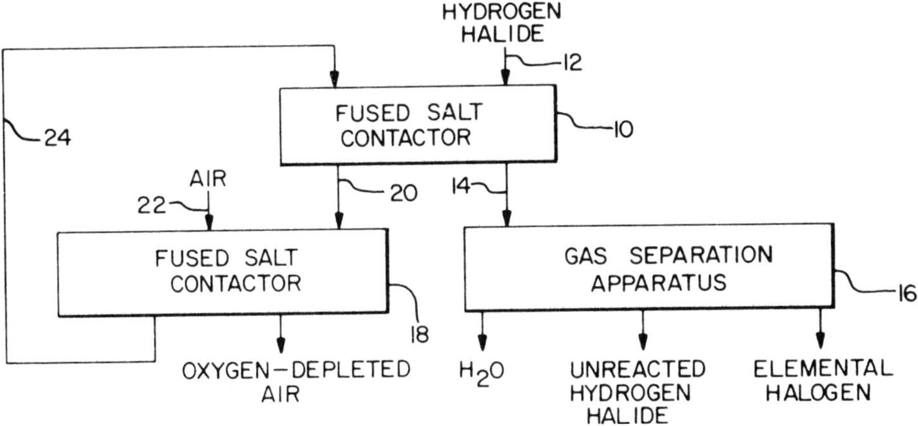

Source: U.S. Patent 4,107,280

When contacted with this mixture, the hydrogen halide is oxidized by reaction with the vanadium compound according to the following general reaction in which X is Cl, Br or I.

$$2HX + V_2O_5 \longrightarrow V_2O_4 + X_2 + H_2O$$

Kinetics do not allow the reaction according to the above equation to proceed to completion. For this reason the effluent gas mixture leaving the salt contactor in a line **14** contains a mixture of steam, unreacted halogen halide gas, and elemental halogen gas. It is an advantage of this process that little or no free oxygen gas is formed in the contactor and, therefore, that components of the effluent gas mixture can easily be separated by conventional gas separation apparatus **16**.

Contact with the hydrogen halide gas causes the vanadium in the molten salt mixture to be reduced to a lower valence state. The vanadium must be regenerated therefrom before the salt mixture can be reused. This regeneration is accomplished by transporting the spent salt mixture to a second fused salt container **18** via a line **20**. The vanadium reacts with oxygen in the gas and this is regenerated to its higher valence state. The regenerated salt is returned to the first fused salt contactor, in a line **24**, and the oxygen-bearing gas either recirculates, if economically advantageous, or is vented to the atmosphere.

Contacting with Molten Cuprous and Cupric Chlorides

According to *H. Riegel and V.A. Strangio; U.S. Patent 4,119,705; October 10, 1978; assigned to The Lummus Company*, hydrogen chloride and oxygen are contacted with a molten mixture of cuprous and cupric chloride in an oxidation reaction zone to enrich the cupric chloride content of the melt, and the melt introduced into a dechlorination zone wherein gaseous chlorine is removed from the melt.

The oxidation reactor is operated at a pressure higher than the dechlorination reactor, and molten salt circulation rates are controlled in a manner such that the cupric chloride content and temperature of the salt introduced into the dechlorination reaction zone are higher than the cupric chloride concentration and temperature of the melt introduced into the oxidation reactor.

Referring to Figure 8.9, an oxygen-containing gas, such as air or oxygen in line **10**, hydrogen chloride in line **11** and a recycle stream, containing oxygen and chlorine in line **12** is combined in line **13** and introduced into an oxidation reactor generally indicated as **14**, containing means for increasing gas-liquid contact, such as a packed bed, schematically indicated as **15**.

Figure 8.9: Flow Diagram for Deacon Process

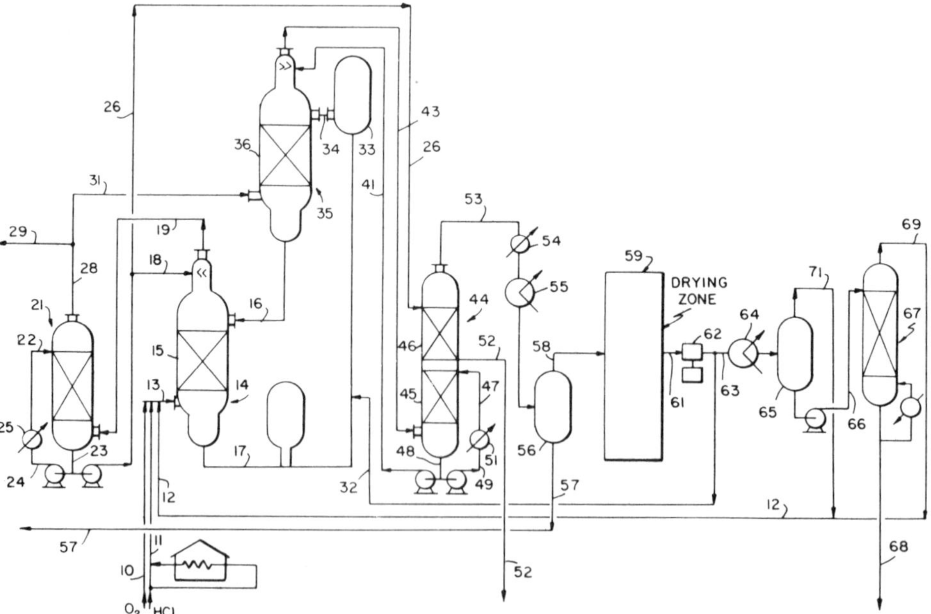

Source: U.S. Patent 4,119,705

A molten salt mixture containing cupric and cuprous chloride, and further including a melting point depressant, in particular, potassium chloride, is introduced into reactor **14** through line **16** to countercurrently contact the gaseous mixture introduced into the reactor through line **13**. As a result of the countercurrent contact between the molten salt mixture and the gas introduced through line **13**, the hydrogen chloride is oxidized to chlorine, and such chlorine values are recovered by the molten salt by enriching the cupric chloride content of such molten salt.

A molten salt mixture, enriched in cupric chloride, is withdrawn from reactor **14** through line **17** for further processing to recover chlorine values therefrom.

A gaseous effluent, containing unreacted oxygen, inerts, such as nitrogen introduced with the oxygen containing gas, equilibrium amounts of hydrogen chloride and some chlorine is contacted in the upper portion of reactor **14** with an aqueous hydrogen chloride quench liquid introduced into the top of the reactor through line **18** to effect cooling of the effluent, with such cooling also resulting in vaporization of the quench liquid. In general, the effluent is cooled to a temperature in the order of from 835° to 450°F as a result of such quenching.

A partially cooled gaseous effluent, containing oxygen, nitrogen, hydrogen chloride, chlorine and water vapor is withdrawn from reactor **14** through line **19** and introduced into a quench cooling tower, schematically indicated as **21**. In quench cooling tower **21**, the gas is directly contacted with an aqueous hydrogen chloride quench liquid introduced into the top of the tower through line **22** to separate hydrogen chloride from the gas and recover the hydrogen chloride as an aqueous hydrogen chloride solution.

A dilute aqueous hydrogen chloride solution, generally containing 8 to 20 wt % of hydrogen chloride is withdrawn from tower **21** through line **23** and a first portion thereof is passed through line **24** including a cooler **25** for subsequent introduction into the quench tower **21** through line **22**. A further portion of the aqueous hydrogen chloride is employed as a quench liquid in line **18** for cooling the gaseous effluent in the top of oxidation reactor **14**. A further portion of the aqueous hydrogen chloride is employed in line **26** for quenching.

A gaseous effluent, essentially free of hydrogen chloride, and containing oxygen, nitrogen, chlorine and water vapor is withdrawn from quench tower **21** through line **28** and a first portion thereof passed through line **29** for purging from the system subsequent to effecting neutralization thereof with a suitable caustic (not shown).

The remaining portion of the gaseous effluent withdrawn from the quench tower **21** is passed through line **31** for subsequent introduction into a dechlorination reactor for effecting stripping of chlorine values from the molten salt.

The molten salt, enriched in cupric chloride, in line **17** is lifted by a suitable lift gas in line **32** into a separation vessel, schematically indicated as **33** for separation of the molten salt from the lift gas.

The separated molten salt and lift gas is passed from separation vessel **33** through line **34** and introduced into a dechlorination reactor, schematically indicated as **35** and containing means for increasing gas-liquid contact, such as a packed bed, schematically indicated as **36**. The molten salt, enriched in cupric chloride, is countercurrently contacted in dechlorination reactor **35** with a stripping gas in line **31**. As a result of such contact, chlorine values are stripped from the melt as gaseous chlorine thereby reducing the cupric chloride content of the molten salt.

Cyanogen Chloride and Gaseous Chlorine

R. Miller; U.S. Patents 4,100,263; July 11, 1978; and 4,175,116; November 20, 1979; assigned to Ciba-Geigy Corporation describes a process for preparing cyanogen chloride by reacting chlorine and hydrogen cyanide in a manganous chloride-containing aqueous medium to form gaseous cyanogen chloride and dissolved hydrogen chloride, and the subsequent conversion of the hydrogen chloride to chlorine employing manganese dioxide and nitric acid as cyclic reagents. The process proceeds according to the following reaction sequence.

(1) $HCN + Cl_2 \xrightarrow[\text{Solution}]{MnCl_2} HCl + CNCl$

(2) $MnO_2 + 4HCl \longrightarrow MnCl_2 + 2H_2O + Cl_2$

(3) $MnCl_2 + MnO_2 + 4HNO_3 \longrightarrow 2Mn(NO_3)_2 + Cl_2 + 2H_2O$

(4) $2HCl + 2HNO_3 + MnO_2 \longrightarrow Mn(NO_3)_2 + Cl_2 + 2H_2O$

(5) $Mn(NO_3)_2 \longrightarrow MnO_2 + 2NO_2$

(6) $2NO_2 + H_2O + \frac{1}{2}O_2 \longrightarrow 2HNO_3$

This sequence and the various recycle streams are diagrammatically depicted in Figure 8.10 which represents a flow sheet of the total process.

Figure 8.10: Flow Sheet of Process for Preparing Cyanogen Chloride and Gaseous Chlorine

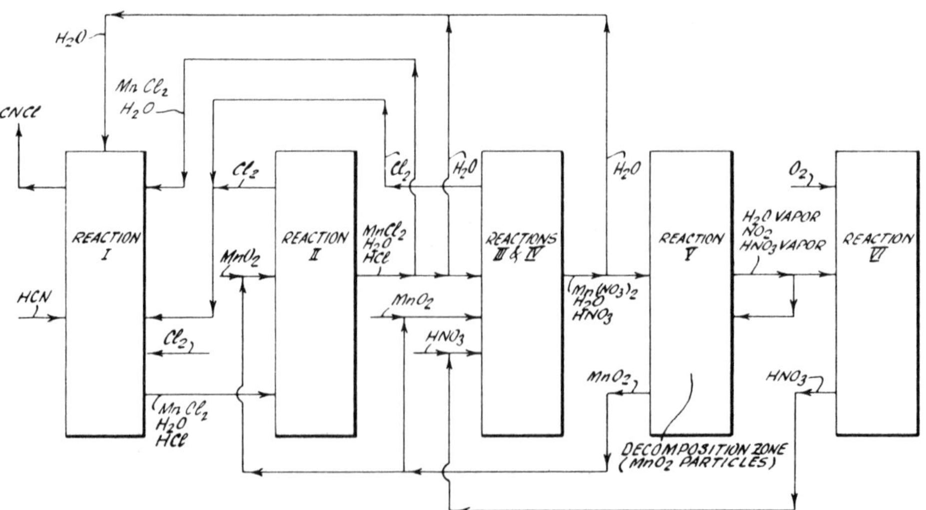

Source: U.S. Patent 4,100,263

Example: The following example presents a reaction scheme for reactions (2), (3) and (4).

The cyanogen chloride reactor was operated so that the effluent leaving the reactor contained 7.5 parts of HCl dissolved in 80 parts of water along with 20 parts of $MnCl_2$. A small amount of free chlorine was also present in the effluent solution. The effluent flowed to a reaction zone which was also fed with a 100% excess of recycled MnO_2.

The resultant slurry was then boiled at a temperature of 103°C, whereupon the HCl reacted with the MnO_2 to form water and chlorine in accordance with reaction (2). For each 6 parts of HCl that reacted, 3.58 parts of MnO_2 was converted to 5.18 parts of $MnCl_2$. The chlorine formed in the reaction along with the dissolved chlorine contained in the effluent vaporized and the gaseous chlorine was returned to the CNCl reactor.

Approximately 80% of the HCl reacted quickly to form a concentrated $MnCl_2$ solution (approximately 25 wt %). The reaction was conducted under sufficient pressure so that the evolved chlorine (2.92 parts for each 6 parts of HCl which reacted) could flow back to the CNCl reactor. At that point, the bulk of the solution (approximately 80%) was pumped through a heat exchanger where it was cooled to about 15°C and then was pumped into the CNCl reactor at a level between the two scrubbing zones. The quantity of $MnCl_2$ recycled to the CNCl reactor was equal to the quantity contained in the liquid effluent that left the reactor.

The remainder of the $MnCl_2$ solution was pumped to an evaporator in which sufficient water was evaporated to form a solution containing about 50% $MnCl_2$. The vapor was condensed, cooled and recycled to the top of the CNCl reactor.

The concentrated 50% $MnCl_2$ solution was then pumped to a second reaction zone in which it was contacted with 100% excess of solid MnO_2 and a threefold excess of nitric acid whose concentration was close to the azeotropic composition.

In this reaction zone, which was an elongated column fitted with a bottom reboiler, the slurry entered close to the top and the evolved chlorine left from the top for recycle to the CNCl reactor. By operating so that the aqueous phase boiled at the pressure required to enable the evolved chlorine to flow into the CNCl reactor, chlorine formation reaction (3) was necessarily carried out at a temperature in excess of 120°C the maximum boiling point of the HNO_3-H_2O system at atmospheric pressure. At this elevated temperature, the speed of the reaction was quicker than at lower temperatures.

The water contained in the $MnCl_2$ solution fed to the reaction zone plus the water formed in the reaction diluted the nitric acid. By using an excess of concentrated nitric acid, this water has a minor effect on the acid concentration. Under these conditions, the desired reaction went rapidly and substantially to completion. For each 5.18 parts of $MnCl_2$ plus an accompanying 0.31 parts of HCl, it was seen that 3.94 parts of MnO_2 and 10.86 parts of HNO_3 reacted.

The mixture formed on completion of the reaction was composed of 28.53 parts water, 32.58 parts nitric acid containing 15.44 parts dissolved manganous nitrate with the excess 7.52 parts manganese dioxide suspended in the liquid. The bulk of the manganese dioxide was permitted to settle out and thereafter was recycled to the reaction zone for reaction (3). The supernatant liquid substantially free from suspended MnO_2 was passed to a fractional distillation unit. Water was fractionally distilled out of the mixture to form a solution containing about 57% nitric acid in which the manganous nitrate was dissolved.

The solution was then pumped at a controlled rate into a fluidized bed of particles of manganese dioxide which were maintained above 200°C by immersed electric heating elements. For this experiment, steam was utilized as the initial fluidizing medium. The gas leaving the top of the reactor was cooled by allowing it to flow through two condensers connected in series but separated by a gas-liquid separator.

A second gas-liquid separator was connected to the outlet of the second condenser which, in turn, was vented to a scrubber. The first condenser was cooled with ordinary cooling water, while the second was cooled with ice water. Substantially all of the water vapor and nitric acid condensed in the first condenser and collected in the first gas-liquid separator. Liquid NO_2 and a small amount of dilute nitric acid condensed in the second condenser and flowed into the second gas-liquid separator. A small amount of brown-colored gas (NO_2) flowed into the scrubber.

The condensate was not processed further. However, it was noted that the amount of solids in the fluidized bed gradually increased, The experiment was ended before bed overflow occurred. The gases exiting from the fluidized bed did not contain any solids, thereby indicating a quantitative conversion of manganous nitrate.

CHLORINE RECOVERY

Recovery from Chemical Plant Waste Streams

A process is disclosed by *R.W. Lynch; U.S. Patent 4,196,140; April 1, 1980; assigned to Olin Corporation* for recovering recoverable chlorine from chemical plant waste streams.

An aqueous stream containing recoverable chlorine is reacted with an alkali metal hydroxide, such as sodium hydroxide, to form a slurry of solid particles of alkaline earth metal hydroxide, such as calcium hydroxide, suspended in a liquid. The calcium hydroxide is filtered or otherwise separated from the liquid. The liquid is admixed with an organic alcohol to form an organic-aqueous solution. A halogenating agent, such as chlorine, is reacted with the organic-aqueous solution to form a solution of organic hypochlorite in an organic phase and an aqueous phase.

The solution of organic hypochlorite is phase separated to form aqueous and organic phases. The aqueous phase containing sodium chloride may be recycled for use as a reactant in a chlor-alkali electrolytic cell.

The organic phase containing organic hypochlorite may be used as a chlorinating agent or may be treated with an acid, such as hydrochloric acid, to reclaim free chlorine.

Example: About 198 parts of a solution which analyzed as 9.6% $Ca(OCl)_2$, 0.98% $CaCl_2$, 18.16% NaCl, and the remainder as H_2O was admixed with about 26 parts of an aqueous solution of about 50% sodium hydroxide. The mixture was stirred for about 15 min and filtered at atmospheric pressure through a glass pad filter for about 3 hr.

About 158 parts of the resulting filtrate which analyzed as 8.5% NaOCl, 0.8% NaOH, 0% $CaCl_2$, 16.0% NaCl, and the remainder as H_2O were admixed with about 21 parts tertiary butyl alcohol and about 230 parts of 1,1,2-trichloro-1,2,2-trifluoroethane in a glass flask. Molecular chlorine gas was bubbled through the solution for about 15 min at a temperature of about 25°C. An organic phase (about 248 parts) and an aqueous phase (about 150 parts) were formed. The organic phase, after separation from the aqueous phase, was analyzed and found to contain 18.5 parts of tertiary butyl hypochlorite.

The percent conversion of sodium hypochlorite, defined as 100 times (the ratio of chemical equivalents of sodium hypochlorite used divided by the chemical equivalents of sodium hypochlorite originally present), was about 46%.

The percent yield of tertiary butyl hypochlorite, defined as 100 times (the ratio of chemical equivalents of tertiary butyl hypochlorite formed divided by the chemical equivalents of sodium hypochlorite used), was about 47%.

Recovery of Chlorine and Hydrogen Chloride from a Combustion Gas

W.H. Prahl; U.S. Patent 4,157,380; June 5, 1979 describes a process of removing chlorine (Cl_2) and hydrogen chloride (HCl) from a combustion gas, such as a combustion gas formed from incinerating chlorine-containing organic materials, which comprises the following.

(1) Lowering the temperature of the combustion gas below the melting point of cupric chloride or mixture thereof with other salts;

(2) Contacting the cooled combustion gas of step (1), in the presence of oxygen, with copper of lower than a divalent oxidation state, such as a cuprous compound or its equivalent, in a quantity sufficient to absorb chlorine and hydrogen chloride present in gas, thereby converting the cuprous compound to cupric chloride (in order to absorb substantially all HCl and Cl_2 it is desirable to employ a stoichiometric excess of cuprous compound); and

(3) Contacting cupric chloride with a reducing agent, thereby converting the cupric chloride to cuprous chloride or an equivalent compound of lower than divalent oxidation state and the reducing agent to a chlorinated product.

The products formed from the reducing agent in step (3) can be removed; and cuprous chloride formed in step (3) can be reused in step (2).

Example: Figure 8.11 is a schematic representation of the equipment that can be used in this example. 500 kg/hr of a Cl-waste containing 68% chlorine is introduced through line **2-31** into burner **2-3**, and is burned with 2,750 kg/hr of air entering through line **2-32**. The temperature of the combustion gas is lowered to about 600°C by introducing through line **2-33**, recycled flue gas of about 200°C. A checkerwork of firebricks promotes mixing and prevents radiation from reaching the boiler.

Figure 8.11: Recovery of Chlorine and Hydrogen Chloride from a Combustion Gas

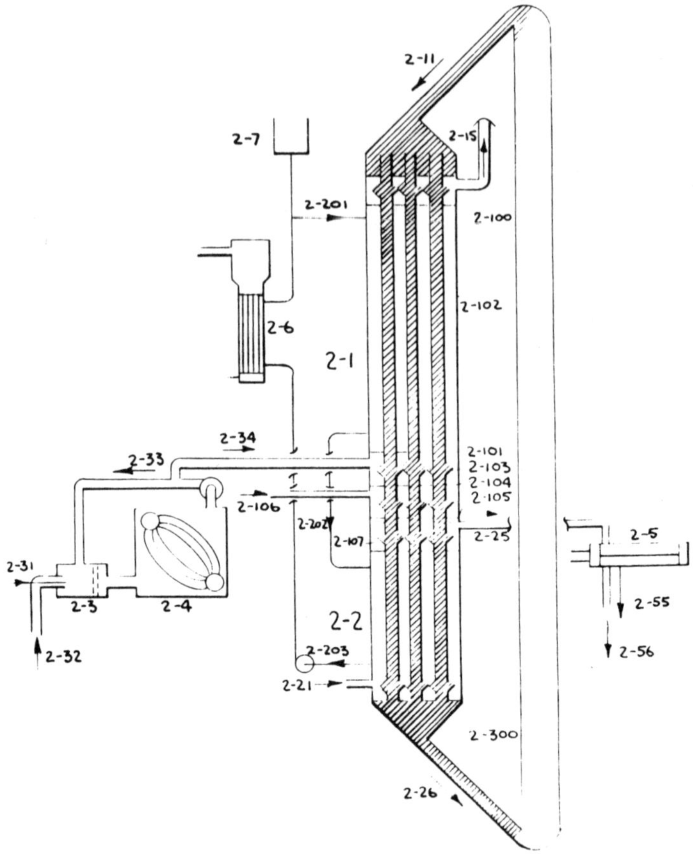

Source: U.S. Patent 4,157,380

The combustion gas passes a steam boiler, symbolized by **2-4**, in which it generates about 1,800 kg/hr of low pressure steam, thereby being cooled to about 200°C. The mixture leaving the boiler through line **2-34** contains about 158 kg/hr oxygen, 2,118 kg/hr nitrogen, 132 kg/hr water, 497 kg/hr carbon dioxide, and 350 kg/hr hydrogen chloride.

It enters through line **2-34** the upper part, **2-1**, of a moving bed reactor. **2-1** contains about 7 m^3 of a reaction mass prepared by impregnating alumina with about 20 wt % of copper chloride. In certain cases it may be advantageous to incorporate, in addition to the copper chloride, alkali salts, such as potassium chloride, rare earth chlorides, etc., in order to prevent evaporation of copper, enhance the reactivity, etc. In general such additions are of little overall usefulness. Evaporation of copper chloride is best prevented by operating at the lowest possible temperature and by having combustion gas and regenerating gas passing the reaction mass in opposite directions. The reaction mass is contained in about 400 4" pipes **2-100**, of about 2.5 m length, rolled or welded on both ends into tube sheets, **2-101**, and surrounded by a shell **2-102** which contains a heat transfer liquid (e.g., Dowtherm).

The combustion gas, freed of chlorine and hydrogen chloride, leaves the system through stack **2-15**. It contains approximately, 81 kg/hr oxygen, 2,113 kg/hr nitrogen, 219 kg/hr water, and 497 kg/hr carbon dioxide.

The reaction mass moves in the course of about 15 min from the top of the bottom of tubes **2-100**. In its downward path it enters next a chamber formed by tube sheets **2-104** and **2-105**. An inert gas, for instance nitrogen, or water vapor, enters this chamber through line **2-106**, and flows slowly upward and downward, forming a barrier between combustion gas and regenerating gas.

The reaction mass enters through funnels **2-107** the lower part **2-2** of the moving bed reactor system, essentially identical with the upper part, except that the pipes are 2 m long. About 20 kg/hr of methane enter through line **2-21**, move upward in the pipes, regenerate the reaction mass, and are thereby converted into about 184 kg/hr of carbon tetrachloride and 175 kg/hr of hydrogen chloride.

This mixture leaves reactor part **2-2** through line **2-25**, and goes to condenser **2-5**. Carbon tetrachloride is condensed and goes through line **2-55** for purification, while gaseous hydrogen chloride leaves through line **2-56** for further use. The regenerated reaction mass leaves **2-2** through line **2-26**, and is elevated back to the top of the system, entering through line **2-11**. The elevation is effected by any of the conventional means of moving granular solids. Shown here is a mechanical elevator **2-300**. Another preferred method would be an air conveyor.

Regeneration by methane is endothermic, and the lower part **2-2** of the reactor has to be heated. That is preferably done by circulating the heat transfer fluid successively through both parts of the reactor system. Cool heat transfer liquid enters the shell part **2-1** through line **2-201**, takes up the heat of absorption in **2-1**, moves on through line **2-202** into the lower part, **2-2** and gives off part of the heat to the tubes in **2-2**. It leaves **2-2** through line and pump **2-203** and passes through a cooling device, where the rest of the excess heat is taken up, for instance, by producing steam in boiler **2-6**. Surge tank **2-7** keeps the cooling system full and under pressure.

The regeneration of the reaction mass by means of hydrocarbons undergoing a substitution reaction yields about one-half of the original chlorine content of the Cl-waste in form of gaseous hydrogen chloride. Although under certain circumstances this may be acceptable, it will be in general more desirable to recover the whole chlorine content of the Cl-waste in form of chlorinated organic product. This is achieved by separating the hydrogen chloride from the chlorinated product and recycling the former to the absorption system.

Phase Change Apparatus

A.E. Bennett; U.S. Patent 4,128,409; December 5, 1978; assigned to Tioxide

Group Limited, England describes a process for the recovery of chlorine from chlorine-containing gas mixtures in a multistream heat exchanger wherein the chlorine-containing mixtures are separated from other gases which would solidify or liquify in the process, either before or after compression of the mixture, and is then cooled below the dew-point of chlorine in one stream of the heat exchanger. The liquid and gaseous phases thus produced are then separated and repassed separately through the heat exchanger, the chlorine passing to recovery and the gas phase to recooling before being repassed once more through the heat exchanger prior to discharge. A predetermined mixture of gaseous and liquid phases is also recycled separately through the heat exchanger before recompression and return via the heat exchanger to the phase-separation apparatus.

As shown in Figure 8.12, the chlorine-containing gas feed **19** is first compressed in a compressor **20** and then gases which, if present in the gas mixture in substantial quantity, would form solids or which would liquify with chlorine in the subsequent process steps are removed in a purification step shown generally as block **21**.

Compressed and purified chlorine-containing tail gas passes through pipe **1** to the multistream heat exchanger **2** and continues separately through channel **3** in the heat exchanger wherein it is in efficient heat exchange contact with channels **4**, **5**, **6** and **7** through which other gas streams pass, as described below.

During passage through **3** the gas is cooled and liquid chlorine is formed. The mixture passes to phase separator **8** wherein liquid and gas phases separate.

Liquid chlorine from the phase separator passes to channel **4** wherein it is heated and vaporized and passed to storage (not shown).

Figure 8.12: Schematic Diagram for Phase-Change Recovery Process

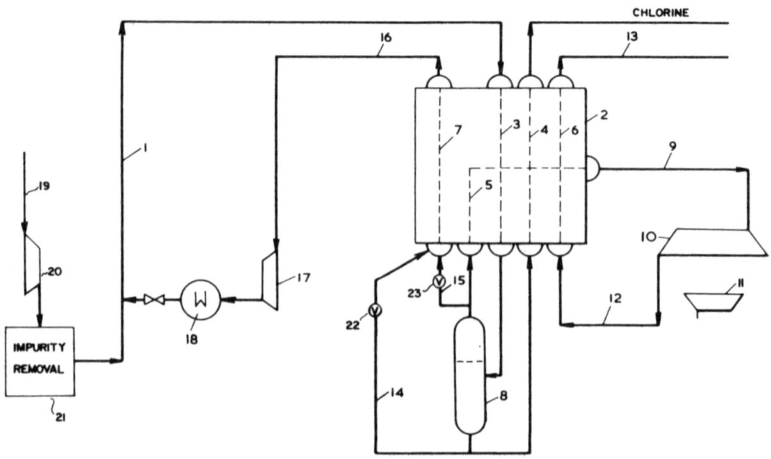

Source: U.S. Patent 4,128,409

Cooled and compressed gas from the phase separator passes through channel **5** and is heated before being withdrawn through pipe **9** to turbo-expander **10** wherein the gas is allowed to expand and is cooled. As the gas expands work is carried out and turbine wheel **11** is rotated and may be used to drive a gas blower (not shown).

The cooled gas passes via pipe **12** to channel **6** of the heat exchanger before being discharged through pipe **13**. Part of the liquid and gas in the phase separator are withdrawn through pipes **14** and **15**, mixed, expanded in expansion valves **22** and **23** and supplied to channel **7** of the heat exchanger before being withdrawn via pipe **16**, compressed in compressor **17**, cooled in cooler **18** and mixed with incoming compressed and purified tail gas in pipe **1** prior to return to the heat exchanger through channel **3**.

Example: Tail gas from the chloride process for the production of pigmentary titanium dioxide has the following properties: a pressure of 1.035 bars (abs); a temperature of 40°C; an analysis of 66.15% Cl_2, 26.05% O_2, and 7.8% inert gases.

The gas is compressed to 7.0 bars absolute in a centrifugal compressor and is cooled in a multiple-stream heat exchanger to -90°C at which temperature approximately 99% of its chlorine content is condensed to the liquid phase. The mixture is then transferred to a phase separator.

The greater part of the liquid phase (consisting mainly of chlorine) is removed and its pressure reduced to about 1.25 bars absolute before being passed through the heat exchanger. The resulting warmed gaseous chlorine is then passed to storage after further compression.

The greater part of the gas phase from the phase separator (mainly oxygen, inert gases and some chlorine) is recirculated through the heat exchanger and is withdrawn therefrom at about -32°C before passing to a turbo-expander wherein the pressure of the gas is reduced to about 1.4 bars absolute and its temperature is reduced to about -97°C. The expansion of the gas in the turbo-expander may provide useful work, for example, in the operation of a gas blower. The cold gas from the turbo-expander is again recirculated through the heat-exchanger before discharge therefrom.

A mixture of liquid chlorine and cold gas phase from the phase separator is produced, the ratio of the two components being 1.94:1.0, and the pressure of the mixture being about 1.9 bars absolute. The resulting cold gas mixture is recirculated through the heat exchanger, recovered thereform, compressed, cooled, mixed with fresh incoming compressed tail gas and supplied to the multi-stream heat exchanger as previously described.

It is calculated that on a commercial scale operation the operating costs of the process may be less than 50% of those entailed in the process of British Patent 1,274,710 and an even greater reduction in such costs can be obtained by the process when compared with the operating costs of the previously used compression/refrigeration processes for the recovery of chlorine.

Absorption by Iron (II) Chloride

E. Zirngiebl; U.S. Patent 4,082,631; April 4, 1978; assigned to Bayer AG, Germany discloses a process for the absorption of chlorine from a chlorine-containing gas, comprising contacting the chlorine-containing gas with an aqueous solution containing about 1 to 400 g/ℓ of iron(II) chloride and at least one of copper(I) and copper(II) ions, the chlorine being absorbed with conversion of the iron(II) chloride into iron(III) chloride, thereafter electrolyzing the iron(III) chloride to form chlorine and iron(II) chloride and recycling the iron(II) chloride for the absorption of additional chlorine.

Example: 0.2 m^3/hr of gas, containing 5 vol % of chlorine and 95 vol % of air, was treated with an absorption liquid containing approximately 127 g of $FeCl_3$, 36 g of $FeCl_2$ and 0.5 g of dissolved Cu per liter in an absorption tower **2** as shown in Figure 8.13. The average residence time in the absorption tower was between 1 and 2 min.

Figure 8.13: Process Flow Sheet for Chlorine Recovery by Absorption

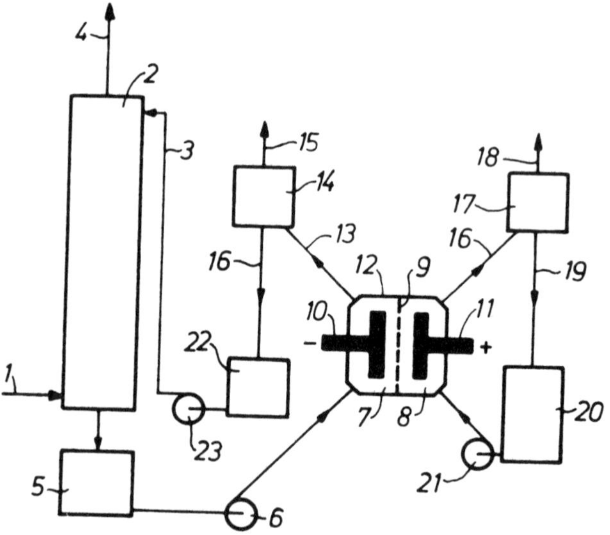

Source: U.S. Patent 4,082,631

The oxidized absorption liquid, containing 163 g/ℓ of Fe(III), 353 g/ℓ of Cl^-, 45.6 g/ℓ of HCl and 0.5 g/ℓ of Cu^{2+}, was delivered to the cathode compartment of the electrolysis cell **12** through the catholyte pump **6**. A membrane of a fluorine plastic with grafted-on sulfonic acid groups having ion exchanger properties (Nafion) was used as the diaphragm **9**. A graphite plate was used as the cathode **10**, while a similar plate of activated titanium (platinum metal coating approximately 1 μ thick) was used as the anode.

The anode side was initially charged with the same solution which was also introduced into the cathode compartment. The ratio by volume of the catholyte solution to the anolyte solution amounted to approximately 3:2. Electrolysis itself was carried out at about 25 A (corresponding to 36 A/dm^2). After an electrolysis time of about 1 hr, during which about 53 g of Fe(II) were formed on the cathode side and the corresponding quantity of chloride on the anode side, the catholyte solution was returned to the absorption tower through **13**, **14**, **22** and **23**. The voltage of the electrolysis cell amounted to approximately 7 volts. There was no significant evolution of hydrogen on the cathode side. The anode gas consisted of highly pure chlorine.

Chlorine and Cyanuric Acid Values from Polychloroisocyanuric Acids

A process for recovering chlorine values from an aqueous solution of a chloroisocyanuric acid compound is disclosed by *M.C. Fullington and L.C. Hirdler; U.S. Patent 4,138,559; February 6, 1979; assigned to Olin Corporation*. After reacting a mineral acid with the aqueous solution to form an acidified reaction mixture containing dissolved chlorine, the reaction mixture is fed to a stripping column which employs an inert gas to remove the dissolved chlorine.

The stripping column is maintained to provide a continuous liquid phase and a noncontinuous gas phase. Chlorine gas is readily recovered in a vessel such as a scrubber. Cyanuric acid may be subsequently recovered from the chlorine-depleted solution.

Example 1: An aqueous slurry of trichloroisocyanuric acid was prepared by a process in which an aqueous slurry of monosodium cyanurate was reacted with hypochlorous acid and chlorine gas. After filtration and recovery of the product, a filtrate containing 0.8% of trichloroisocyanuric acid and 6.5% of sodium chloride was heated in a heat exchanger to a temperature of 35°C.

Following heating, the aqueous solution was pumped to a reaction vessel. Hydrochloric acid (32%) was introduced and the aqueous solution acidified. In the reaction vessel, some chlorine gas was released and this gas was piped to a scrubber containing a solution of sodium hydroxide to produce a solution of sodium hypochlorite. The acidified solution,containing dissolved cyanuric acid, dissolved chlorine, hypochlorous acid, and sodium chloride was fed to a 10' stripping column packed with ceramic Intalox saddles.

After the acidified solution reached a level of about 4', air heated to a temperature of 50°C, was fed through an air sparge at the lower end of the column at a rate to provide a ratio of liquid to gas of 15:1. During continuous stripping of the acidified solution of trichloroisocyanuric acid, aerated liquid occupied about 100% of the height of the stripping column. The liquid level in the column was controlled by limiting the rate of flow of the chlorine-stripped solution of cyanuric acid from the bottom of the column.

Air containing chlorine was removed from the top of the column and fed to the scrubber. The chlorine-stripped solution removed from the bottom of the column had an available chlorine concentration of 20 ppm and its temperature was 35°C. This recovered solution containing dissolved cyanuric acid was transferred from the column to a jacketed reactor where its pH was adjusted to 4.4±0.5 by the addition of a solution of sodium hydroxide. Ethylene glycol solution, at a temperature of -5°C, was circulated through the jacket of the reactor to cool the cyanuric acid solution to 10°C.

During cooling, the solution was agitated and recirculated to keep cyanuric acid suspended in the solution. The cooled solution was pumped through a 150 μ

bag filter to separate the cyanuric acid crystals from the solution. During the chlorine stripping and cyanuric acid recovery, this procedure prevented the scaling of packing materials and equipment internals by precipitated cyanuric acid.

Example 2: *(Comparative)* – The filtrate of the same composition and temperature as that of Example 1 was acidified by the procedure of Example 1. The acidified solution and air was fed to the 10' stripping column used in Example 1 at the same liquid to gas ratio. The stripping column was operated in the conventional manner where the liquid was not retained in the packed section of the tower to provide a liquid level. The aqueous solution removed from the bottom of the column had an available concentration in the range of 300 to 1,000 ppm, indicating that inefficient and inadequate stripping of the dissolved chlorine had taken place.

CHLORINE FOR WATER PURIFICATION

Membrane Cell for Electrolysis of Sodium Chloride

A chlorine generator for the production of chlorine and hydrogen, with the chlorine being used for the treatment of swimming pools, sewage treatment facilities and drinking water is provided by *D. Yates; U.S. Patent 4,097,356; June 27, 1978*. The generator is characterized by an ion-permeable perfluorosulfonic acid membrane separating an anode compartment containing sodium chloride and cathode compartment.

Referring to Figure 8.14, the chlorine generator is shown in conjunction with a swimming pool for the chlorination of the water in the pool. A swimming pool **18** is in fluid communication via a pipe **20** with a pump **22**. The pump **22** removes water from the swimming pool **18** and circulates this water through a pipe **26** to a filtration system **28**.

Additionally, a line **24** is connected to the pipe **20** such that water is withdrawn from the pipe and thereafter split into two streams, with the first stream flowing through a chlorine aspirator **32** and the second stream flowing through a hydrogen aspirator **34**.

The line **24** may be connected to the pipe such that water may be withdrawn from the pipe and circulated to a chlorine sensor **30** of a chlorine generator **10**. The chlorine sensor measures the amount of chlorine existing within the water flowing through the line **24** and hence is indicative of the chlorine concentration of the water in the swimming pool. The chlorine sensor, which may be any conventional chlorine sensor, is electrically connected to a power supply **78** of the chlorine generator **10** such that whenever the chlorine concentration in the water of the swimming pool is below a predetermined level the power supply is activated thereby causing the chlorine generator to generate chlorine.

The chlorine aspirator **32** and the hydrogen aspirator **34** are connected to a common line **12** such that water flowing through the aspirators **32** and **34** flows to a venturi **14** contained within a return line **16**. The line **16** is also connected to a filtration system **28** as the return line to the swimming pool. Accordingly, the water returning to the swimming pool through the return line **16** has been substantially filtered by the filtration system **28** in addition to being chlorinated by the chlorine generator **10**.

Figure 8.14: Chlorine Generator Shown with Swimming Pool

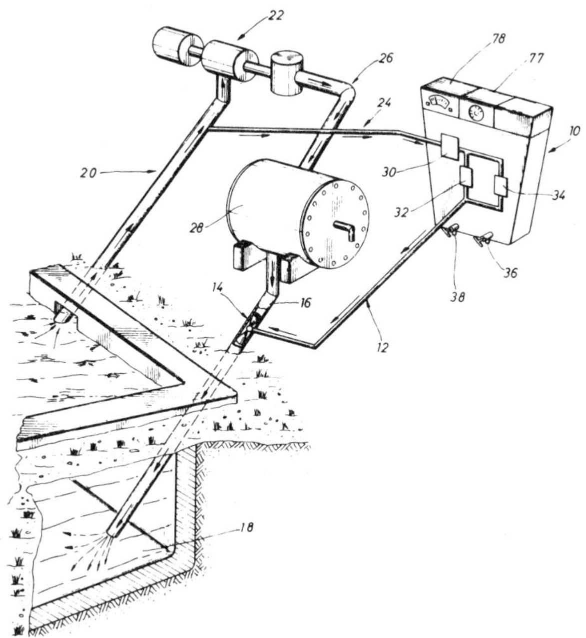

Source: U.S. Patent 4,097,356

As the water enters the swimming pool, the hydrogen gas tends to bubble to the surface of the pool where it is dissipated in the atmosphere. However, the chlorine is retained within the water due to its solubility therein and is circulated throughout the pool, thereby adjusting the chlorine content of the water in the swimming pool.

Water-Cooled Device for Generating Chlorine and Ozone

A compact chlorine generation device developed by *T. Themy; U.S. Patent 4,171,256; October 16, 1979* generates chlorine and usually ozone. The device is water cooled. A power transformer has a housing thereabout in heat conductive contact therewith. A highly heat conductive metal frame is attached in direct efficient heat conductive relation with the housing.

The frame has a conduit therethrough generally adjacent to the housing. A metal block has a semiconductor power rectifier held in a cavity therein and in thermal contact therewith. The block is attached in direct heat conductive relation with the frame. A coolant is flowed through the conduit thereby cooling the transformer and the rectifier. A chlorine generating electric cell has an anode and a cathode which are connected to the respective dc output poles of the rectifier.

The coolant which exits the conduit is led to the cell wherein it serves as the electrolyte. Outflow liquid from the cell and having an enhanced chlorine level dissolved therein is directed to a reservoir which may be a swimming pool, a health spa, a purified water supply, or the like. The chlorine content of the reservoir is thereby increased to provide a sterilizing and disinfecting action.

Through utilizing such an apparatus, it has been found that a transformer normally rated at 70 A can be used with up to 500 A current (at the same 15 volts) of normal operation, without any damage to the transformer. Similarly, the rectifier circuit can handle much larger currents at constant voltage without any harm thereto.

The overall apparatus is quite compact since no fins are needed for air cooling. Further, in view of the relatively high rate of liquid flow, the temperature of the aqueous liquid flowing therethrough does not vary significantly, generally no more than 1°C and in any event no more than about 5°C, at most. This is important as ozone production is adversely affected by high temperatures. The overall system is quite efficient, utilizing the normal pump found on a swimming pool or the like to supply the water flow, thus, keeping down cost of the apparatus.

On-Site Electrolytic Generation of Hypochlorite

A. Pellegri and P.M. Spaziante; U.S. Patent 4,172,773; October 30, 1979; assigned to Oronzio de Nora Impianti Electrochimici, SpA, Italy provide a method and apparatus for halogenating water for sterilization purposes by producing the desired active halogen concentration with a low salt content directly in the stream of water to be treated.

As shown in Figure 8.15, the apparatus comprises an electrolysis cell **1** consisting of a compartment **2** containing a cathode **3** made of steel or nickel or hastelloy or titanium or other electrically conducting material exhibiting a low overvoltage to hydrogen evolution and a compartment **4** separated from compartment **2** by a porous anode **5** made of a sintered valve metal activated with nonpassivatable and electrocatalytic material resistant to the anodic environment.

Figure 8.15: On-Site Electrolytic Generation of Hypochlorite

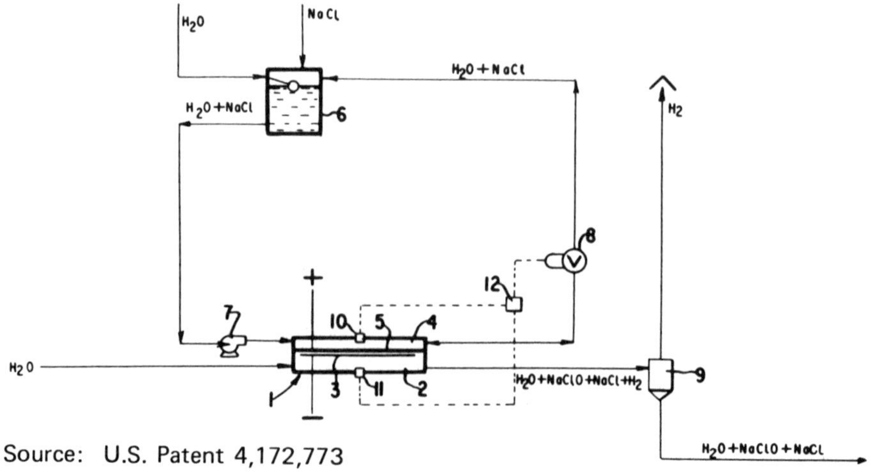

Source: U.S. Patent 4,172,773

The brine circuit is comprised of a saturation tank **6** where salt and the brine solution are contacted, a circulating pump **7**, preferably upstream to the electrolytic cell **1**, a second valve **8** downstream to the electrolytic cell and a suitable direct current supply.

The electrolyte leaving the compartment **2** after electrolysis contains hypochlorite and the hydrogen which is produced at the cathode. A suitable gas separation tank **9** allows the venting of gaseous hydrogen from the effluent hypochlorite solution. The automatic control of the optimal pressure differential through the porous anode **5** may be effected by two pressure gauges **10** and **11** in the brine compartment **4** and in the water compartment **2**, respectively. Pneumatic or electric signals from the two gauges are suitably operated by a control device which, through an actuating system, acts on control valve **8**.

Electrolytic Generation of Halogen Biocides

N.W. Stillman; U.S. Patent 4,100,052; July 11, 1978; assigned to Diamond Shamrock Corporation discloses an electrolytic cell for the generation of low cost halogen biocidally active agent from an aqueous solution having a low halogen salt content for use in the treatment of sewage or other liquid affluents, especially those of fresh-water swimming pools or fresh-water cooling towers. The electrolytic cell is used in line with pumps generally associated with the distribution of waters for swimming pools or cooling towers.

Example: Water in a swimming pool of 12,000 gal capacity was treated to adjust the concentration of NaCl and residual chlorine to 1.11 g/ℓ and 3.48 ppm, respectively. An electrolytic cell **10**, as shown in Figure 8.16, containing an anode center electrode plate **18**, five bipolar electrode plates **20** and a cathode end electrode plate **16** on each side to yield an anode area of 200 in^2 (1,290 cm^2) was connected to a rectifier capable of delivering 6 A at 25 volts.

Figure 8.16: Perspective View of Cell for Generating a Biocidal Agent

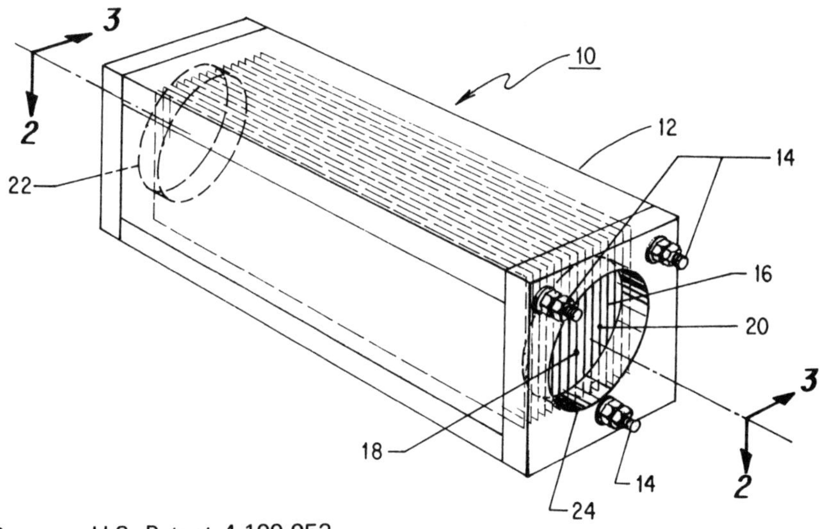

Source: U.S. Patent 4,100,052

A pump capable of delivering 50 gal/min (189 ℓ/min) was installed on the chlorinator. The chlorinator was placed in continuous operation; the results are shown below.

Day	Chlorine Concentration (ppm) Pool	Cell	Remarks
0	3.48	3.64	-
1	0.80	1.75	23 volts, 3.25 amps
5	1.20	1.84	Current efficiency = 28.2%
6	0.40	0.92	-
7	0.20	0.70	-
8	<0.10	*	*
11	0.36	0.70	3.25 amps, NaCl concentration = 1.06 g/ℓ

*Hot weather caused a decline in the residual chlorine. Dry chlorine was added to the pool.

Swimming Pool Chlorinator Requiring Minimal Attention

L. Persson, B. Hansen and E. Eklund; U.S. Patent 4,136,005; January 23, 1979; assigned to AG Licento, Switzerland provide an electrolytic chlorinator which is particularly suited for residential swimming pools or other installations having a small chlorine demand and which accordingly is a practical alternative to chlorination by the addition of chlorine tablets or hypochlorite compounds.

The chlorinator is capable of safe and unattended operation for extended periods (months) and requires substantially no operator manipulation other than infrequent replenishment of a supply of the material to be electrolyzed (salt in solid form) and setting of the desired chlorine output.

As shown in Figure 8.17, the chlorinator illustrated by way of example is connected to a recirculating conduit **11** for a swimming pool **12** containing a body of water to be chlorinated.

The recirculating conduit **11** has an intermittently operating recirculating pump **13**, between the pressure and the suction sides of which the chlorinator is connected by way of an intake conduit **14** and a return conduit **15**. The chlorinating and recirculating system of the pool **12** also comprises valves, filters and other ancillary devices which have been omitted for clarity. The chlorinator is adapted to feed pure chlorine to the pool water during operating periods of variable duration and frequency of repetition; the operating periods preferably coincide with the operating periods of the recirculating pump **13**. The main parts of the chlorinator are as follows.

(a) An electrolytic cell **20** in which an anolyte in the form of concentrated brine (aqueous sodium chloride solution) is decomposed electrolytically so that gaseous chlorine, hydrogen gas and waste liquid containing sodium hydroxide are formed;

(b) A brine source **21** which contains a charge of solid sodium chloride and feeds concentrated brine to the electrolytic cell;

(c) A water metering device **22** which feeds controlled amounts of water to the electrolytic cell **20** and to the brine source

21 and which controls the brine feed to the electrolytic cell by controlling the water feed to the brine source;

(d) A float-controlled mixing valve **23** which feeds water to the metering device **22** and which receives the gaseous chlorine produced in the electrolytic cell and mixes it with a portion of a water stream diverted from and returned to the recirculating conduit **11**;

(e) An eductor or jet pump **24** which is connected between the conduits **14** and **15** and combines the portion of the water stream chlorinated in the mixing valve **23** with the remainder of the stream diverted from the recirculating conduit; and

(f) An electric current supply and control device **25** which during the operating periods feeds direct current of selected amperage to the electrodes of the electrolytic cell and which controls the metering device **22** and, moreover, has certain monitoring and safety functions.

Figure 8.17: Swimming Pool Chlorinator Requiring Minimal Attention

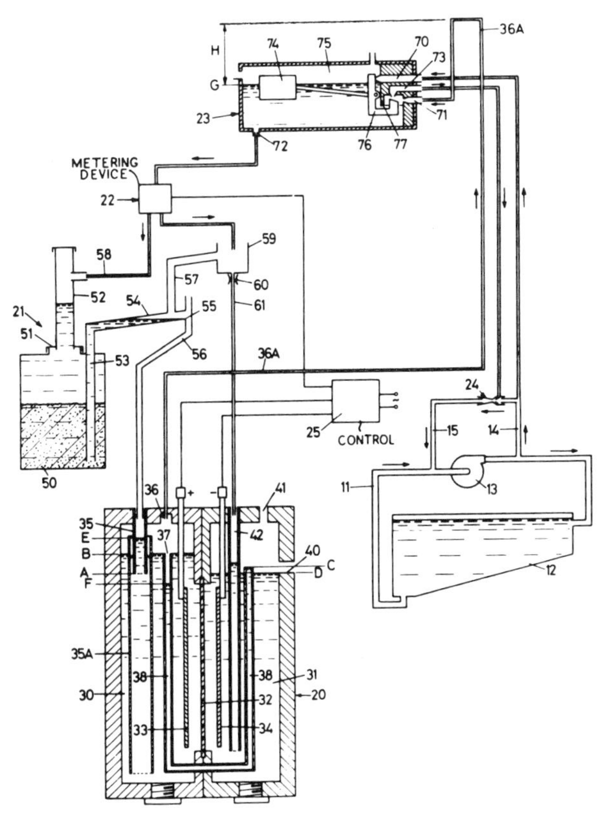

Source: U.S. Patent 4,136,005

Caustic Chlorine Filter Bag

A.E. Ingram; U.S. Patent 4,172,788; October 30, 1979 describes a bag filter adapted chemically for use with caustic chlorine that is strong enough to need service about every six months rather than about every six weeks.

The filter is made of monofilament polypropylene, a comparatively fragile material but chemically adapted for the purposes, in the shape of a pillow case with a long edge opening and oppositely, adjacent an end thereof, a conduit opening. All edges are reinforced with a strip of felted polypropylene, a comparatively heavy and strong material and also chemically adapted for use with caustic chlorine.

The filter is additionally reinforced with pads of the same felted material where supported in a container. In use, a flat filter leaf with a conduit is inserted into the filter through the opening with the conduit opening coinciding with the conduit, and the long edge opening tightly sewn together. Caustic chlorine is introduced into the filter through the conduit opening and filter leaf, and escapes into the container through the filter free of salt. Both the monofilament and felted polypropylene define minute interstices that are impervious to salt but not to chlorine.

ADDITIONAL PROCESSES

Desalinization and Chemical Extraction

D.D. Childress; U.S. Patent 4,176,023; November 27, 1979; assigned to Desal-Chem, Inc. describes a combined desalinization and extraction process for brine water having a salinity of 7½ to 9%. The brine water is introduced to a concentrator basically similar to a shell-and-tube type heat exchanger vertically arranged with upper and lower chambers above and below the tube section and communicating with each other through the tubes.

A heating element in the lower chamber causes the brine water to be heated until it reaches its boiling temperature. Vapors are removed from the upper chamber and are externally compressed so as to create a partial vacuum in the upper chamber. The compressed vapors are passed from the compressor to the concentrator into the spaces on the outside of the tubes where the vapors are condensed as liquid water. The condensed freshwater on the outside of the tubes is removed.

The remaining brine water within the tubes, which is concentrated at 28% salinity, is conducted to a plurality of electrolytic cells having positively charged anodes and negatively charged cathodes. The concentrated brine is electrolyzed with low voltage-direct current to release chlorine gas, caustic alkali containing primarily sodium hydroxide, hydrogen gas, and an inert material containing calcium, nitrogen, and magnesium oxide. The chlorine gas is conducted to a mist extractor-separator to remove any impurities and then compressed to form liquid chlorine. The hydrogen gas is conducted to a mist extraction separator to remove any impurities.

Example: A barrel of brine water is equal to approximately 159 liters of brine water. The concentrator has an outside diameter of 1.3 m and a height of 2 m.

A quantity of 300 barrels of brine introduced to the concentrator will yield 225 barrels of freshwater and 75 barrels of concentrated brine.

The 75 barrels of concentrated solution are then passed through the chlorine cells, which have a capacity of approximately 2,700 kg. The cell stack has outside dimensions approximately 1.4 m high, 1.7 m long and 1 m wide. With the brine salinity at approximately 28%, the cell efficiency is at its peak; however, assuming, for the sake of this example, 80% efficiency, the resulting products are 6,375 kg of chlorine gas, and 5,010 kg of caustic alkali, and a small amount of hydrogen gas.

Solar Powered Gas Generation

H.D. Brown; U.S. Patent 4,080,271; March 21, 1978 describes the utilization of solar energy by means of a reflector, a boiler and a turbogenerator whereby solar heat is concentrated and electric power is produced, utilization of the exhaust heat from the power cycle in the distillation of seawater, and utilization of the electric power in a plurality of electrolytic cells whereby hydrogen and oxygen are extracted from the distilled water and hydrogen and chlorine are extracted from the seawater.

Referring to the Figure 8.18, solar heat **1** is the energy input, and a reflector **2** concentrates the solar heat upon a boiler **3**. A boiler feed pump **4** supplies fluid from a boiler fluid reservoir **5** to the boiler, wherein the fluid is heated to a high temperature which causes it to vaporize. The expanded vapor, at a high pressure, enters a turbine **6**, causing the turbine blades and shaft to rotate, and exits from the turbine to a primary heat exchanger **7**, wherein it is cooled and condensed, being returned thence to reservoir **5**.

Seawater, the incoming raw material, is piped from an external source into a seawater reservoir **8** and is fed thence at ambient temperature into a secondary heat exchanger **9**, wherein it is heated and whence it is fed to the primary heat exchanger, wherein it is further heated to a high temperature which causes it to evaporate, leaving in the primary heat exchanger a residue comprising an aqueous solution of sodium chloride, which is periodically drained from the primary heat exchanger through a valve **10** to a residue chamber **11**.

Water vapor from the primary heat exchanger is returned to the secondary heat exchanger, wherein it is cooled and condensed as distilled water, flowing thence to a distilled water reservoir **12**.

An alternator **13**, mechanically driven by the turbine, generates an electrical potential in the form of an alternating electric current at a suitable output voltage. The electric current is conducted in parallel circuits to a primary transformer **14** and to a secondary transformer **15**, in each of which the voltage is stepped down to a lower magnitude suitable for utilization as hereinafter described.

The alternating currents leaving the transformers are conducted in separate circuits respectively to a primary rectifier **16** and to a secondary rectifier **17**, wherein they are rectified to direct currents, which are conducted thence in separate circuits respectively to a primary gas generator **18**, comprising a plurality of electrolytic cells, and to a secondary gas generator **19**, comprising a further plurality of electrolytic cells.

Distilled water from reservoir **12** is electrolyzed in the primary gas generator, thus producing hydrogen and oxygen, which are separately and respectively piped to a hydrogen reservoir **20** and to an oxygen reservoir **21**.

Figure 8.18: Solar Powered Gas Generation

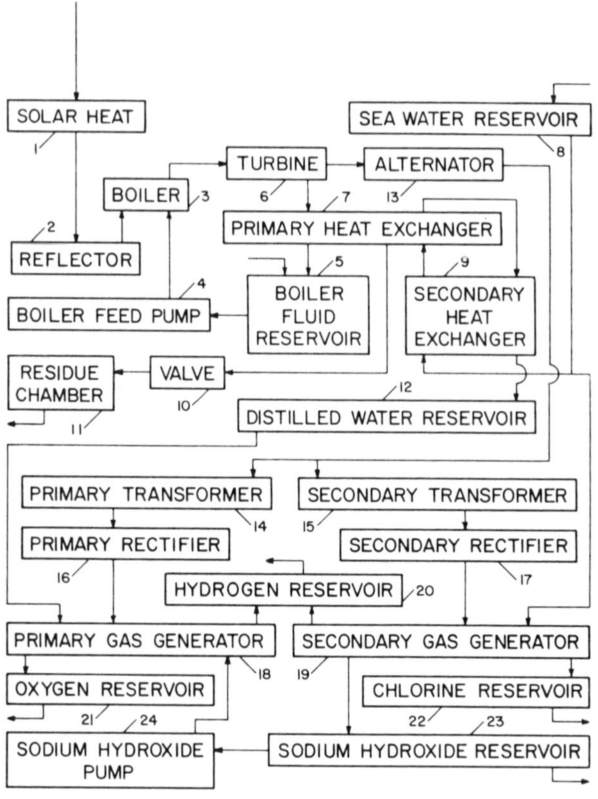

Source: U.S. Patent 4,080,271

Seawater from reservoir **8** is electrolyzed in the secondary gas generator, thus producing chlorine, a further quantity of hydrogen, and a solution of sodium hydroxide in water, which are separately and respectively piped to a chlorine reservoir **22**, to hydrogen reservoir **20** and to a sodium hydroxide reservoir **23**.

Integrated System for Producing Chlorine Dioxide and Chlorine

In the process developed by *G. Cowley, R. Swindells and M. Kostanecki; U.S. Patent 4,086,329; April 25, 1978; assigned to Erco Industries Limited, Canada*, chlorine, including at least part of the by-product chlorine from the chlorine dioxide producing reaction, is reacted in a known exothermic reaction with sulfur dioxide, in accordance with the equation as follows.

(1) $$SO_2 + Cl_2 + 2H_2O \longrightarrow H_2SO_4 + 2HCl$$

The mixture of sulfuric acid and hydrochloric acid is fed to the chlorine dioxide generator, the hydrogen ions and chloride ions contained in the mixture utilized to provide hydrogen ions, chloride ions and sulfate ions to the chlorine dioxide-producing reaction.

Due to the inefficiencies introduced to the chlorine dioxide-producing process by the competing reaction according to equation (2), imbalances result from feed of the products of equation (1) as the sole source of the hydrogen, chloride and sulfate ion requirements, and these must be compensated for if a stable steady state chlorine dioxide-forming procedure is to be maintained.

$$(2) \qquad ClO_3^- + 6H^+ + 5Cl^- \longrightarrow 3Cl_2 + 3H_2O$$

The disclosed process achieves such compensation while at the same time allowing variation in the quantity of sodium sulfate produced per mol of chlorine dioxide.

Referring to Figure 8.19, there is illustrated a chlorine dioxide producing system **10** including a generator **12**. In the generator, there is present a chlorine dioxide-producing reaction medium containing sodium, chloride, chlorate, sulfate and hydrogen ionic species.

Figure 8.19: Integrated System for Producing Chlorine Dioxide and Chlorine

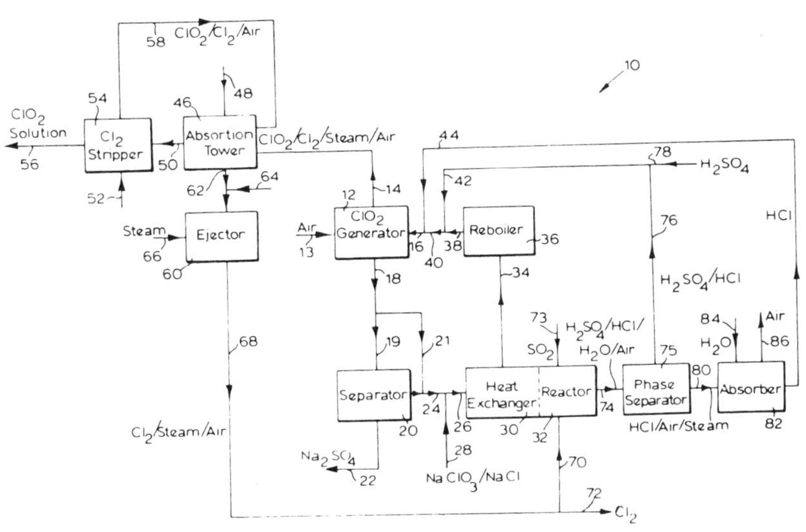

Source: U.S. Patent 4,086,329

The reaction medium is maintained at its boiling point while the vessel is maintained under a vacuum. The temperature and vacuum may vary widely, typically from 25° to 90°C at a pressure of 20 to 500 mm Hg abs. Air is bled into the generator by line **13** to provide the absolute pressure in the generator.

The boiling temperature of the reaction medium causes water to be evaporated therefrom which dilutes the chlorine dioxide and chlorine formed from the reaction medium. The resultant gaseous mixture is removed from the reaction zone by line **14**. When after initial start-up the reaction medium becomes saturated with sodium sulfate, anhydrous neutral sodium sulfate precipitates from the reaction medium.

The generator **12** is run in a substantially continuous manner with the spent chemicals of the reaction medium being replenished continuously, in this embodiment, by an aqueous feed solution in line **16** containing sodium chlorate, sodium chloride, hydrochloric acid and sulfuric acid, and the liquid input being balanced by liquid and steam output so that the liquid level in the generator remains substantially constant.

The precipitated sodium sulfate is removed, continuously or intermittently, from the generator, as a slurry with some spent reaction medium by line **18**. The removed slurry is partly passed by line **19** to a separator **20** wherein the crystalline sodium sulfate is separated from the mother liquor and is recovered therefrom by line **22** as one of the products of the system **10** for use in pulp mill operations. The remainder of the slurry in line **19** is recycled to the generator in a recirculation loop.

Recovery of Acidic Gases

H.B.H. Cooper; U.S. Patent 4,069,117; January 17, 1978 discloses a cyclic and regenerative process for the removal and recovery of at least one acid gas from a mixture of gases including carbon dioxide comprising the following.

(a) Electrolyzing an aqueous alkali metal chloride in a two-compartment cell including a cationic permselective ion-exchange membrane between the anode and cathode compartments to produce chlorine, hydrogen and aqueous alkali metal hydroxide;

(b) Producing hydrochloric acid from the hydrogen and the chlorine;

(c) Treating the mixture with the aqueous alkali metal hydroxide to form aqueous alkali metal carbonate/bicarbonate and to form aqueous alkali metal salt from at least one acid gas;

(d) Reacting the hydrochloric acid with the aqueous alkali metal salt to form aqueous alkali metal chloride and liberate the acid gas for recovery; and

(e) Cycling the aqueous alkali metal chloride to the electrolyzing for production of hydrogen, chlorine and aqueous alkali metal hydroxide.

Electrowinning Metal from Chloride Solution

F. Grontoft; U.S. Patent 4,155,821; May 22, 1979; assigned to Falconbridge Nickel Mines Limited describes a method for electrowinning metal from metal chloride-bearing electrolyte in an electrolytic cell wherein diffusion of chlorine from anolyte to catholyte is limited, thereby limiting concentration of chlorine gas around the cell so as to be within acceptable limits.

Catholyte is withdrawn through a catholyte overflow duct in the cell to establish a catholyte level and the anode is maintained in an electrolyte-permeable diaphragm which is attached to a hood extending above the electrolyte level. Chlorine gas generated at the anode is withdrawn from the hood by applying suction via a suction duct located in the hood above the catholyte level and electrolyte is drawn through the diaphragm into the anolyte and upwardly inside the hood. Anolyte is withdrawn therefrom under the suction as an overflow via an anolyte outlet duct in the hood above the catholyte level.

Example: A cell for the electrowinning of nickel from a chloride-bearing electrolyte contained 36 nickel cathodes, each between 2 of 37 graphite anodes. The average electrode spacing from cathode center to anode surface was about 75 mm. The anode diaphragms were made of tightly woven cloth of artificial fiber, the anode hoods of thermosetting plastic sealed to the graphite and the diaphragms were attached to the hoods below the catholyte level.

The electrolyte feed was supplied to the cell at a rate of 1.5 m^3/hr, a pH of 2.4 and a temperature of 62°C, with the following composition in grams per liter: 67.0 Ni, 88.0 Cl, 50.3 SO_4, 15.0 H_3BO_3, and 28.0 Na.

Current was supplied at 8.9 kA, a cathode current density of 172 A/m^2, and a potential of 3 V. Current efficiency was 98.5% and nickel production from the cell was 9.6 kg/hr. The mean height of the anolyte overflow levels in the hoods above the electrolyte overflow level in the cell was 45 mm, and the total suction set at a control point remote from the cell was 70 mm wg.

The specific gravity of the electrolyte was 1.18. Under these conditions electrolyte was drawn through the diaphragm and anolyte overflowed from the hoods through the outlet duct at a rate of 1.1 m^3/hr, a pH of 1.6, a temperature of 60°C, and with a nickel content of 61 g/ℓ. At the same time electrolyte overflowed from the open cell through the overflow duct at a rate of 0.4 m^3/hr, a pH of 2.5, a temperature of 60°C, and with a nickel content of 60 g/ℓ. There was no smell of chlorine in the atmosphere surrounding the cell.

Liquid Chlorine Electrochemical System

A gas phase free, liquid chlorine electrochemical system and apparatus are described by *T.G. Hart; U.S. Patent 4,086,393; April 25, 1978; assigned to Energy Development Associates*. The housing containing the electrodes is filled with sufficient electrolyte so that the pressure therein is sufficient to liquify the chlorine as it is generated.

To operate the embodiment of Figure 8.20, cell **1**, reservoirs **2** and **3** and all interconnections and projections are completely filled with, e.g., a 40% aqueous zinc chloride electrolyte and all gas vented. Additional electrolyte is then pumped into the system to raise the pressure to about 6.3 kg/cm^2, as is indicated by gauge **17**. Circulation pump **14** is adjusted to the flow rate appropriate for charge and the charge current flow is started between electrodes **4** and **5**. Chlorine initially generated at electrode **4** goes directly into solution in the electrolyte.

The system pressure steadily rises due to foreign gas generation at chlorine electrode **4** and the foreign gas accumulates in gas space **18**, making space for itself by compressing the electrolyte slightly. Typically, by the time the electrolyte is saturated with chlorine, the system pressure has risen to about 8 kg/cm². After electrolyte saturation, the chlorine liquifies as it is generated, assuming that the system temperature is about 30°C, which is above the chlorine hydrate formation temperature. The circulation of pump **14** prompts the liquid chlorine **20** to settle at the bottom of reservoir **2**.

Figure 8.20: Liquid Chlorine Electrochemical System

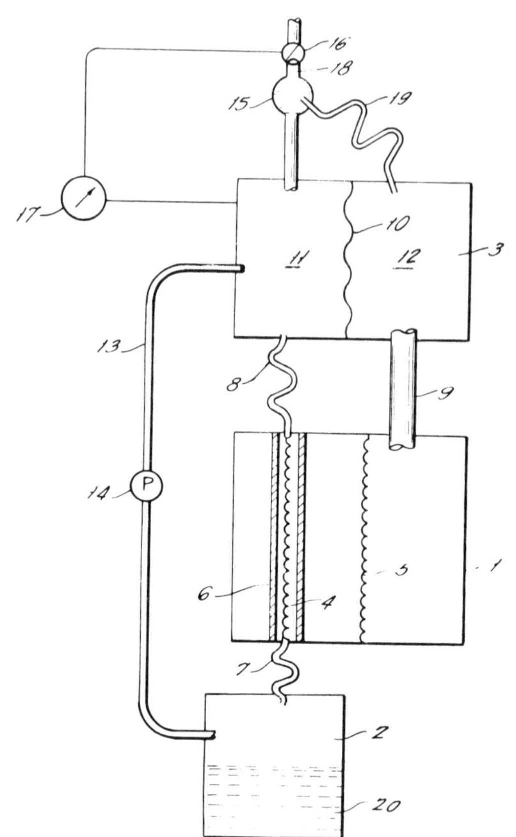

Source: U.S. Patent 4,086,393

The system pressure will rise due to continued generation of foreign gas and when the pressure reaches about 10.1 kg/cm², valve **16** is opened to vent gas out of gas space **18** and reduce system pressure to about 8 kg/cm². Valve **16** can connect gas space **18** with another gas space which contains, e.g., a 20% hydrogen, 80% nitrogen mixture at about 8 kg/cm² and a device for catalyzing a reaction between chlorine and hydrogen and oxygen and hydrogen.

When charging is completed, electrodes **4** and **5** are disconnected from the power source, pump **14** is deactivated, and the system stands idle awaiting discharge.

Chlorine within separator **6** at the end of charging is relatively quickly dissipated, partly by movement through separator **6** into the other compartment, and mainly by conversion to chloride at electrode **4**. The loss of chlorine is small because of the relatively small volume of free space inside separator **6** and within electrode **4** which is, e.g., a titanium wire-spiral coated with ruthenized oxide. Replenishment of the chlorine inside separator **6** by diffusion through conduits **7** and **8** is relatively slow because of the maximum length and minimum cross-section thereof. Therefore, the self discharge of this embodiment is small.

In order to discharge, pump **14** is adjusted to the appropriate discharge rate and electrodes **4** and **5** are connected to the load. In this zinc/chlorine example, zinc is depleted at electrode **5** and chlorine is depleted at electrode **4**. The voltage loss at chlorine electrode **4** depends partly on how efficiently dissolved chlorine is brought into contact with the electrode surface. Therefore, one purpose of pump **14** is to maintain sufficient electrolyte turbulence to insure a consistently high chlorine level at the surface of electrode **4**.

COMPANY INDEX

The company names listed below are given exactly as they appear in the patents, despite name changes, mergers and acquisitions which have, at times, resulted in the revision of a company name.

INVENTOR INDEX

U.S. PATENT NUMBER INDEX

Copies of U.S. patents are easily obtained
from the U.S. Patent Office at 50¢ a copy.

NOTICE